Complexity Characteristics of Natural Language

Complexity Characteristics of Natural Language

Guest Editors

Stanisław Drożdż
Jarosław Kwapień
Tomasz Stanisz

MDPI

Basel • Beijing • Wuhan • Barcelona • Belgrade • Novi Sad • Cluj • Manchester

Guest Editors
Stanisław Drożdż
Complex Systems Theory Department
Polish Academy of Sciences
Kraków
Poland

Jarosław Kwapień
Complex Systems Theory Department
Polish Academy of Sciences
Kraków
Poland

Tomasz Stanisz
Complex Systems Theory Department
Polish Academy of Sciences
Kraków
Poland

Editorial Office
MDPI AG
Grosspeteranlage 5
4052 Basel, Switzerland

This is a reprint of the Special Issue, published open access by the journal *Entropy* (ISSN 1099-4300), freely accessible at: https://www.mdpi.com/journal/entropy/special_issues/HF50GG8ME9.

For citation purposes, cite each article independently as indicated on the article page online and as indicated below:

Lastname, A.A.; Lastname, B.B. Article Title. *Journal Name* **Year**, *Volume Number*, Page Range.

ISBN 978-3-7258-7002-8 (Hbk)
ISBN 978-3-7258-7003-5 (PDF)
https://doi.org/10.3390/books978-3-7258-7003-5

Contents

Preface

This Reprint addresses natural language as a complex system shaped by interacting structural, statistical, and causal mechanisms. Its scope spans multiple linguistic scales and data sources, linking theory with empirical analysis. The aim is to consolidate recent progress and motivate causal, multiscale approaches to language complexity. The Reprint is written to serve researchers in linguistics, complexity science, information theory, and artificial intelligence who seek a unified and forward-looking perspective.

Stanisław Drożdż, Jarosław Kwapień, and Tomasz Stanisz
Guest Editors

entropy

MDPI

Editorial

Editorial: Complexity Characteristics of Natural Language

Stanisław Drożdż [1,2,*], Jarosław Kwapień [1] and Tomasz Stanisz [1]

[1] Complex Systems Theory Department, Institute of Nuclear Physics, Polish Academy of Sciences, 31-342 Kraków, Poland
[2] Faculty of Computer Science and Mathematics, Cracow University of Technology, 31-155 Kraków, Poland
* Correspondence: stanislaw.drozdz@ifj.edu.pl

The contributions collected in this Special Issue testify to the continued vitality and conceptual breadth of research at the intersection of natural language, complexity science, and information theory. Taken together, the ten papers presented here advance a shared perspective that natural language is not merely a vehicle for meaning, but rather a multiscale complex system whose statistical, dynamical, and structural properties encode deep regularities of human cognition, social interaction, and cultural evolution.

Several contributions deepen our quantitative understanding of scaling, correlations, and variability in language. Detailed analyses of punctuation and sentence-length variability reveal that punctuation is not a superficial stylistic device but rather a key organizer of linguistic correlations, capable of inducing multifractal structure and long-range dependence in texts. Studies of experimental literary works demonstrate that even highly unconventional narratives retain or sometimes even amplify these complexity signatures, and in some cases show remarkable translation invariance, pointing to constraints that transcend individual languages.

Other papers broaden the empirical scope of linguistic complexity by examining language across spatial, temporal, grammatical, and genealogical scales. Large-scale analyses of social media data show how rank diversity and scaling laws depend jointly on grammar, geography, and time, revealing both universal patterns and context-sensitive deviations. At a macroscopic level, the study of language families uncovers Zipf-like regularities that may reflect long-term "entropic" processes of diversification and contact, connecting linguistic evolution to ideas from statistical physics.

A third thematic strand addresses causality and structure in linguistic systems. Some contributions move beyond correlation by explicitly asking what causes what in language change and use. By combining information-theoretic complexity measures with causal inference techniques, evidence is provided demonstrating that changes in morphological complexity can precede and drive changes in syntactic organization, rather than the reverse. Complementary work critically reassesses long-standing hypotheses about social drivers of complexity, showing—using improved data and modeling—that the proportion of non-native speakers does not straightforwardly reduce linguistic complexity.

This Special Issue also connects classical complexity questions with contemporary NLP and machine learning. Some contributions explicitly engage with large language models (LLMs), either as analytical tools or as objects of study. Novel approaches combine linguistic structure (e.g., relational triples) with LLMs to address tasks such as objectivity detection in informal texts. Other contributions show that, despite their impressive fluency, current LLM-generated texts lack certain hallmark properties of natural language—most notably robust long-range dependence—highlighting a gap between human language production and present-day architectures. Work on informativity and predictability further clarifies how information-theoretic concepts should be interpreted when applied to noisy corpus data and probabilistic models, while a forward-looking analysis discusses both the promise and the limitations of LLMs for causal reasoning in text.

Despite decades of progress, several gaps have persisted in the study of language complexity. In particular, many earlier studies relied on static correlations, leaving causal relationships underspecified. Results were often fragmented across linguistic levels (characters, words, syntax, discourse) and data types (literary texts, spoken language, social media). The rapid rise of LLMs created an urgent need to reassess classical complexity measures in light of generative models that can mimic surface statistics without penetrating deeper dependencies. This Special Issue directly addresses these gaps by (i) integrating causal inference with information-theoretic and algorithmic complexity measures, (ii) systematically comparing multiple linguistic scales and data sources, and (iii) explicitly contrasting human-generated and machine-generated language. In doing so, the collected works show that complexity is not a monolithic property but an emergent outcome of interacting constraints operating across time, scale, and representation.

Looking ahead, the interaction between complexity science and large language models is likely to define the next phase of research. Several directions stand out as particularly suggestive. Among these, empirical evidence that natural language exhibits long-range dependence and multifractal organization indicates that future LLM architectures must incorporate richer, more persistent memory mechanisms. Understanding which complexity signatures matter most for meaning, coherence, and creativity remains an open challenge. As models become capable of producing fluent explanations, distinguishing genuine causal understanding from plausible-sounding narratives will be of crucial importance. Complexity-based diagnostics may help us to evaluate whether LLMs capture causal structure or merely approximate correlations. Extending complexity analyses to low-resource languages, multimodal communication, and hybrid human–AI texts will test the universality of observed laws and reveal where current models fall short. Rather than treating LLMs as black boxes, future work should use insights from information theory, scaling laws, and algorithmic complexity to design principled benchmarks that go beyond accuracy and perplexity.

In conclusion, this Special Issue demonstrates that the study of natural language complexity remains both theoretically rich and practically relevant. By combining rigorous quantitative methods with emerging AI technologies, the field is well positioned to deepen our understanding of language as a complex adaptive system and to guide the development of language models that more faithfully reflect its structure and dynamics.

Author Contributions: Writing—original draft preparation, S.D.; writing—review and editing, S.D., J.K., and T.S. All authors have read and agreed to the published version of the manuscript.

Conflicts of Interest: The authors declare no conflicts of interest.

https://doi.org/10.3390/e28010098

entropy

MDPI

Article

Multifractal Hopscotch in *Hopscotch* by Julio Cortázar

Jakub Dec [1], Michał Dolina [1], Stanisław Drożdż [1,2,*], Jarosław Kwapień [2,*] and Tomasz Stanisz [2]

[1] Faculty of Computer Science and Telecommunications, Cracow University of Technology, 31-155 Kraków, Poland; jakub.dec@pk.edu.pl (J.D.); michal.dolina@pk.edu.pl (M.D.)

[2] Complex Systems Theory Department, Institute of Nuclear Physics, Polish Academy of Sciences, 31-342 Kraków, Poland; tomasz.stanisz@ifj.edu.pl

* Correspondence: stanislaw.drozdz@ifj.edu.pl (S.D.); jaroslaw.kwapien@ifj.edu.pl (J.K.)

Abstract: Punctuation is the main factor introducing correlations in natural language written texts and it crucially impacts their overall effectiveness, expressiveness, and readability. Punctuation marks at the end of sentences are of particular importance as their distribution can determine various complexity features of written natural language. Here, the sentence length variability (SLV) time series representing *Hopscotch* by Julio Cortázar are subjected to quantitative analysis with an attempt to identify their distribution type, long-memory effects, and potential multiscale patterns. The analyzed novel is an important and innovative piece of literature whose essential property is freedom of movement between its building blocks given to a reader by the author. The statistical consequences of this freedom are closely investigated in both the original, Spanish version of the novel, and its translations into English and Polish. Clear evidence of rich multifractality in the SLV dynamics, with a left-sided asymmetry, however, is observed in all three language versions as well as in the versions with differently ordered chapters.

Keywords: natural language; punctuation; sentence length variability; multiscaling; discrete Weibull distribution

1. Introduction

One of the perspectives on research on natural language is the one derived from complexity science [1,2]—as multiple traits identify natural language as a complex system [3]. Numerous methods widely used in studying complex systems, based on concepts originating in information theory [4–6], time series analysis [7–11], network science [12–18] and the theory of power-law probability distributions [19–22], have been used to grasp the various quantitative characteristics of natural language. Understanding the mechanisms behind such characteristics has the potential to improve the methods of natural language processing and generation. This is especially relevant today, given the heightened focus on large language models (LLMs) [23,24], incorporated in generative AI systems like OpenAI's ChatGPT, Microsoft's Bing Chat, and Google's Bard. One of the language properties recently investigated with statistical methods is the usage of punctuation in written language—it has been demonstrated that the distribution of word counts between consecutive punctuation marks in literary texts generally follows a discrete Weibull distribution [3,25,26]. What is interesting and significant, however, is the fact that for different languages the two parameters of this Weibull distribution are, to a large extent, selectively different and specific to a given language. Consistent with this fact, the punctuation distributions of texts after translation into another language are still governed by the Weibull distribution but with the parameters of the target language [3,25]. Interestingly, the patterns of sentence-ending punctuation marks, such as periods, exhibit much more flexibility and are less strictly bound by this distribution. These intervals, equivalent to sentence lengths, can therefore display a more diverse range of patterns.

Among the works considered important in the world literature, a book interesting from the aforementioned quantitative perspective is *Hopscotch* by Julio Cortázar. At its core, this novel challenges traditional narrative structures and invites readers to engage with the text in a non-linear manner. The novel presents multiple reading options: one can either follow the conventional order of chapters or choose to explore the narrative via a suggested "hopscotch" order, jumping between chapters in a non-sequential fashion [27,28]. This experimental approach allows readers to experience the story in a unique, personalised way reflecting the author's desire to break away from the conventional storytelling. As a consequence of this free-form structure of the novel, the order of sentences and smaller textual constituents depends on a chosen sequence of chapters, which implies a strong impact on long-range correlations, both the linear and nonlinear ones. Since the novel comprises an eclectic mix of styles, including passages of stream-of-consciousness and experimental language play, the overall richness and depth of the narrative results in a convoluted, multilevel linguistic construct that, if transformed into numbers, is expected to reveal substantial complexity with self-similarity being one of its likely manifestations.

2. Materials and Methods

In its original Spanish version [29], *Hopscotch* consists of 155 chapters that can be read in the printed order, the "hopscotch" order recommended by the author, in which the number of the subsequent chapter is explicitly given at the end of the previous one, or in any other order formed by permutating the chapters that is chosen by a reader. Because of such alternatives, all foreign-language translations preserve the structure of the novel and the author recommendations. To illustrate this point, along with the Spanish text, there are two translations considered as well: the English one [30] and the Polish one [31]. We transform the text in each language version into a sentence length variability (SLV) time series by counting words between each two consecutive sentence-ending punctuation marks (these can be periods, interrogation marks, exclamation marks, and other equivalent marks). The related statistical data are collected in Table 1. The SLV data reflect the evolution of the size of the main functional building blocks throughout the whole novel, which has been shown to be substantially informative [25,32].

Table 1. Essential statistics on the SLV data considered in this study.

	Spanish	English	Polish
Minimum sentence length (words)	1	1	1
Maximum sentence length (words)	850	860	681
Average sentence length (words)	16.6	17.4	13.9
Minimum sentence length (characters)	1	1	1
Maximum sentence length (characters)	4037	4058	3730
Average sentence length (characters)	77.3	77.7	74.9
Minimum chapter length (sentences)	4	3	3
Maximum chapter length (sentences)	1710	1720	1873
Average chapter length (sentences)	108.0	109.3	125.5

Punctuation may be viewed as a process of interrupting an otherwise ceaseless stream of words that improves understanding of the message and allows a reader to have necessary breaks. These functions make occurrences of the punctuation marks suitable for the survival analysis [33]. It has already been documented in the literature that the inter-mark distances develop distributions that can be modelled by the discrete Weibull distribution [3,25]. This distribution is given by the following probability mass function [34]:

$$f(k) = (1-p)^{k^\beta} - (1-p)^{(k+1)^\beta}, \quad p \in (0,1), \quad \beta > 0 \tag{1}$$

and the following cumulative distribution function:

$$\mathcal{F}(k) = 1 - (1-p)^{k^{\beta}}. \quad (2)$$

The latter describes the probability that the random variable takes on a value greater than k. The discrete Weibull distribution can be considered as a generalisation of the geometric distribution with probability p to which it conforms if $\beta = 1$, i.e., if the probability $\mathcal{F}(k)$ is constant in time. If $\beta > 1$, the probability increases with time while the opposite is true for $\beta < 1$. The discrete Weibull distribution is used in various fields, including survival analysis, weather forecasting, and study of textual data [33,35,36]. In the case of punctuation, $f(k)$ is the probability that a mark will occur exactly after k words.

Self-similarity or a lack of characteristics scale is among the most significant properties of natural complex systems. From an empirical perspective, this property can manifest itself in a non-trivial temporal organisation of measurement outcomes formed as a time series. In particular, complexity is often associated with a cascade-like hierarchy of data points that reveal multiscaling, which makes practical methods of identifying this sort of organisation a valuable tool in complex systems research [37]. Our experience so far allows us to consider the multifractal detrended fluctuation analysis (MFDFA) as the most reliable method in this respect [38,39]. This method constitutes a multiscale generalisation of an earlier, commonly used detrended fluctuation analysis (DFA) [40].

Here, the essential steps of the MFDFA algorithm are briefly sketched. Let $U = \{u_i\}_{i=1}^{T}$ be a time series of T consecutive measurements of some observable u. We partition it into M_s non-overlapping windows of length s starting from both ends of U, which gives us $2M_s$ such windows. In each window, we eliminate potential non-stationarity of the signal by applying a detrending procedure to an integrated signal (the so-called profile) $X = \{x_i\}_{i=1}^{s}$ whose elements have the following form:

$$x_i = \sum_{j=1}^{i} u_j. \quad (3)$$

The detrending is carried out with the help of a polynomial $P^{(m)}$ of order m (we use $m = 2$ throughout this study) that is best-fitted to X in each window $\nu = 0, \ldots, 2M_s - 1$ and the variance of the resulting detrended signal is then calculated:

$$f^2(\nu, s) = \frac{1}{s} \sum_{i=1}^{s} (x_i - P^{(m)}(i))^2. \quad (4)$$

A family of fluctuation functions of order q is defined in the next step based on the average variance across all the windows:

$$F_q(s) = \left\{ \frac{1}{2M_s} \sum_{\nu=0}^{2M_s-1} \left[f^2(\nu, s) \right]^{q/2} \right\}^{1/q}, \quad (5)$$

where q is a real number. The functions $F_q(s)$ have to be calculated for a number of different values of scale s and index q. Typically, minimum s is chosen above the length of the longest sequence of constant values of U and maximum s is $T/5$. In contrast, there is no typical range of q. This index can be viewed as being related to the moments of the signal, so the extreme values cannot be chosen as too large for a time series with heavy tails.

If the fluctuation functions depend on s as power laws:

$$F_q(s) \sim s^{h(q)} \quad (6)$$

for a number of different choices of q, it indicates that the time series under study is either monofractal (when $h(q)$ is constant in q) or multifractal otherwise. The function $h(q)$ is called the generalised Hurst exponent, because for $q = 2$, $h(q) = H$, where H is the standard Hurst exponent [41,42]. In terms of visualisation, fractal $F_q(s)$ form linear plots on double logarithmic plots. A convenient way of presenting the multifractal property of data is singularity spectrum $f(\alpha)$. It can be derived from $h(q)$ by the following formulae:

$$\alpha = h(q) + qh'(q),$$
$$f(\alpha) = q[\alpha - h(q)] + 1, \tag{7}$$

where α is a measure of data-point singularity equivalent to the Hölder exponent and $f(\alpha)$ is the Legendre transform of $h(q)$. The function $f(\alpha)$ can be interpreted geometrically as a fractal dimension of the subset of the whole data set with the Hölder exponent equal to α [43]. For a monofractal time series the pair $(\alpha, f(\alpha))$ is a single point, while for a multifractal one it is a parabola with shoulders pointing down. The usefulness of this representation comes from a fact that the broader the singularity spectrum is, the richer is the multifractality of the time series, which can be viewed as a measure of time series complexity. Sometimes the parabola of $f(\alpha)$ is distorted and asymmetric, which indicates that the data points of different amplitude have different scaling properties [44–47]. Alternatively, the scaling type may be expressed in terms of the multifractal spectrum $\tau(q)$ defined by

$$\tau(q) = qh(q) - 1. \tag{8}$$

For monofractal time series, $\tau(q)$ depends linearly on q (because $h(q)$ is constant then), while it is nonlinear for multifractal ones, which allows one for an easy detection of multiscaling.

3. Results and Discussion

3.1. Linear Correlations

It is instructive to look into data structure before any further step is made. In the left column of Figure 1, time series of SLV in three different language versions of the text are shown, each of them representing an unsigned process with a heavy tail. The SLV time series representing the text with two alternative chapter orders, the one recommended by the author and a completely random one, are shown in the right column (top and middle panels) of the same figure. In both cases, there is a clear data clustering observed (long sentences tend to group with the long ones, while small sentences tend to group with the small ones), the time series are characterised by memory. This by no means differs from other pieces of literature where temporal memory is naturally present. It stems from the fact that sentence lengths are often connected with writing style and certain features of narrative, which usually last for some paragraphs or pages (for instance, longer sentences are often used in descriptions and slow-paced parts of texts, while short sentences are more typical for fast-paced parts and dialogues). This property vanishes if the sentences are shuffled instead of the chapters. Plots of the Pearson autocorrelation function (ACF) for the two authorial chapter orders are presented in Figure 2. ACF shows statistically meaningful values over a range of up to 200–400 consecutive sentences, after which it reaches noise level in both chapter orders. A trace of power-law decay of ACF can be noticed because of the double-log scale of the plots. Roughly the same picture is obtained if a random permutation of chapters is considered. This means that there is no difference between the proposed chapter orders and any other order that can be preferred by a reader (it must be noted, however, that it is impossible to consider all possible orderings as we deal with $155! \approx 10^{273}$ possible permutations here). As it might be expected, the sentence-level randomisation kills any genuine structure of SLV.

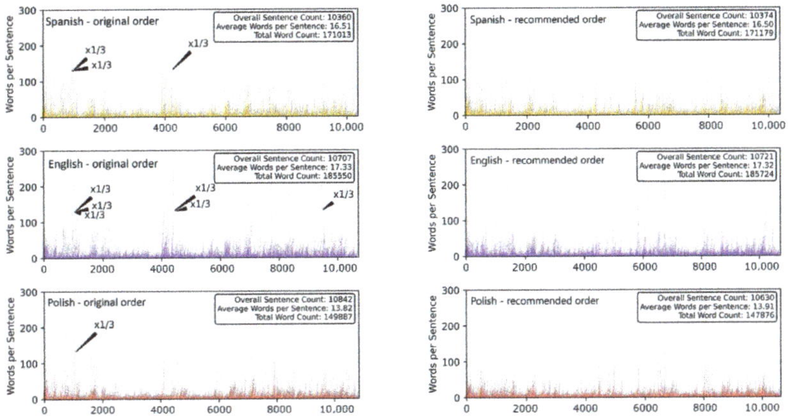

Figure 1. Time series of sentence length variability (SLV) measured in words for the original Spanish text of *Hopscotch* (**top**) and its translations into English (**middle**) and Polish (**bottom**). In order to ensure sufficient readability and preserve the vertical scale in all panels, the bars representing a few sentences with excessive length in each text have been rescaled by a factor of 1/3 (black arrows). Time series of SLV are presented in alternative orders of chapters: the printed order (**left**) and the order recommended by the author (**right**).

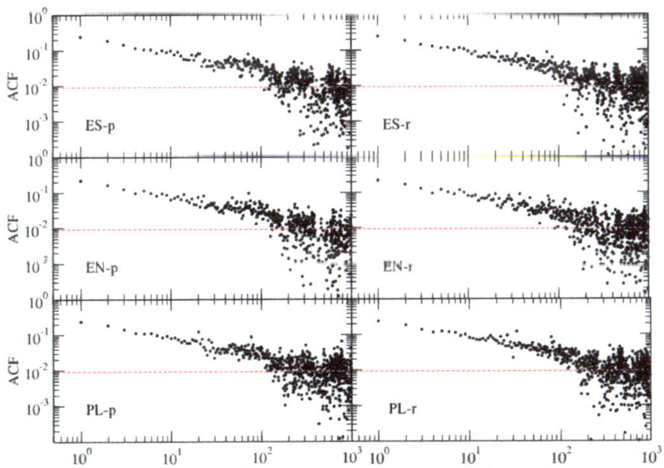

Figure 2. Autocorrelation function for SLV time series for two chapter orders: the printed order ("-p", **left**) and the recommended order ("-r", **right**), and for three languages: Spanish (ES, **top**), English (EN, **middle**), and Polish (PL, **bottom**). Noise level is indicated by a red dashed line in each panel. Note the double logarithmic scale of the plots.

3.2. The Weibull Analysis

For all three language versions of *Hopscotch*, histograms for the respective SLV time series have been calculated; they are shown in Figure 3 (dark colours) together with the best-fitted discrete Weibull distribution defined by Equation (1). It is evident that there is no agreement between theory and the empirical distribution in each case—the result that matches results of similar analyses of different texts already reported in the literature [3,25]. A much better agreement between the two can be seen for a different type of time series (punctuation mark distance variability, PMDV), in which distances between consecutive punctuation marks have been considered without restricting the analysis to sentence endings. Thus, if also such marks as comma, colon, semicolon, dash, etc. are included in the study, the discrete Weibull distribution becomes a better model for the data (see light-colour histograms in Figure 3). For all the language versions, the values of β for the PMDV time series exceed 1, which can be viewed as the normal case for written language, in which the probability of using a punctuation mark after a given number of words since the previous instance of writing a mark (the hazard function) increases with this number. It distinguishes *Hopscotch* from *Finnegans Wake* by James Joyce, where an abnormal value $\beta < 1$ has been found [48]. In contrast, for the SLV data, $\beta \approx 1$ has been obtained for two out of three languages; it corresponds to the exponential distribution of such data. This is better seen if the plots are semi-logarithmic (see the insets in Figure 3), where the exponential decay of the histogram tails and the corresponding fits are represented by straight lines.

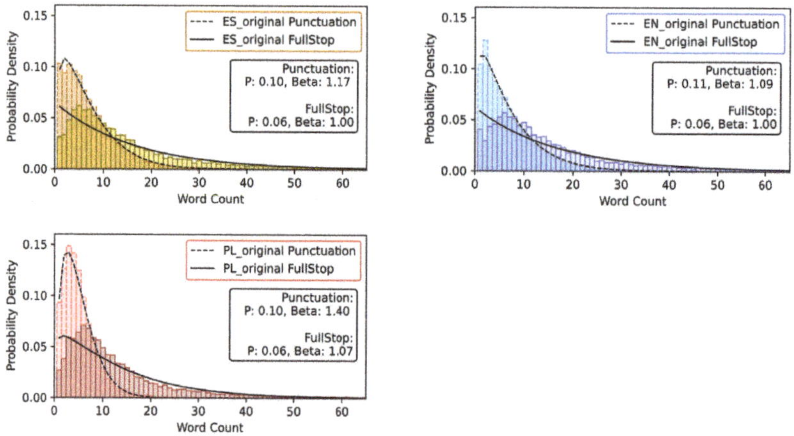

Figure 3. Histograms of the SLV time series (dark colour) for three different language versions of *Hopscotch*: Spanish (**top left**), English (**top right**), and Polish (**bottom**) together with the histograms of the distances (in words) between any two consecutive punctuation marks (both the sentence-ending ones and the intra-sentence ones, light colour). Discrete Weibull distributions that have been least-square fitted to the histograms of both types are also shown in each panel (dashed and solid lines, respectively). In each case, values of the p and β parameters of the fits are explicitly given in legend boxes. (Insets) Histogram of the SLV time series with the fitted discrete Weibull distribution presented on the half-logarithmic scale.

3.3. Generalised Hurst Exponents

The classic Hurst analysis of time series allows one to quantify linear autocorrelation, as it expresses how the observed value span depends on the length of observation [41]. In its generalised version $h(q)$, it allows to detect differences in memory effects between fluctuations of different magnitudes [49]. It naturally enters the multiscale analysis through Equation (6). $h(q)$ is a non-increasing function of its argument with the special case being a constant function $h(q) = H$ for all qs. In a practical situation, the range of considered values of q must be limited and dependent on the probability distribution function of the underlying stochastic process. Typically, the reference is that the qth moment of this distribution has to exist. In the present case, the related probability mass function is close to exponential, therefore no specific theoretical limit for q has to be applied. Because of this freedom, the parameter q has been limited to $-7 \leq q \leq 7$, which is a reasonable compromise between the range width, which should be as large as possible, and the available sample size. Figure 4 shows the generalised Hurst exponents for the three languages and the two authorial chapter orders. In each case $h(q)$ is a decreasing function which indicates that the time series are heterogeneous in their scaling behaviour for different qs. The larger the variability of $\Delta h = h(q_{\max}) - h(q_{\min})$ is, the more heterogeneous they are. However, as $0.37 \leq \Delta h \leq 0.47$ the difference between the cases is more quantitative than qualitative. It is not surprising, because, on the one hand, by changing the order of chapters, one effectively preserves the autocorrelation range up to the average number of sentences per chapter and, on the other hand, good translations into foreign languages are expected to preserve the structure of the original language version as closely as possible.

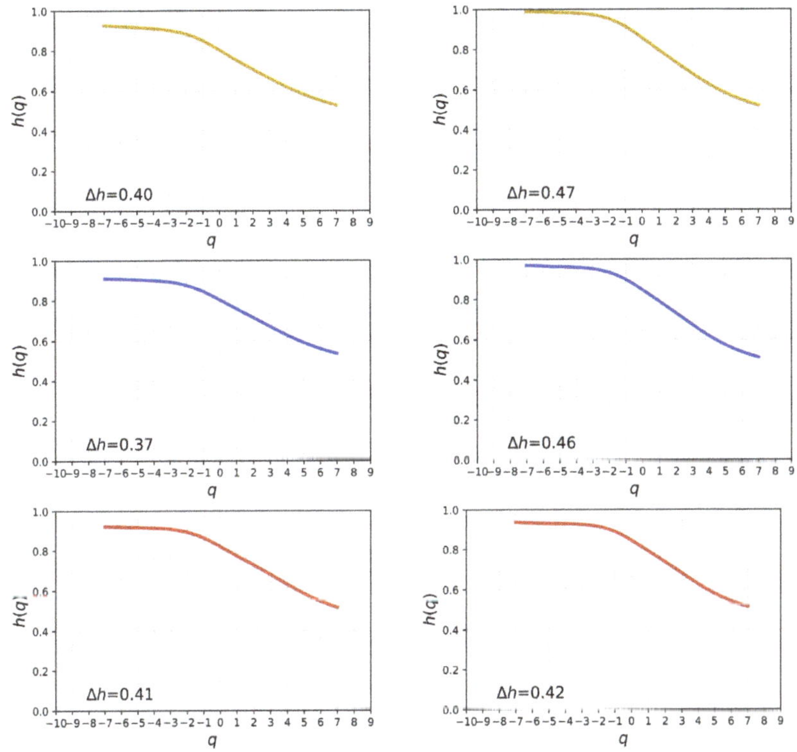

Figure 4. The generalised Hurst exponent $h(q)$ for the SLV time series representing *Hopscotch* in different languages: Spanish (**top**), English (**middle**), and Polish (**bottom**). The printed order (**left**) and the recommended order (**right**) of chapters are shown separately. The difference between extreme values of $h(q)$ for $-7 \leq q \leq 7$ is given by Δh in each case.

3.4. Multifractal Analysis

The family of the generalised Hurst exponents $h(q)$ is closely related to the Hölder exponents α and the singularity spectrum $f(\alpha)$ via Equation (7). From the perspective of the amount of information that one can extract from the data, the $f(\alpha)$ representation is more natural and convenient than $h(q)$, especially if a fractal analysis is carried out. However, both $h(q)$ and $f(\alpha)$ are of little use without inspecting of the fluctuation functions $F_q(s)$ first. This is because they carry information about the possible existence of fractal scaling over a sufficient range of scales s, which is a crucial characteristic of the data. Therefore, Figures 5–7 show the family of $F_q(s)$ calculated in the range $-7 \leq q \leq 7$ for different languages and different chapter orders.

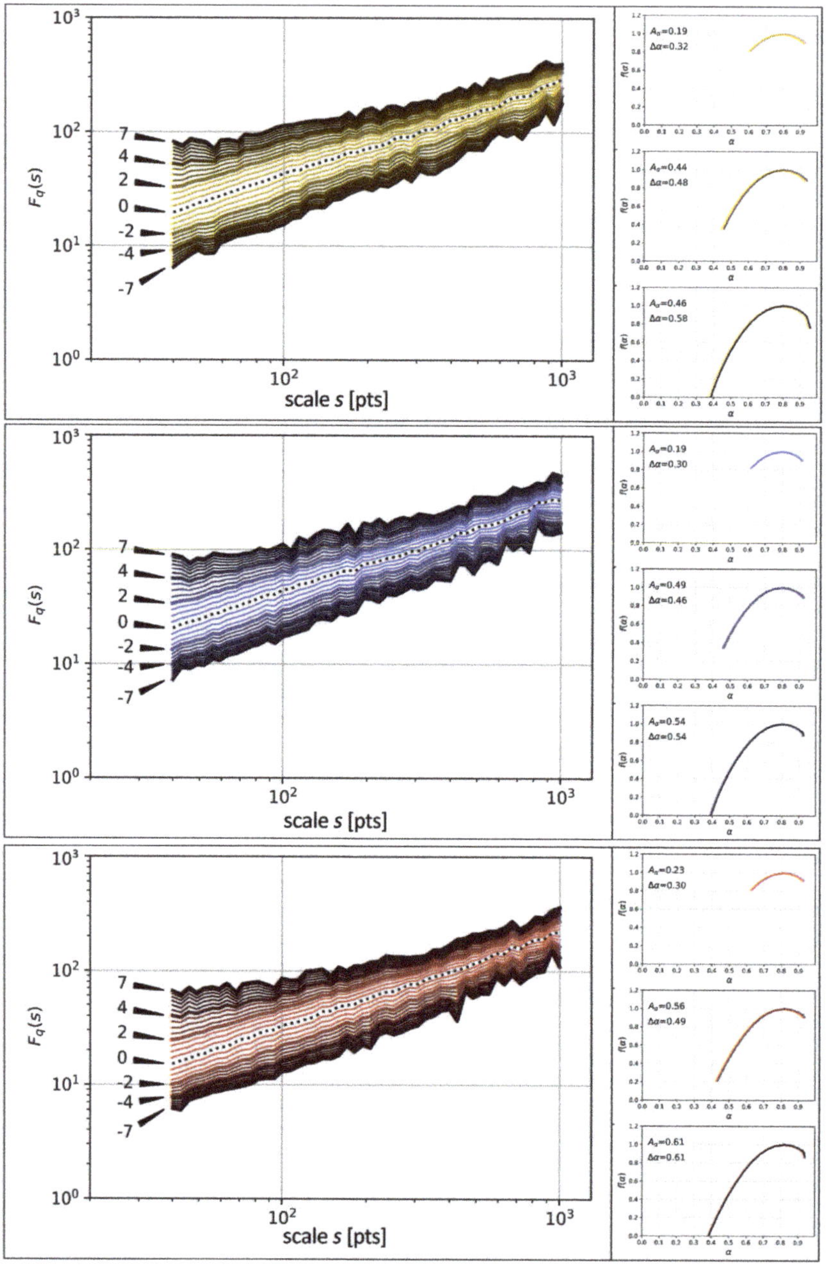

Figure 5. (Main) q-dependent fluctuation functions $F_q(s)$ calculated for time series of SLV in the Spanish original text of *Hopscotch* (**top**) and its two translations into English (**middle**) and Polish (**bottom**). The printed order of chapters is considered. The plots of $F_q(s)$ for particular values of q are indicated by arrows. (Side) Singularity spectra $f(\alpha)$ associated with the exponents $h(q)$ calculated from the scaling regions of the respective functions $F_q(s)$ for three intervals: $q \in [-2, 2]$ (**top right**), $q \in [-4, 4]$ (**middle right**), and $q \in [-7, 7]$ (**bottom right**). In each case the asymmetry coefficient A_α and the singularity spectrum width $\Delta\alpha$ are also given.

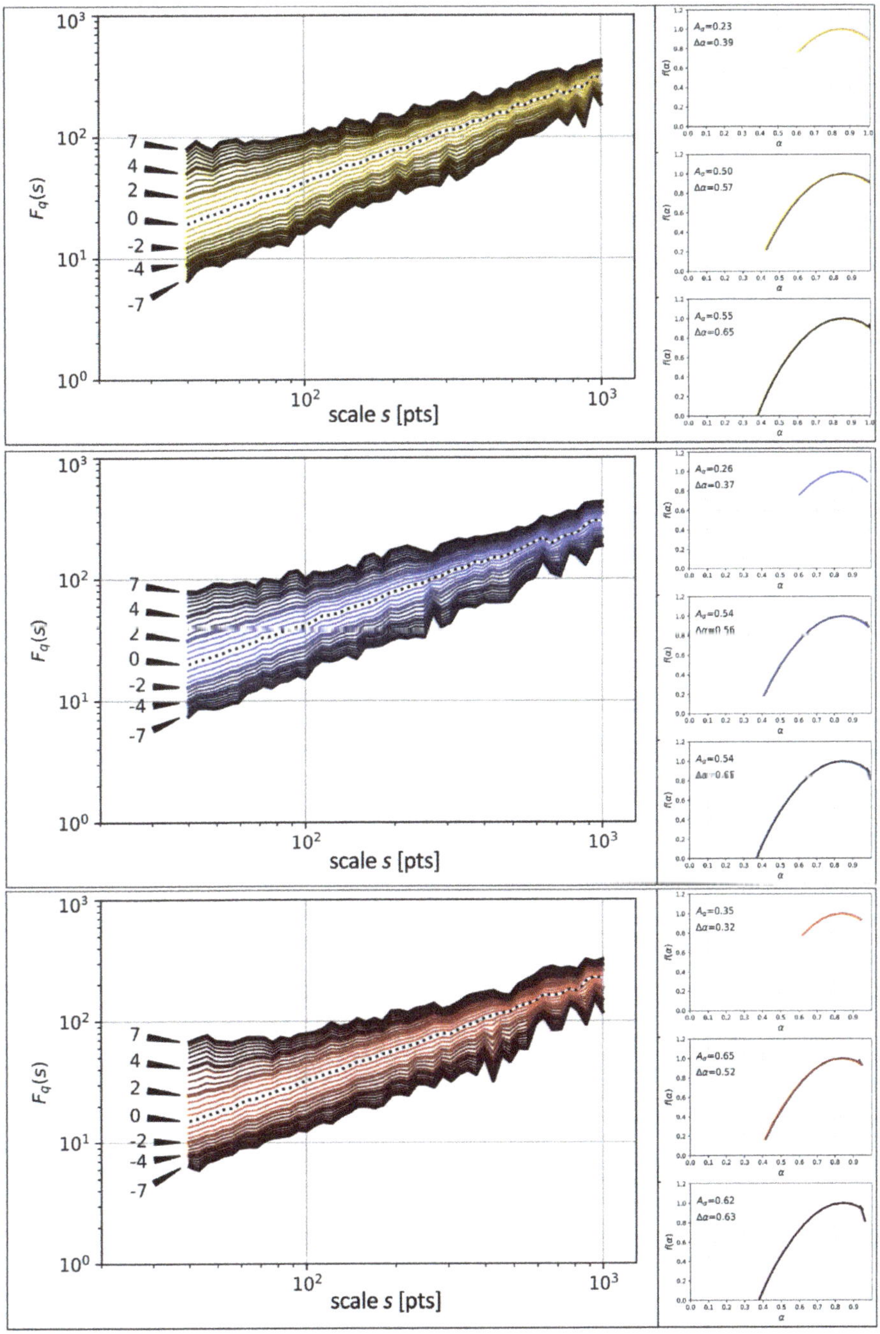

Figure 6. The same functions as in Figure 5 for the recommended order of chapters.

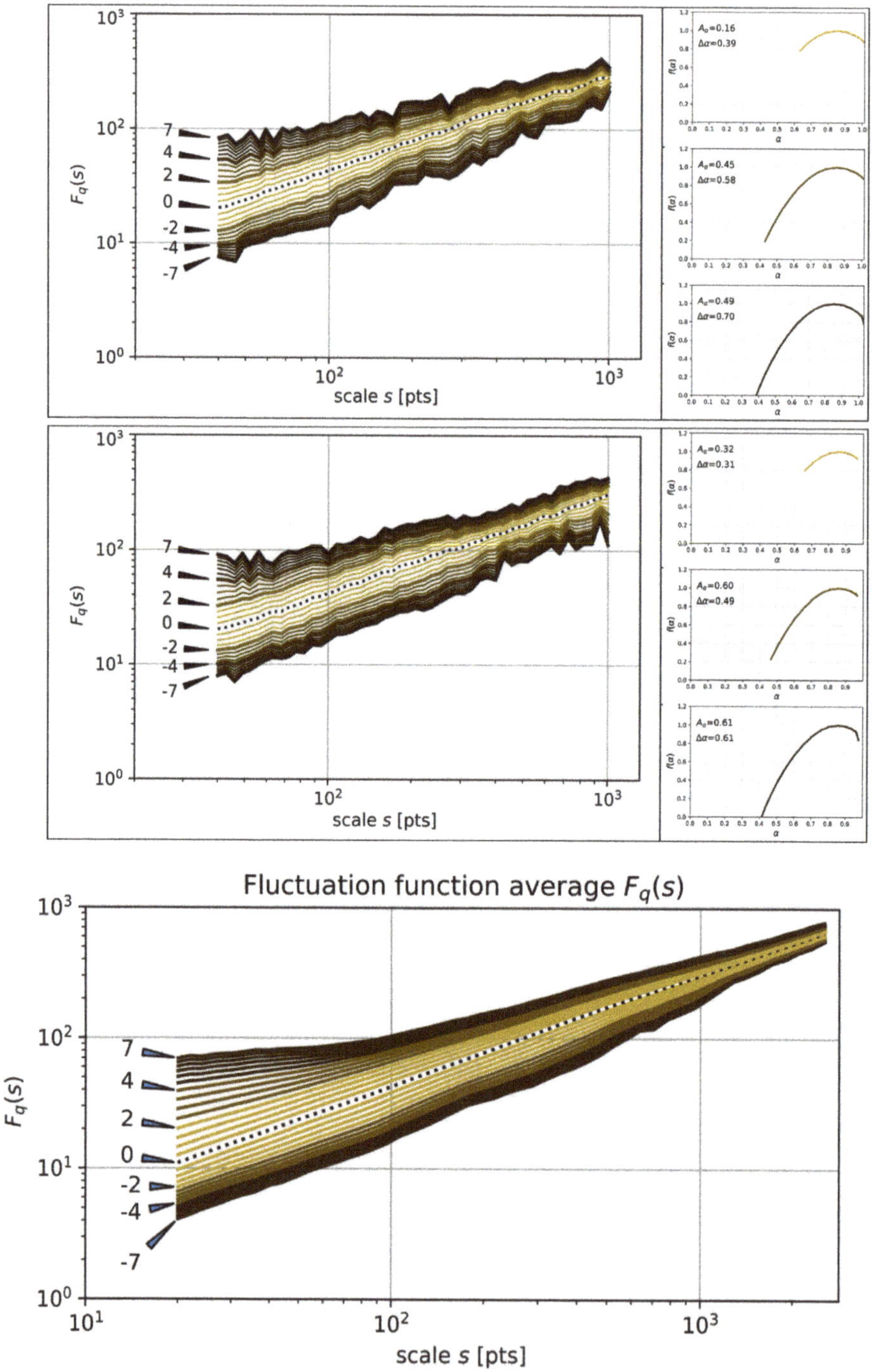

Figure 7. The same functions as in Figure 5 for the Spanish version of *Hopscotch* with two random permutations of chapters (**top** and **middle**) and the average taken over a set of 1000 sample permutations (**bottom**).

Let one start with the printed order of chapters; the corresponding fluctuation functions for different values of q are displayed in Figure 5 (main panels) for the three languages considered in this work. All the functions show a power-law dependence (approximately straight lines on double logarithmic plots) over some range of scales, whose right limit reaches typically 300–400 consecutive sentences for both positive and negative qs. The singularity spectra derived from the scaling exponents $h(q)$ for different intervals of q are shown in the side panels of each main panel of Figure 5. The spectra have their maximum at $\alpha \approx 0.8$, which points out to a strong persistence in SLV. By going through these side panels from top to bottom, the range of the values of q gradually extends from $-2 \leq q \leq 2$ to $-7 \leq q \leq 7$. The rationale behind considering different spans of q is that there is a certain degree of asymmetry in the density of the lines related to the functions $F_q(s)$ between positive and negative qs in the main panels. Indeed, if one looks at the values of the asymmetry index A_α for the $f(\alpha)$ spectra, it assumes a minimum value for the most narrow range of q and it increases with the increasing $|q|$. This observation is valid for all the three languages even though particular values of A_α can differ from each other for the equivalent intervals of q in different languages. The asymmetry is left-sided, which means that the multiscaling comes predominantly from large fluctuations in SLV. This type of asymmetry is more often seen in empirical data than their opposite counterpart [46]. As regards the widths $\Delta\alpha$ of the $f(\alpha)$ spectra, they increase with $|q|$, which is a normal behaviour in such a context, but their value indicates that the SLV time series under study show the multifractal scaling even for the most narrow range of q. This result has been expected, since one of our past studies focused on books with a stream-of-consciousness narrative [32].

Now, the order of chapters can be altered to agree with the one recommended by Cortázar. This is his "hopscotch" order, which allows a reader for viewing the book's plot from a different perspective. The respective results in terms of $F_q(s)$ and $f(\alpha)$ are documented in Figure 6, whose structure is the same as the structure of Figure 5. By altering the order, one destroys the temporal structure of the SLV time series for scales above the average length of a chapter expressed in the number of sentences, while the structure on shorter time scales remains largely unaltered. This leads to a visible significant distortion of the behaviour of $F_q(s)$ for the scales $s \approx 10^3$ with respect to the printed order of chapters, which tends to be inclining towards monofractality there (a narrow beam of almost parallel lines). In contrast, for the scales inside the chapters, the changes are less evident. By looking at the singularity spectra obtained for the recommended order, one observes a systematic increase in both the asymmetry A_α and the width $\Delta\alpha$ if compared to the results for the printed order. This effect suggests an existence of a richer variety of the singularity strengths expressed by the Hölder exponents α in the considered time series. The origin of this behaviour remains unclear at the present stage of research, however. Together with the increased left–right asymmetry of the spectra, a slight shift of the maximum of $f(\alpha)$ in the direction of larger values of α can be identified for the recommended order ($\alpha \approx 0.85$). This seems to be a systematic effect for the three languages. Finally, the change in A_α is related to this shift, because together with the unchanged value of $\alpha_{\min}$ they cause the elongation of the left branch of $f(\alpha)$, while the length of the right branch remains roughly the same in both situations despite the fact that $\alpha_{\max}$ is larger for the recommended order than it is for the printed order.

It is interesting to investigate whether such effects can also be observed for any random permutation of chapters. Figure 7 shows the results of the multifractal analysis of the SLV data corresponding to such permutations. The results for two individual permutations are shown there together with the average results for 1000 different random permutations. Only the results for the original Spanish version of the text are shown; the remaining two language versions are characterised by quantitatively similar results. Randomisation of the chapter order brings results that do not differ much from the results for the recommended order: for both individual permutations, the spectrum width $\Delta\alpha$ is elevated with respect to its value for the printed order (0.65 and 0.70 vs. 0.56) but it is comparable to the value for the recommended order (0.65 and 0.70 vs. 0.65). Regarding the asymmetry index A_α, the situation is not so evident as in the case of $\Delta\alpha$, because the results for one of the random permutations and for the printed order are close to each other (0.49 vs. 0.46), even though a larger discrepancy is also possible (0.61 vs. 0.46). These numbers refer to the maximum span of q ($-7 \leq q \leq 7$). It is also instructive to compare the results for individual chapter orders with the average over many random permutations (see Figure 7 (bottom)).

The significance of the results reported above has also been tested against the null hypothesis of no correlation. This has been performed by a conventional surrogate testing [50], in which the SLV time series are randomised at the individual sentence level, which destroys memory. Results show that in this case the singularity spectra are shifted to smaller values of α and the maximum of $f(\alpha)$ is located near $\alpha = 0.5$. Even though $\Delta\alpha$ for such surrogate data remains as high as 0.2, the inherent properties of the MFDFA procedure suggest that $f(\alpha)$ has been broadened spuriously by the finite-size effects related to the heavy-tailed probability distribution functions of the SLV data [51,52]. Another surrogate testing based on the Fourier-phase randomisation of time series, which destroys all nonlinear correlations [53], brings the expected results, i.e., the singularity spectra become point-like and located exactly at $\alpha \approx 0.5$. Both types of surrogate tests confirm the statistical validity of the presented outcomes.

4. Summary

The time series of SLV representing Julio Cortázar's novel *Hopscotch* in its original Spanish version and in two foreign-language translations were analysed by means of the autocorrelation function, the Weibull analysis, and the MFDFA. Long-range memory effects were observed in terms of a power-law decay of ACF that extended beyond the scales defined by the average chapter length for the printed order of chapters and for other possible chapter orders, including the one recommended by the author. The fluctuations of the SLV time series were distributed in agreement with an exponential distribution, at least to a certain extent (small fluctuations somehow deviated from this pattern). For a comparison, the time series of PMDV, where all punctuation marks were considered, were modelled by the discrete Weibull distribution with $\beta > 1$—the value that did not differ from its standard values for written texts. Finally, by using the MFDFA formalism, the fractal properties of the SLV time series were studied. They happened to be multifractal for both the printed and recommended orders, as well as for all the considered random orders. The main difference between the results obtained for different chapter orderings was the observation that all the conceptual orders (i.e., the non-printed ones) showed an enhanced variety of singularities (i.e., broader singularity spectra) present in the time series and expressed by the Hölder exponents. They also developed stronger left-sided asymmetry as compared with their counterparts for the printed order of chapters, the effect produced by a subtle but noticeable shift of the maxima of the spectra towards larger values of the Hölder exponent. It has to be stressed that multifractal structures in literary works are a rare property that have been identified only in a small fraction of texts [3,32]. Interestingly, this and the other considered properties seem to be invariant under translation into foreign languages if a translator pays sufficient attention to the structure of the original (see also [21,25,26]).

The results obtained in this analysis indicate that the printed order of chapters in *Hopscotch* cannot be distinguished statistically if compared to either the recommended order or a generic order obtained via a random permutation of chapters. Also, we do not observe any qualitative difference between the recommended order and generic orders, which leads us to a conclusion that neither the printed order nor the recommended order introduce significant temporal correlations on scales above the average number of sentences in a chapter, at least such correlations that could be identified with the methodology applied here. From this perspective, each chapter constitutes a largely independent block whose position in the narrative can be arbitrarily changed without changing the statistical properties of SLV.

If the results of this study are compared with the earlier results for *Finnegans Wake* by J. Joyce, one observes that *Hopscotch* is more conventional not only in terms of literary style but also in terms of the statistical properties [3,32,48]. Its role in the world's literature stems from its innovative construction including the nonlinear and reader-engaging narrative rather than the structural and lingual complexity that is so striking in *Finnegans Wake* or *Ulysses*. Nevertheless, the results of this work provide an inspiration for more systematic research of the most prominent and unique literary works, as these are the ones that are potent to catalyze opening new perspectives in linguistics, including those in the field of large language models. In the latter context, it is important that the methodology presented here allows for the identification and quantification of various types of correlations in written texts, including long-range ones. In the terminology of nonlinear dynamics this means a reduction in the effective dimensionality of the corresponding 'phase space' and therefore a reduction in the actual number of parameters involved. In LLMs, nowadays commonly based on neural networks, such facts can be exploited to significantly, and perhaps even crucially, reduce the number of applied parameters and thus increase their efficiency in terms of time and energy consumption.

A related prospective direction of future research, somehow parallel yet substantially distinct from the multifractal approach discussed here, is a dynamically oriented analysis of SLV, in which one looks for low-dimensional attractors in a reconstructed phase-space of "text-writing dynamics". Apart from some early attempts [11,54], this direction still remains largely unexplored.

Author Contributions: Conceptualization, J.D., M.D., S.D. and T.S.; Methodology, S.D. and J.K.; Software, J.D., M.D., J.K. and T.S.; Validation, J.D., M.D., S.D., J.K. and T.S.; Formal analysis, J.D., M.D., S.D. and J.K.; Investigation, J.D., M.D., S.D., J.K. and T.S.; Resources, J.D., M.D. and T.S.; Data curation, J.D., M.D. and T.S.; Writing—original draft, J.K.; Writing—review & editing, S.D., J.K. and T.S.; Visualization, J.D. and M.D.; Supervision, S.D. All authors have read and agreed to the published version of the manuscript.

Funding: This research received no external funding.

Institutional Review Board Statement: Not applicable.

Data Availability Statement: The data presented in this study are available on request from the corresponding author. The data are not publicly available due to copyright.

Conflicts of Interest: The authors declare no conflicts of interest.

References

1. Hébert-Dufresne, L.; Allard, A.; Garland, J.; Hobson, E.A.; Zaman, L. The path of complexity. *NPJ Complex.* **2024**, *1*, 4. [CrossRef]
2. Kwapień, J.; Drożdż, S. Physical approach to complex systems. *Phys. Rep.* **2012**, *515*, 115–226. [CrossRef]
3. Stanisz, T.; Drożdż, S.; Kwapień, J. Complex systems approach to natural language. *Phys. Rep.* **2024**, *1053*, 1–84. [CrossRef]
4. Dębowski, L. *Information Theory Meets Power Laws: Stochastic Processes and Language Models*; Wiley: Hoboken, NJ, USA, 2020. [CrossRef]
5. Takahira, R.; Tanaka-Ishii, K.; Dębowski, L. Entropy Rate Estimates for Natural Language—A New Extrapolation of Compressed Large-Scale Corpora. *Entropy* **2016**, *18*, 364. [CrossRef]
6. Montemurro, M.A.; Zanette, D.H. Universal Entropy of Word Ordering Across Linguistic Families. *PLoS ONE* **2011**, *6*, e19875. [CrossRef]
7. Alvarez-Lacalle, E.; Dorow, B.; Eckmann, J.P.; Moses, E. Hierarchical structures induce long-range dynamical correlations in written texts. *Proc. Natl. Acad. Sci. USA* **2006**, *103*, 7956–7961. [CrossRef]
8. Liu, J.; Gunn, E.; Youssef, F.; Tharayil, J.; Lansford, W.; Zeng, Y. Fractality in Chinese prose. *Digit. Scholarsh. Humanit.* **2023**, *38*, 604–620. [CrossRef]
9. Sánchez, D.; Zunino, L.; Gregorio, J.D.; Toral, R.; Mirasso, C. Ordinal analysis of lexical patterns. *Chaos* **2023**, *33*, 033121. [CrossRef]
10. Pawłowski, A. Time-Series analysis in linguistics: Application of the ARIMA method to cases of spoken Polish. *J. Quant. Linguist.* **1997**, *4*, 203–221. [CrossRef]
11. Kosmidis, K.; Kalampokis, A.; Argyrakis, P. Language time series analysis. *Physica A* **2006**, *370*, 808–816. [CrossRef]
12. Cancho, R.F.I.; Solé, R.V. The small world of human language. *Proc. R. Soc. Lond. Ser. B Biol. Sci.* **2001**, *268*, 2261–2265. [CrossRef]

13. Liu, H.; Xu, C. Can syntactic networks indicate morphological complexity of a language? *Europhys. Lett.* **2011**, *93*, 28005. [CrossRef]
14. Cong, J.; Liu, H. Approaching human language with complex networks. *Phys. Life Rev.* **2014**, *11*, 598–618. [CrossRef] [PubMed]
15. Wachs-Lopes, G.A.; Rodrigues, P.S. Analyzing natural human language from the point of view of dynamic of a complex network. *Expert Syst. Appl.* **2016**, *45*, 8–22. [CrossRef]
16. Kulig, A.; Kwapień, J.; Stanisz, T.; Drożdż, S. In narrative texts punctuation marks obey the same statistics as words. *Inf. Sci.* **2017**, *375*, 98–113. [CrossRef]
17. Stanisz, T.; Kwapień, J.; Drożdż, S. Linguistic data mining with complex networks: A stylometric-oriented approach. *Inf. Sci.* **2019**, *482*, 301–320. [CrossRef]
18. Raducha, T.; Gubiec, T. Predicting language diversity with complex networks. *PLoS ONE* **2018**, *13*, e0196593. [CrossRef] [PubMed]
19. Naranan, S.; Balasubrahmanyan, V. Models for power law relations in linguistics and information science. *J. Quant. Linguist.* **1998**, *5*, 35–61. [CrossRef]
20. Newman, M. Power laws, Pareto distributions and Zipf's law. *Contemp. Phys.* **2005**, *46*, 323–351. [CrossRef]
21. Ausloos, M. Punctuation effects in english and esperanto texts. *Physica A* **2010**, *389*, 2835–2840. [CrossRef]
22. Piantadosi, S.T. Zipf's word frequency law in natural language: A critical review and future directions. *Psychon. Bull. Rev.* **2014**, *21*, 1112–1130. [CrossRef] [PubMed]
23. Shanahan, M.; McDonell, K.; Reynolds, L. Role play with large language models. *Nature* **2023**, *623*, 493–498. [CrossRef] [PubMed]
24. Zhao, W.X.; Zhou, K.; Li, J.; Tang, T.; Wang, X.; Hou, Y.; Min, Y.; Zhang, B.; Zhang, J.; Dong, Z.; et al. A Survey of Large Language Models. *arXiv* **2023**, arXiv:2303.18223v13.
25. Stanisz, T.; Drożdż, S.; Kwapień, J. Universal versus system-specific features of punctuation usage patterns in major Western languages. *Chaos Solitons Fractals* **2023**, *168*, 113183. [CrossRef]
26. Dec, J.; Dolina, M.; Drożdż, S.; Kwapień, J.; Liu, J.; Stanisz, T. Quantifying patterns of punctuation in modern Chinese prose. 2024, *submitted*.
27. Bernstein, J. Rayuela, Chapter 34: A structural reading. *Hispanófila* **1974**, *52*, 61–70. [CrossRef]
28. Simpkins, S. "The Infinite Game": Cortázar's "Hopscotch". *J. Midwest Mod. Lang. Assoc.* **1990**, *23*, 61–74. [CrossRef]
29. Cortázar, J. *Rayuela*; Sudamericana: Buenos Aires, Argentina, 1963.
30. Cortázar, J. *Hopscotch*; Rabassa, G., Translator; Pantheon Books: New York, NY, USA, 1966.
31. Cortázar, J. *Gra w Klasy*; Chądzyńska, Z., Translator; Wydawnictwo Literackie: Kraków. Poland, 1985.
32. Drożdż, S.; Oświęcimka, P.; Kulig, A.; Kwapień, J.; Bazarnik, K.; Grabska-Gradzińska, I.; Rybicki, J.; Stanuszek, M. Quantifying origin and character of long-range correlations in narrative texts. *Inf. Sci.* **2016**, *331*, 32–44. [CrossRef]
33. Miller, R. *Survival Analysis*; John Wiley & Sons: Hoboken, NJ, USA, 1997.
34. Nakagawa, T.; Osaki, S. The discrete Weibull distribution. *IEEE Trans. Reliab.* **1975**, *R-24*, 300–301. [CrossRef]
35. Johnson, N.L.; Kotz, S.; Balakrishnan, N. *Continuous Univariate Distributions*; Wiley-Interscience: Hoboken, NJ, USA, 1994.
36. Altmann, E.G.; Pierrehumbert, J.B.; Motter, A.E. Beyond Word Frequency: Bursts, Lulls, and Scaling in the Temporal Distributions of Words. *PLoS ONE* **2009**, *4*, e7678. [CrossRef]
37. Jimenez, J. Intermittency and cascades. *J. Fluid Mech.* **2000**, *409*, 99–120. [CrossRef]
38. Kantelhardt, J.W.; Zschiegner, S.A.; Koscielny-Bunde, E.; Havlin, S.; Bunde, A.; Stanley, H. Multifractal detrended fluctuation analysis of nonstationary time series. *Physica A* **2002**, *316*, 87–114. [CrossRef]
39. Oświęcimka, P.; Kwapień, J.; Drożdż, S. Wavelet versus detrended fluctuation analysis of multifractal structures. *Phys. Rev. E* **2006**, *74*, 016103. [CrossRef] [PubMed]
40. Peng, C.K.; Buldyrev, S.V.; Havlin, S.; Simons, M.; Stanley, H.E.; Goldberger, A.L. Mosaic organization of DNA nucleotides. *Phys. Rev. E* **1994**, *49*, 1685. [CrossRef]
41. Hurst, H.E. The long-term storage capacity of reservoir. *Trans. Am. Soc. Civ. Eng.* **1951**, *116*, 2447. [CrossRef]
42. Heneghan, C.; McDarby, G. Establishing the relation between detrended fluctuation analysis and power spectral density analysis for stochastic processes. *Phys. Rev. E* **2000**, *62*, 6103–6110. [CrossRef] [PubMed]
43. Halsey, T.C.; Jensen, M.H.; Kadanoff, L.P.; Procaccia, I.; Shraimant, B.I. Fractal measures and their singularities: The characterization of strange sets. *Phys. Rev. A* **1986**, *33*, 1141–1151. [CrossRef]
44. Ohashi, K.; Amaral, L.A.; Natelson, B.H.; Yamamoto, Y. Asymmetrical singularities in real-world signals. *Phys. Rev. E* **2003**, *68*, 065204. [CrossRef]
45. Cao, G.; Cao, J.; Xu, L. Asymmetric multifractal scaling behavior in the Chinese stock market: Based on asymmetric MF-DFA. *Physica A* **2013**, *392*, 797–807. [CrossRef]
46. Drożdż, S.; Oświęcimka, P. Detecting and interpreting distortions in hierarchical organization of complex time series. *Phys. Rev. E* **2015**, *91*, 030902. [CrossRef]
47. Gómez-Gómez, J.; Carmona-Cabezas, R.; Ariza-Villaverde, A.B.; Gutiérrez de Ravé, E.; Jiménez-Hornero, F.J. Multifractal detrended fluctuation analysis of temperature in Spain (1960–2019). *Physica A* **2021**, *578*, 126118. [CrossRef]
48. Stanisz, T.; Drożdż, S.; Kwapień, J. Statistics of punctuation in experimental literature—The remarkable case of *Finnegans Wake* by James Joyce. *Chaos* **2024**, 34, 083124. [CrossRef] [PubMed]
49. Monjo, R.; Meseguer-Ruiz, O. Review: Fractal Geometry in Precipitation. *Atmosphere* **2024**, *15*, 135. [CrossRef]

50. Theiler, J.; Prichard, D. Constrained-realization Monte-Carlo method for hypothesis testing. *Physica D* **1996**, *94*, 221–235. [CrossRef]
51. Drożdż, S.; Kwapień, J.; Oświęcimka, P.; Rak, R. Quantitative features of multifractal subtleties in time series. *Europhys. Lett.* **2009**, *88*, 60003. [CrossRef]
52. Kwapień, J.; Blasiak, P.; Drożdż, S.; Oświęcimka, P. Genuine multifractality in time series is due to temporal correlations. *Phys. Rev. E* **2023**, *107*, 034139. [CrossRef]
53. Schreiber, T.; Schmitz, A. Improved Surrogate Data for Nonlinearity Tests. *Phys. Rev. Lett.* **1996**, *77*, 635–638. [CrossRef]
54. Ausloos, M. Measuring complexity with multifractals in texts. Translation effects. *Chaos Solitons Fractals* **2012**, *45*, 1349–1357. [CrossRef]

entropy

MDPI

Article

Language Statistics at Different Spatial, Temporal, and Grammatical Scales

Fernanda Sánchez-Puig [1,2,3], Rogelio Lozano-Aranda [1,2], Dante Pérez-Méndez [2,4], Ewan Colman [2,5], Alfredo J. Morales-Guzmán [6], Pedro Juan Rivera Torres [2], Carlos Pineda [7] and Carlos Gershenson [2,8,9,10,*]

1 Facultad de Ciencias, Universidad Nacional Autónoma de México, Mexico City 04510, Mexico; fernanda@ciencias.unam.mx (F.S.-P.); rogelio97@ciencias.unam.mx (R.L.-A.)
2 Centro de Ciencias de la Complejidad, Universidad Nacional Autónoma de México, Mexico City 04510, Mexico; dante.perez.m@gmail.com (D.P.-M.); ecolman@exseed.ed.ac.uk (E.C.); pjr62@cam.ac.uk (P.J.R.T.)
3 Instituto de Fisica Interdisciplinar y Sistemas Complejos, Universidad de las Islas Baleares, 07122 Palma de Mallorca, Spain
4 Posgrado en Ciencias de la Computación, Universidad Nacional Autónoma de México, Mexico City 04510, Mexico
5 Roslin Institute, University of Edinburgh, Midlothian EH8 9YL, UK
6 MIT Media Lab, Cambridge, MA 02139, USA; alfredo.mo.gu@gmail.com
7 Instituto de Física, Universidad Nacional Autónoma de México, Mexico City 04510, Mexico; carlospgmat03@gmail.com
8 School of Systems Science and Industrial Engineering, Binghamton University, Binghamton, NY 13902, USA
9 Instituto de Investigaciones en Matemáticas Aplicadas y Sistemas, Universidad Nacional Autónoma de México, Mexico City 04510, Mexico
10 Santa Fe Institute, Santa Fe, NM 87501, USA
* Correspondence: cgg@binghamton.edu

Abstract: In recent decades, the field of statistical linguistics has made significant strides, which have been fueled by the availability of data. Leveraging Twitter data, this paper explores the English and Spanish languages, investigating their rank diversity across different scales: temporal intervals (ranging from 3 to 96 h), spatial radii (spanning 3 km to over 3000 km), and grammatical word ngrams (ranging from 1-grams to 5-grams). The analysis focuses on word ngrams, examining a time period of 1 year (2014) and eight different countries. Our findings highlight the relevance of all three scales with the most substantial changes observed at the grammatical level. Specifically, at the monogram level, rank diversity curves exhibit remarkable similarity across languages, countries, and temporal or spatial scales. However, as the grammatical scale expands, variations in rank diversity become more pronounced and influenced by temporal, spatial, linguistic, and national factors. Additionally, we investigate the statistical characteristics of Twitter-specific tokens, including emojis, hashtags, and user mentions, revealing a sigmoid pattern in their rank diversity function. These insights contribute to quantifying universal language statistics while also identifying potential sources of variation.

Keywords: Twitter; rank diversity; geolocalization; ngrams; language models; statistics; scaling laws; complexity

1. Introduction

Statistical linguistics has emerged as a significant area of study in the past century [1], with Zipf's law standing as a cornerstone in understanding the statistical distribution of rank–frequency distributions within languages. This law explains a fundamental relationship between the rank order and frequency of occurrence of words, asserting that the frequency of a word is inversely proportional to its rank when ranked by frequency.

In this context, the comparison between random text models and real texts has been pivotal, revealing that genuine texts comprehensively span the entire lexical spectrum irrespective of word length, thus underscoring the profound meaningfulness embedded within Zipf's law [2–6]. Concurrently, random text models have been posited as explanatory constructs for Zipf's law [7].

Several studies have been possible only recently because of the availability of large amounts of data (see the next Section). When enough data are available, not only statistical studies are possible for a single moment but also the dynamics of language, i.e., how language statistics change in time, are feasible as well. In this work, we investigate language usage dynamics across temporal, spatial, and grammatical scales using geolocated Twitter data.

Our aim is to determine the significance of each scale in language statistics, evaluating how changes in these scales influence rank dynamics.

Analyzing language change over time intervals enhances our understanding of language variations and its influencing factors. This method is valuable for examining geographical, grammatical, and temporal language variants, revealing how languages differ across different regions and periods.

Our research focuses on the differences of language usage at different time scales. By examining these temporal variations, we aim to uncover factors that drive linguistic change, contributing to a deeper understanding of language dynamics.

Smaller regions may exhibit less word diversity due to limited statistical sampling, and our analysis seeks to observe these differences across various scales. Despite a common language, conversation topics vary daily and regionally, and our statistical analyses effectively characterize these differences.

The work is mostly concerned with understanding what factors contribute to rank diversity, and the conclusions are mostly about this.

To our knowledge, this is the first study of its kind. Discovering the answers to our questions should be interesting, even if the results are not surprising. We believe that the importance of rank diversity (and other measures) will only become clear after multiple studies are conducted to evaluate its usefulness. However, we cannot assume its usefulness first and then interpret the data accordingly.

2. Related Work

Some studies have delved into the origins and evolution of language [8–11], leveraging the accessibility of vast datasets such as digitized books and the proliferation of social media. These datasets facilitate an in-depth exploration of language dynamics and evolutionary patterns over time [12–14].

Almodaresi et al. [15] have characterized various lexical distributions across different analytical levels, demonstrating their adherence to a log-normal distribution at both user and country levels.

Prior research has scrutinized language variations across time scales spanning years and centuries [12,16], offering valuable insights into diverse fields including lexicography, grammar evolution, collective memory, technology adoption, censorship, historical epidemiology, and the lifespan of commonly used words and phrases.

Other authors have conducted statistical studies showing the differences between Spanish and English. In [17], researchers analyzed large text samples from both languages and tested various methods to identify the language of a short text sample. The Kolmogorov–Smirnov Goodness-of-Fit test was found to be the most reliable, with a significance level of 0.0077 for a sample of around 107 characters (about 21 words). This research, as well as ours, supports the idea that each language has distinct statistical characteristics that enable identification from a short text sample.

Twitter data have been extensively utilized for analyzing language in various contexts such as sentiment analysis, topic analysis, (mis)information dissemination, and activity patterns [14,18–22]. Additionally, it has been instrumental in understanding global synchronization patterns [21], spreading mechanisms [20], and political polarization [23]. Leveraging metadata encompassing location, time, and text, we can analyze dynamics and geography across multiple scales. Previous research has identified variations in text usage and interaction mechanisms across different countries or cultures, including differences in the usage of URLs, hashtags, mentions, replies, and retweets [24,25]. Moreover, studies have shown that Twitter hashtags exhibit patterns consistent with the geographical landscape, reflecting the virtual space's representation of the physical world [26]. Similar investigations have been conducted on characterizing rank dynamics in the Chinese microblog Sina Weibo [27], focusing on metrics such as the time spent by hashtags on the list, the timing of their appearance, rank diversity, and ranking trajectories.

Our previous studies [26,28] have revealed a consistent pattern of changes in word usage across various languages, which we quantified using a metric called *rank diversity*. This metric, calculated by examining a corpus of words ranked by frequency within specific time intervals, captures the variability in the occupancy of word ranks over time. When a rank consistently features the same word, rank diversity is low, whereas if different words occupy the same rank across time intervals, rank diversity is high. By plotting rank diversity against rank, we can analyze temporal changes in word usage. Our previous research [28–30] has shown that the relationship between rank diversity and rank can be approximated by a sigmoid function with similar parameters observed across different languages.

Twitter data present unique characteristics that make it a compelling subject of study. Unlike books or longer written texts, tweets are limited to a small number of characters, making them an intriguing medium for exploring language usage and its statistical differences from longer texts. Additionally, Twitter provides a finer temporal resolution compared to physically published materials, and its geotagging feature allows for the examination of geographical variations in language use at very fine scales. Moreover, the social network nature of Twitter, including user interactions such as mentions and retweets, as well as trending topics, creates a distinct language ecosystem. Finally, Twitter offers a vast dataset suitable for conducting comprehensive statistical analyses.

In this study, we conducted an analysis of over 20 million geolocated tweets originating from eight different countries. Our investigation focused on examining rank diversity across a range of spatial, temporal, and grammatical scales. The primary aim was to evaluate how variations in these scales impact rank diversity and to elucidate their implications for language dynamics.

While previous research has highlighted the statistical similarities between punctuation marks and words, as discussed in [31] and related literature, our study took a different approach. Since Twitter users, especially from the time when there was a 144-character limit, tended to disregard punctuation marks. Still, this should not make a difference in our analysis, as we are not analyzing rank distributions (that could be described with Zipf's law) but rather rank dynamics. Thus, we concentrated solely on analyzing linguistic patterns involving Ngrams. This decision allowed us to delve deeper into the nuances of language dynamics across different scales without the confounding influence of punctuation. Moreover, this allowed us to compare directly results with a previous analysis of the Google Ngrams dataset, which also lacks punctuation marks.

Statistical language modeling has been used for information retrieval by treating each document as a language sample and a query as a generation process [32]. This way, language models can be used as probability distributions where documents are ranked based on the probability of generating the query from their language models.

3. Methods

3.1. Rank Diversity

We define rank diversity $d(k)$ as the number of words occupying a given rank k during the period of time of the study, which is divided by the number T of time intervals considered.

Therefore, rank diversity is given by

$$d(k) = \frac{|X(k)|}{T}, \tag{1}$$

where $|X(k)|$ is the cardinality (i.e., number of unique words) that appear at rank k during all time intervals.

Rank dynamics have been studied fields of sports [33] and other systems like natural, social, economic, and infrastructural systems. The authors of [34] precisely measured sampling effects on different measures of rank dynamics.

If a rank is occupied by a single word at all times, then rank diversity is minimal. While if in each time interval we have a different word in a given rank, then the rank diversity is maximum.

Rank diversity is an intuitive and straightforward measure of diversity within a dataset, providing direct insights into the variety and spread of elements. It is computationally efficient and robust, making it suitable for large datasets. In contrast, the Kullback–Leibler (KL) divergence is excellent for comparing probability distributions but can be less intuitive when choosing the base of the logarithm and the units of measurement, and it is sensitive to outliers. Rank correlation coefficients measure the similarity between two rankings but do not address diversity within a single dataset. Therefore, while KL divergence and rank correlation coefficients have their strengths, rank diversity offers unique benefits, especially in contexts focused on understanding the richness and variability within single datasets, such as linguistic studies.

3.2. Data

We used geolocated Twitter data to examine shifts in language usage patterns. These tweets were gathered using the Twitter Streaming Application Programming Interface (API), which serves as an intermediary mechanism facilitating data transfers between systems. Among the vast volume of collected tweets, only a subset is geolocalized. Our dataset comprises over 20 million geolocated tweets originating from Argentina, Canada, Colombia, India, Mexico, South Africa, Spain, and the United Kingdom, which are all posted in 2014. Notably, this was during a period when tweets were restricted to 140 characters, and threaded conversations were not as prevalent. We computed rank diversity metrics using various time intervals (Δt). It is worth mentioning that mentions and hashtags were included in the dataset and treated as separate words. These geolocated tweets contain precise latitude and longitude coordinates indicating their geographical origin at the time of creation.

Ferrer and Cancho showed in [35] that the word frequency spectrum typically follows a power law with an exponent around $\beta = 2$, although this can vary significantly. This variation reflects how we balance effective communication (maximizing information transfer). The more we prioritize effective communication, the higher the exponent. The range of exponents goes from about $\beta = 1.6$ (lower cost) to $\beta = 2.4$ (higher cost). We performed the analysis of Zipf's exponents in Figure 1.

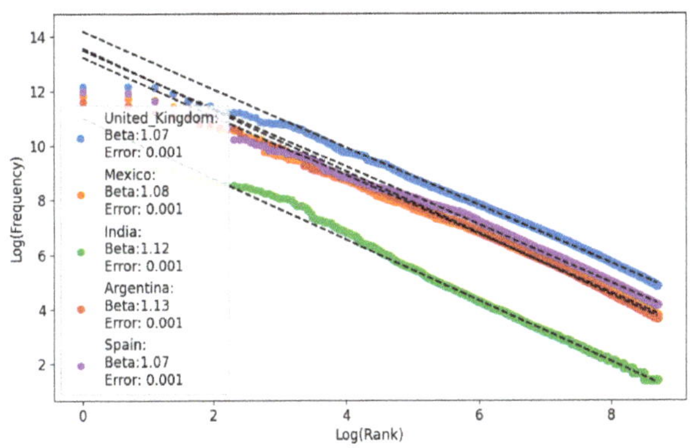

Figure 1. Zipf analysis of monogram data. Errors less than 0.05 show a good fit with Zipf's law, although models with more variables can produce better fits [30]. In texts with beta exponents between $1 < \beta < 2$, the variety of lexical units employed is restricted, and there are many repetitions. Dotted lines show the lineal regression of Zipf's beta exponent.

3.3. Grammatical Scale

The grammatical scale, as defined in our study, pertains to the length of n-gram blocks utilized, where n ranges from 1 to 5 [16]. Single words are considered monograms or 1-grams, while sets of five words are termed 5-grams, and so forth.

The use of sequences of 5-grams has a compatibility reason: since the Google n-grams dataset includes up to 5-grams, our current study enables comparisons with previous research (ours and from others) that have used this dataset.

These n-gram blocks are constructed by partitioning sentences into words and assembling sets of contiguous words of length n with overlaps. For instance, the sentence "This is a sentence" yields three bigrams: ("This is", "is a", and "a sentence").

In our previous research, we investigated how the grammatical scale impacts the rank dynamics of words using the Google Books n-gram dataset [29]. Our findings revealed that changes in the grammatical scale have a more pronounced effect on language statistics (such as rank diversity, change probability, rank entropy, and rank complexity) compared to alterations in the language itself (across English, Spanish, French, German, Italian, and Russian).

3.4. Temporal Scale

To delineate the temporal scale, it is essential to recall that we define rank diversity $d(k)$ as the count of words occupying a particular rank k across all time intervals, which is divided by the total number of time intervals T considered. By altering the time interval Δt, we can compute the rank diversity for various values of T. It is worth noting that if the same dataset is utilized, as the temporal scale Δt increases, T will decrease.

The temporal scale undergoes variation by doubling its value at each time interval, i.e., $3, 6, 9, 12, 24, 48, 96$ h. Evans and Larsen-Freeman [36] conducted a study on language acquisition across different temporal scales, yielding noteworthy findings.

To elucidate the differences of word ranks (specifically 1-grams, although the principle applies to any grammatical scale), we provide examples of selected Spanish words in Figure 2. For instance, in this scenario, $d(k = 1)$ is calculated by dividing the count of unique series (representing words graphically) that intersect the line denoting point $k = 1$ at some time by the total number of 3 h intervals within a year.

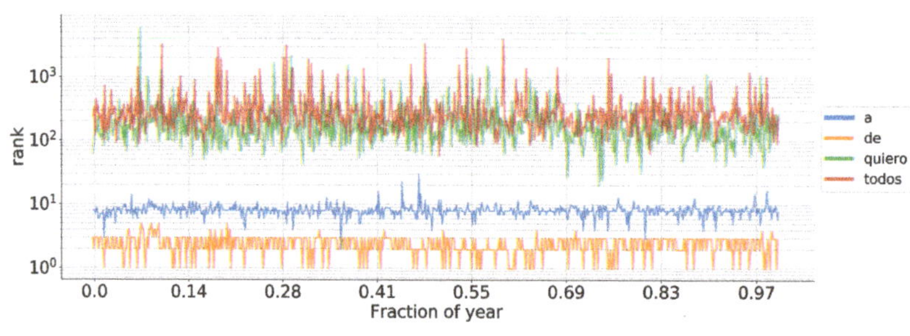

Figure 2. Rank development of some Spanish words during 2014 using a temporal scale of 24 h. Note that if a word has a high rank, its trajectory takes a wider set of possible ranks compared to words with lower ranks [28].

In Figure 3, we present the sigmoid curves for 1, 2, 3, 4, and 5-grams for a 3 h time interval, which is the minimum time unit. We used a statistically significant sample size to ensure comparability, making sure that the smallest set is sufficiently large. Shorter time (and distance) intervals would produce too few data with the Twitter dataset we worked with.

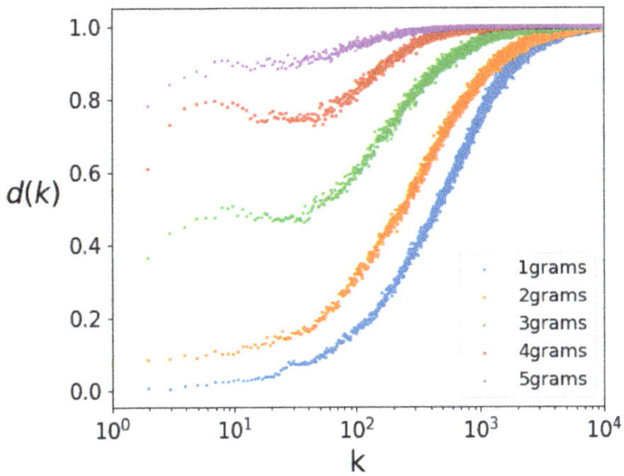

Figure 3. Sigmoid curves for 1, 2, 3, 4, and 5-grams showing that Zipf distribution is valid for a period of 3 h (smallest temporal interval).

3.5. Spatial Scale

We investigate the impact of the spatial scale using tweets exclusively from Mexico, Spain, Argentina, and the United Kingdom, as these countries possess sufficient geolocalized data to achieve statistical significance across various spatial scales (approximately 3.9, 3.7, 4.6, and 5.6 million tweets from Mexico, Spain, Argentina, and the United Kingdom, respectively).

Initially, we established a 3 km radius circle centered on the capital city's geographical midpoint (Mexico City, Madrid, Buenos Aires, and London). Subsequently, we expanded the circle's radius exponentially by powers of two, i.e., 6 km, 12 km, 24 km, 48 km, 98 km, and so forth, until encompassing the entire country.

Our decision to designate the distance from an urban hub (the capital city of each country) rather than the country's geographic center aims to ensure a minimum population density, guaranteeing a sufficient number of tweets within a confined area. Initiating from the geographic center might yield areas lacking adequate data for statistical significance or comparability across countries. Additionally, countries were demarcated using polygons to exclude data from neighboring countries even if they fell within the considered radii.

To mitigate potential biases, we maintain an identical number of tweets across all spatial scales subsequent to the analysis of the smallest scale for each country. For instance, the quantity of tweets within the 3 km radius circle for Mexico totaled 309,792. Consequently, for all subsequent spatial scales, denoted by expanding-radius circles, we randomly sample 309,792 tweets without replacement to ensure uniform analysis conditions.

We conducted sampling to ensure it is statistically significant and comparable, so the smallest set is sufficiently large. Figure 4 illustrates how geographic distribution was modeled as a expanding circle considering the capital city as center of the circle and using Spain as an example.

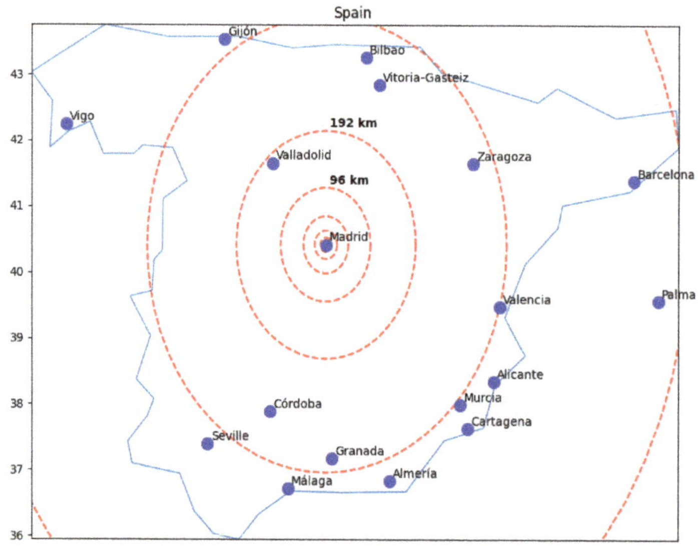

Figure 4. Concentric circles illustrating the geographic distribution for spatial scale in Spain.

3.6. Relevance of Scales

Since we are examining three distinct scales, namely grammatical, temporal, and spatial, to discern any unclear differences in the behavior of rank diversity across these scales, we generated a total of $I \cdot J \cdot S$ rank diversity curves for each country. Here, I, J, and S denote the number of different values that a particular scale can adopt, representing grammatical, temporal, and spatial scales, respectively. Consequently, we generated rank diversity curves corresponding to every unique combination of values from the considered scales. For instance, in the case of Mexico, this equates to $5 \cdot 6 \cdot 11 = 330$ possible combinations.

To obtain a quantitative summary of a rank diversity curve's behavior, which measures how rapidly rank diversity increases relative to rank, and thereby simplify the system's description and reduce observed complexity, we employed estimates of μ. μ represents a parameter of the sigmoid curve indicating the rank value at which the rank diversity curves reach $1/2$.

The sigmoid is the cumulative of a Gaussian distribution, i.e.,

$$\Phi_{\mu,\sigma}(\log_{10} k) = \frac{1}{\sigma\sqrt{2\pi}} \int_{-\infty}^{\log_{10} k} e^{-\frac{(y-\mu)^2}{2\sigma^2}} \mathrm{d}y, \tag{2}$$

and is given as a function of $\log k$ [28].

The fundamental concept for gauging the significance of scale changes on rank diversity behavior lies in recognizing that a lower value of μ signifies a swifter increase in rank diversity concerning rank. It is crucial to note that rank diversity, being a statistical metric, undergoes further averaging with μ, necessitating caution in result interpretation.

To assess the relative importance of a scale concerning the extent to which a transition between two distinct values of that scale impacts variations in rank diversity behavior, we used the following average:

$$\eta(s) = \frac{\sum_i^I \sum_j^J \sigma_{i,j}^s}{I \cdot J}, \tag{3}$$

where $\sigma^s_{i,j}$ corresponds to the standard deviation of estimated values of μ associated to the scale s given fixed i and j values of the two remaining scales, i.e., if

$$\bar{\mu}_{i,j} = \sum_s \frac{\mu^s_{i,j}}{N}, \tag{4}$$

then

$$\sigma^s_{i,j} = \sqrt{\frac{\sum_s(\mu^s_{i,j} - \bar{\mu}_{i,j})^2}{N-1}}. \tag{5}$$

In essence, $\eta(s)$ aims to encapsulate the average dispersion of μ within the scale s, enabling an objective comparison of different scales to determine which one wields the greatest influence on modifying the rate of rank diversity increase across different scale values. In the Results section, we illustrate how $\eta(s)$ effectively quantifies visually discernible trends using graphs of μ plotted against scale values.

This way, high η values indicates a strong association between the scales, meaning that the group means differ significantly. Meanwhile, low η values indicates a weak association, meaning the group means do not differ significantly.

Furthermore, to substantiate the visually observed trends with statistical evidence, we employed a linear regression model to conduct the F-test, evaluating whether at least one of the scales significantly impacts μ and consequently rank diversity. Additionally, the t-test of each coefficient associated with independent variables was utilized to ascertain whether individual scales contribute significantly to explaining the variability of μ within a linear model. In cases involving coefficients representing multiplicative terms in a multiplicative model, the t-test was employed to determine the presence of statistically significant interactions between pairs of scales. Interaction effects between scales are evidenced by observing how the behavior of rank diversity depends on the specific values of other scales as one scale increases. Since such interaction effects may be subtle to discern graphically, a statistical approach serves to validate the hypothesis of their existence.

These models were fitted using the $\log_{10}$ scaled values of temporal and grammatical scales as predictors and μ as a response. In particular, the multiplicative model is

$$Y = \beta_0 + \beta_1 X_1 + \beta_2 X_2 + \beta_3 X_3 + \beta_4 X_1 X_2 + \beta_5 X_1 X_3 + \beta_6 X_2 X_3. \tag{6}$$

The model is simplified into a linear form by excluding terms involving products of predictors (X_i). Here, the coefficients (β_i) represent weights determining the influence of each predictor (i.e., a particular scale) on the response variable, μ. Hypothesis tests are employed to ascertain whether a coefficient significantly deviates from zero, indicating evidence of influence. In cases involving products of two scales, the coefficients quantify the impact of one scale on how the other affects the response. This is evident by factoring a common linear predictor, such as X_1: $Y = \beta_1 X_1 + \beta_4 X_1 X_2 + \beta_5 X_1 X_3 = (\beta_1 + \beta_4 X_2 + \beta_5 X_3) X_1 = \beta_{145}(X_2, X_3) X_1$, where $\beta_{145}(X_2, X_3)$ serves as a multivariate function slope of X_1, delineating how X_2 and X_3 influence the relationship between X_1 and Y. Statistically significant coefficients β_4 and β_5 indicate pairwise interaction between predictors X_2 and X_3 with X_1. Geometrically, β_0 corresponds to the level of the hyperplane best fitting the observations; however, it does not provide further pertinent information in this context. All models were fitted using linear least-squares problems solved via the QR factorization method for numerical stability, leveraging the standard `lm` function in the R programming language.

4. Results and Discussion

The rank diversity of N-grams is computed for tweets originating from eight distinct countries, which were evenly divided between Spanish-speaking nations (Mexico, Argentina, Colombia, Spain) and English-speaking nations (Canada, United Kingdom, India, South Africa). Here, Δt represents the duration between consecutive time "slices" with the total time span denoted as $T\Delta t$. In Figure 5, we present the rank diversity plots for

N-grams, employing a time interval of $\Delta t = 24$ h across varying values of N, ranging from 1 to 5.

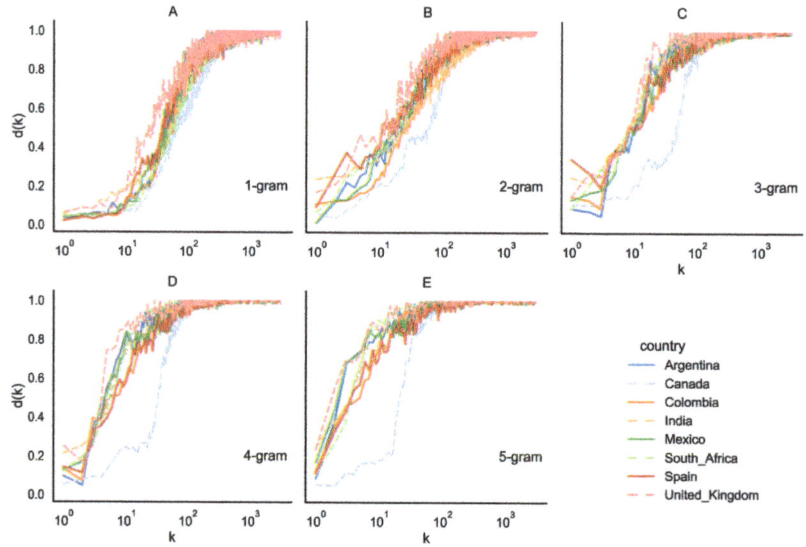

Figure 5. Rank diversity $d(k)$ for eight different countries at different grammatical scales $N = 1, 2, ..., 5$, $\Delta t = 24$ h. Spanish-speaking countries are shown with solid lines, English-speaking countries are shown with dashed lines. (**A**) 1-grams, (**B**) 2-grams, (**C**) 3-grams, (**D**) 4-grams, and (**E**) 5-grams. X-axis is shown in log scale.

Initially, we observe that the sigmoid curve remains a suitable model for describing rank diversity patterns, which is particularly evident when examining shorter time intervals. This consistency holds across all combinations of scales considered. Furthermore, for $N = 1$, the rank diversity profiles exhibit substantial similarity. However, as we increase N, the profiles begin to diverge. This discrepancy implies that 1-grams demonstrate consistent rank diversity regardless of language or country. In contrast, for 2-grams, 3-grams, and 4-grams, distinct trends emerge. Specifically, the curves for Spanish-speaking countries exhibit close alignment, forming a cohesive cluster, while Canada displays a noticeable deviation from this pattern.

In Figure 5, the Spanish-speaking countries appear to be more tightly clustered in general than the other countries studied. Therefore, the rank diversity differences among the Spanish-speaking countries are smaller than the other non-Spanish speaking countries. We calculated this difference to strengthen this argument by using the least squares difference for all countries against all others. In Table 1, we can see that the difference among Spanish speakers ranges from 0.001 to 0.004, while it ranges from 0.005 to 0.21 for other countries. We are unsure if the differences are due to English itself (as India is very similar to Spanish-speaking countries) or their secondary languages that have words in the ranks considered (French in Canada, Hindi and others in India, Afrikaans and others in South Africa). It would be necessary to analyze what percentage of words in different ranges come from other languages and possibly other factors, but this is beyond the scope of the article.

Table 1. Distance between countries using least square difference.

	Mexico	Colombia	Spain	United Kingdom	Canada	South Africa	India
Argentina	−0.003	0.002	−0.001	−0.007	0.014	0.006	0.001
Mexico		0.004	0.002	−0.005	0.017	0.008	0.003
Colombia			−0.002	−0.009	0.012	0.004	−0.001
Spain				−0.007	0.015	0.006	0.002
United Kingdom					0.021	0.013	0.008
Canada						−0.008	−0.013
South Africa							−0.005

This behavior means that there are some other features that make them distinct from the rest.

In English-speaking countries like the UK, Canada, and India, second languages play a significant role, often reflecting the cultural differences. Canada is officially bilingual, with both English and French as official languages. In the UK, the most commonly learned second languages are French, Spanish, and German. India is a highly multilingual country with English serving as an associate official language alongside Hindi. In each region, that should be checked, but this would require an anthropological study, which is way beyond the statistical linguistics scope of this paper.

Sigmoid curves can be likened to the cumulative distribution of a Gaussian with μ denoting the mean (the peak and center of the Gaussian) and σ representing the standard deviation (which determines its width). Within the sigmoid context, μ denotes the x-axis point where the range diversity equals 0.5, while sigma governs the smoothness of the sigmoid (if it were zero, it would resemble a step from zero to one at μ; as it increases, the sigmoid widens). In a previous study ([29]), we conducted a comparative analysis of various grammatical scales across six languages using Google Books data. Typically, σ exhibits a correlation with μ, although it has higher variability. Consequently, we streamlined the range diversity curves, focusing solely on μ for comparisons in this article.

In the subsequent sections, we delve into the impacts of different scales (grammatical, spatial, and temporal) on rank diversity, leveraging estimates of the parameter μ from the sigmoid curves.

4.1. Grammatical Scale

Following the progression of N values in ascending order, Figures 6 and 7 illustrate that as the grammatical scale expands, so does the pace of rank diversity increment. This trend holds consistently across countries, regardless of whether the language is Spanish or English, or the values adopted by the other two scales. Notably, a larger grammatical scale signifies an increase in phrase complexity. At the apex of the scale (5-grams), the rank diversity pertains to blocks comprising five words. The potential combinations of five-word blocks surpass those of four words, which in turn exceed those of three words. Consequently, we observe heightened diversity at the initial ranks compared to lower grammatical scales.

Furthermore, we observe that for 1-grams, μ exhibits similarity between Spanish and English, remaining practically unaffected by the spatial scale. In other words, there are no discernible changes in the pace of rank diversity increment across analyzed areas. However, it does fluctuate in relation to the temporal scale, as evidenced by the findings in the first column of Figure 6. This underscores the significance of employing diverse scales to scrutinize the rank diversity of languages.

Additionally, a visual inspection reveals that alterations in the grammatical scale yield the most significant overall escalation in the pace of rank diversity increment compared to the other two scales. Toward the conclusion of the Results section, we quantitatively validate this observation by comparing average dispersions.

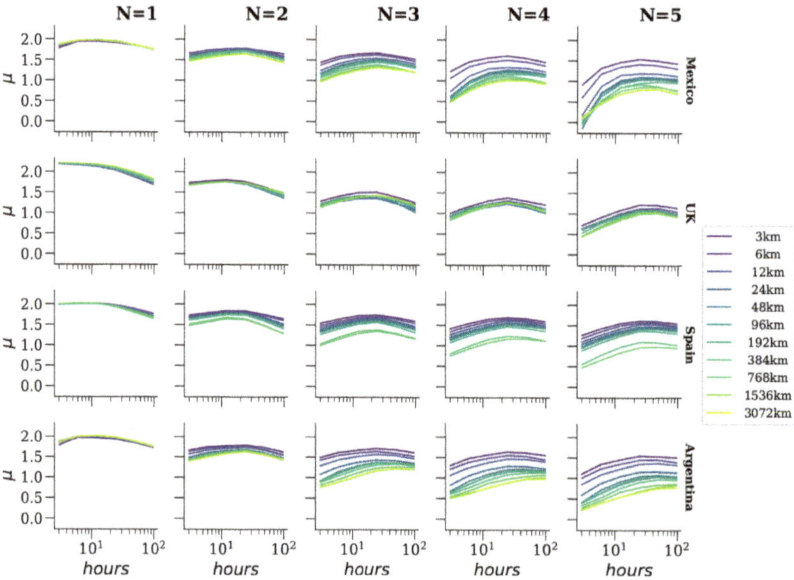

Figure 6. Different estimations of μ. Measure the speed of rank diversity increment vs. different values of the temporal scale. Note that the x-axis is in $\log_{10}$ scale. The N values represent N-grams (grammatical scale). Each row uses Twitter data from the country specified at the right label. Colored lines represent the different spatial scales considered.

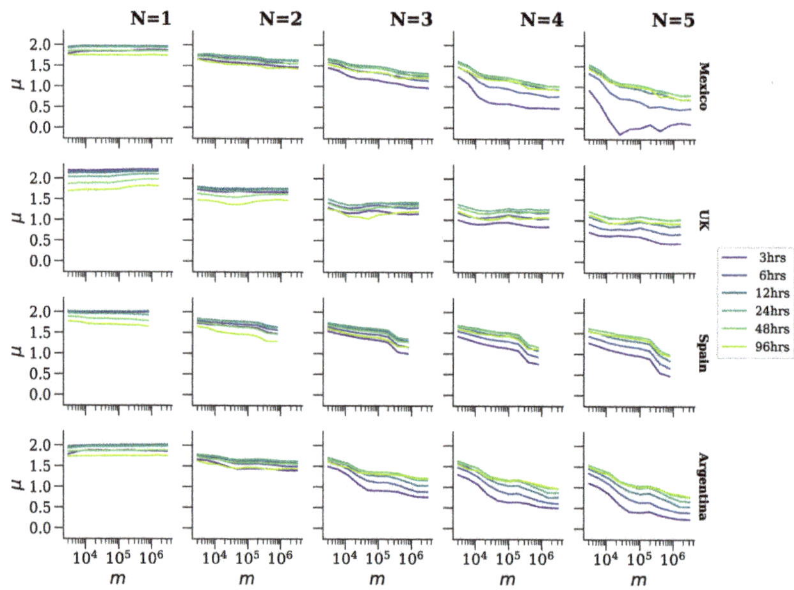

Figure 7. μ estimates vs. different values of the spatial scale. The top and right labels represent the same as in Figure 6. Colored lines here represent the different temporal scales.

4.2. Temporal Scale

In Figure 6, we manipulate the temporal scale along the x-axis to illustrate the correlation between μ and various time intervals Δt. Notably, we observe that the pace of the rank diversity increment does not exhibit a linear increase as observed in the grammatical or spatial scales; instead, it displays a noticeable concave shape. This nonlinear effect stems from the phenomenon whereby the addition of frequencies results in fewer variable positions for the N-grams in the lists constituting the total timespan under analysis. Consequently, μ experiences an increment until reaching a certain time interval, after which it begins to decline. This phenomenon occurs because the denominator in the calculation of rank diversity, which represents the number of possible lists dividing the total timespan, decreases with higher temporal scales. Consequently, the relationship between the speed of rank diversity increment and the temporal scale remains consistent regardless of the country or language. Additionally, it is noteworthy how the shape of the relationship between μ and time alters across various columns, each representing different grammatical scales. This observation suggests that the grammatical scale exerts an influence on the variation of the temporal scale.

4.3. Spatial Scale

In each plot of Figure 6, particularly from the column $N = 3$ onwards, it is evident that with fixed grammatical and temporal scales, the spatial scale influences the speed of rank diversity increments. However, to elucidate this relationship further, Figure 7 plots μ against the spatial scale. Notably, for Spanish-speaking countries, μ decreases with the spatial scale when the grammatical scale exceeds 1, whereas in general, μ remains relatively unchanged against the spatial scale for the United Kingdom. Further investigation is necessary to explore potential explanations, such as whether there exists greater homogeneity in the United Kingdom compared to the other countries and/or whether these results reflect a distinction between English and Spanish. Overall, it is observed that μ also decreases with the grammatical scale.

4.4. Relevance of Scales

The interdependence of different types of scales is evident in their overlapping influences, indicating their lack of independence. A detailed explanation follows.

Now, we address the question of which scale holds the most significance in terms of its impact on the variability of μ and, consequently, on the behavior of rank diversity itself.

To investigate this, we assess the relative importance of these scales using Equations (3)–(5). The results, depicted in Figure 8, confirm that the grammatical scale exhibits the greatest variance relative to the other scales under consideration. Additionally, for all Spanish-speaking countries, both the temporal and spatial scales appear to hold approximately equal importance. However, in the case of the United Kingdom, despite having more data available, the spatial scale seems to possess less significance.

Finally, we present the p-values and the associated estimated F-statistic for the first four terms of model (6) in Table 2.

Table 2. p-values associated with the F-statistic, which represents the significance of the regression, i.e., that at least one scale is contributing to explain the changes in rank diversity. The parameter l is the number of parameters used and H is the number of observations.

Country	Mexico	United Kingdom	Argentina	Spain
$l-1$	3	3	3	3
$H-l$	326	296	326	266
F-statistic	489.6	631.7	546	261
p-value	$<2.2 \times 10^{-16}$	$<2.2 \times 10^{-16}$	$<2.2 \times 10^{-16}$	$<2.2 \times 10^{-16}$

The significance of the observed difference increases as the p-value decreases. Since our results yield a p-value lower than 0.05, we consider it statistically significant. These low p-values indicate that at least one scale is linked to μ, assuming approximate linear correlation.

Our main objective here was to support the hypothesis that changes in certain scales lead to variations in rank diversity behavior. In other words, we aimed to demonstrate that the linear regression model offers a better fit to the data than a model with no independent variables, which implies no influence of the scales on the observed variability.

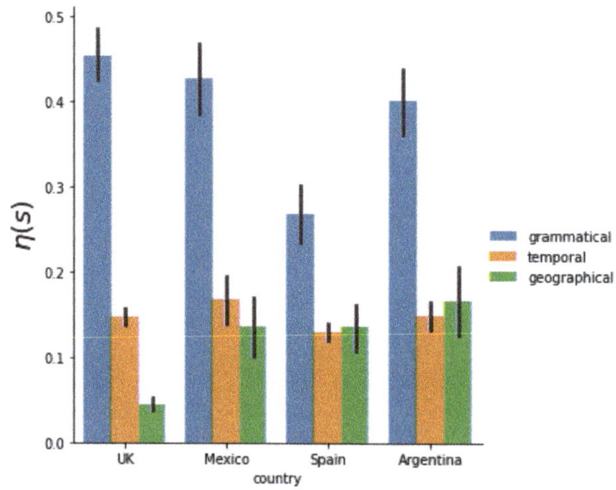

Figure 8. Average dispersion (2) of scale s for each country and a bar chart to compare them.

Specifically, we tested the significance of each associated coefficient to determine whether a particular scale is related to μ. The resulting p-values are detailed in Table 3.

Table 3. Associated p-values to t-tests of the linear coefficients. X_1,X_2, and X_3 represent the grammatical, spatial, and temporal scales, respectively.

p-Value	X_1	X_2	X_3
Mexico	$<2.0 \times 10^{-16}$	$<2.0 \times 10^{-16}$	9.8×10^{-16}
United Kingdom	$<2.0 \times 10^{-16}$	0.125	0.426
Argentina	$<2.0 \times 10^{-16}$	$<2.0 \times 10^{-16}$	$< 2.0 \times 10^{-16}$
Spain	$<2.0 \times 10^{-16}$	$<2.0 \times 10^{-16}$	0.0183

It is noteworthy that for the United Kingdom, the temporal and spatial scales exhibit les significance according to our test compared to the grammatical scale. As illustrated previously in Figure 7, the spatial scale appears to be practically independent of μ. However, in the case of the temporal scale, this suggests that a linear approximation alone is insufficient to adequately capture the relationship between these scales and μ.

Therefore, employing a quadratic model proves to be more effective in elucidating the existence of relationships for the temporal scale. This approach reveals that for this dataset, the temporal relation with μ is nonlinear, as evidenced by the observations depicted in Figure 6.

Alternatively, to assess the statistical significance of interactions between pairs of scales, we can examine the p-values associated with t-tests conducted on the estimated coefficients β_4, β_5, and β_6 in model 6. The results are presented in Table 4.

We observe that all interactions between the grammatical and spatial scales, as well as between the grammatical and temporal scales, are statistically significant. However, the significance is somewhat lower for Argentina. Additionally, there is a notably higher level of interaction between the grammatical and spatial scales, specifically for the United Kingdom.

Table 4. Associated p-values to t-tests of the interaction coefficients. X_1, X_2, and X_3 represent the grammatical, spatial, and temporal scales, respectively.

p-Value	$X_1 * X_2$	$X_1 * X_3$	$X_2 * X_3$
Mexico	$<2.2 \times 10^{-16}$	$<2.2 \times 10^{-16}$	0.57
United Kingdom	$<8.2 \times 10^{-5}$	$<2.2 \times 10^{-16}$	0.12
Argentina	$<2.2 \times 10^{-16}$	$<2.2 \times 10^{-16}$	$<2.6 \times 10^{-4}$
Spain	$<2.2 \times 10^{-16}$	$<2.2 \times 10^{-16}$	0.47

It is interesting that we found minor statistical differences between English and Spanish (compared to other variables). Most of the linguistic studies comparing these two languages focus on other aspects such as phonetics [37]. It would be relevant to make further statistical comparisons across languages.

4.5. Special Tokens

In this section, we concentrate on examining special tokens commonly employed in Twitter discourse: emojis [38,39], hashtags [40], and mentions [41,42]. We investigate the most prevalent occurrences of each within Argentina, Mexico, Spain, and the United Kingdom, along with their respective rank diversities.

An emoji is a pictogram, logogram, ideogram, or smiley utilized in electronic messages and web content. Emojis serve the primary purpose of conveying emotional nuances that may be absent in typed communication. They represent images that can be depicted as encoded characters. Given their widespread usage, emojis play a pivotal role in enhancing our online communication by supplementing text with emotional context akin to body language and facial expressions.

Figure 9 presents a comprehensive overview of the most frequently used emojis, which are arranged in descending order from top to bottom. Notably, the *Smiling Face with Heart-Shaped Eyes* and *Face with Tears of Joy* emojis emerge as the predominant expressions, which are indicative of sentiments like happiness, affection, and amusement. These are followed by a diverse array of symbols representing various emotions, such as love, strength, and positivity.

Interestingly, while some emojis exhibit universal popularity across all countries, others demonstrate regional prevalence (consistent with the results of [38]). For instance, the *Woman Dancing* and *Sun with Face* emojis are predominantly used in Spain, whereas the *Sleeping Face* emoji is more prevalent in Mexico, and the *Unamused Face* emoji is favored in Argentina. Conversely, the *Fire* emoji appears to be exclusive to the United Kingdom.

Furthermore, among the most frequently used Unicode emoji symbols, there are instances of *Emoji Modifier Fitzpatrick*. These emojis can be customized to reflect different skin tones using one of five modifiers. Notably, Argentina's top list includes only the *Light Skin Tone* modifier, while Spain also incorporates the *Medium-Light Skin Tone*. In contrast, both Mexico and the UK feature the *Medium Skin Tone* modifier, with neither country including the *Medium-Dark Skin Tone* or *Dark Skin Tone* in their lists.

A *hashtag* serves as a metadata tag commonly employed on platforms such as Twitter and Instagram. Recognizable by the prefix of the hash symbol #, hashtags function as user-generated labels facilitating the organization and cross-referencing of content based on specific topics or themes. Users leverage hashtags to discover and engage with content relevant to their interests, as the practice enables the aggregation of posts related to a particular hashtag.

Argentina		Mexico		Spain		United Kingdom	
😍	27,906	😍	20,543	😍	21,466	😂	19,726
😂	20,752	🏻	14,564	😂	14,621	😍	18,559
👌	12,239	😂	12,442	🎉	13,037	🏼	15,944
💕	11,523	🏽	9757	🏽	10,799	🏽	14,712
🎉	10,428	😎	7851	🏻	10,743	🇧	9610
😊	8145	👌	6847	💕	9950	🇬	9606
🎶	8133	😊	6090	💪	9926	😊	8283
😘	7594	💕	6079	👌	9326	👌	8261
👏	7196	😘	6006	😘	8776	💕	7111
💪	6624	😁	5567	💃	8643	👍	6807
😭	6284	😭	5551	😊	7581	🏾	6499
💞	6133	💪	5009	🍻	7339	🎉	6457
🙌	5669	🎉	4622	😎	7104	💙	6109
💘	5662	🙌	4054	👏	6926	🙌	6022
🏻	5218	👏	3944	🔝	6891	😘	5609
💙	5217	🏾	3803	🌞	6590	💪	5475
💜	4488	😴	3540	👭	6460	😎	5300
😁	4304	🎶	3297	🇪	6204	🔥	4967
💖	4113	🇲	3255	🇸	6162	😁	4909
😒	4053	🇽	3231	🌊	6156	💜	4378

Figure 9. Comparison of the most frequent emojis and modifiers shown in Argentina, Mexico, Spain and the United Kingdom in geolocalized tweets from the year 2014. The number listed is the frequency of the emojis.

It is pertinent to emphasize that hashtags lack centralized registration or control, allowing any user to create or use them freely. This decentralized nature means that hashtags do not adhere to predefined definitions, and their significance can evolve over time. Consequently, a single hashtag may serve multiple purposes, and the interpretation or accepted meaning of a hashtag is subject to change within the context of online discourse.

Table 5 reveals that the hashtag *#trndnl* emerges as the predominant hashtag across the dataset, which is indicative of discussions surrounding popularity and trending topics. Additionally, notable cities within each country are prominently featured, such as Buenos Aires, Cordoba, and Rosario in Argentina; CDMX/Mexico City, Monterrey, Guadalajara, and Puebla in Mexico; Madrid, Barcelona, and Sevilla in Spain; and London and St Albans in the UK. An intriguing observation is the prevalence of weather-related hashtags exclusively within the European context.

A *mention* is a tweet that contains another person's username anywhere in the body of the tweet. User mentions are identified with the @ symbol within tweets.

Table 6 provides insight into the most frequently mentioned users on Twitter, shedding light on various meeting places such as shopping centers (e.g., @galeriasmx in Mexico), cinemas (e.g., @cinemex in Mexico), and airports (e.g., @heathrowairport in the United Kingdom), as well as renowned companies (e.g., @starbucksuk) and individuals such as politicians (e.g., @mauriciomacri in Argentina) and artists (e.g., @officialmaki in Spain). Notable differences can be discerned across countries, which are indicative of divergent Twitter usage patterns during the period under examination. For instance, Argentina exhibits numerous mentions of artists, Mexico features several references to commercial franchises, Spain includes soccer teams and political parties, while the United Kingdom predominantly mentions the National Rail network.

Table 5. Most frequent hashtags in Argentina, Mexico, Spain and United Kingdom in 2014 geolocalized tweets. The number listed is the frequency of the hashtags.

Argentina		Mexico	
#trndnl	9518	#trndnl	10,836
#buenosaires	5022	#cdmx	6951
#argentina	2372	#mexico	6540
#me	1646	#mexicocity	3631
#cordoba	1502	#job	3508
#rosario	1419	#hiring	3135
#love	1269	#monterrey	2870
#selfie	1088	#endomondo	2626
#friends	1069	#guadalajara	2433
#job	961	#endorphins	2299
#endomondo	939	#friends	2049
#viernes	911	#méxico	1860
#amigos	891	#love	1758
#repost	873	#careerarc	1575
#hiring	868	#photo	1486
#night	807	#jobs	1370
#endorphins	804	#quieremeamame	1333
#domingo	774	#puebla	1215
#sabado	727	#selfie	1183
#carlosrivera	719	#travel	1069
Spain		**United Kingdom**	
#trndnl	19,328	#nowplaying	52,085
#madrid	14,552	#london	35,934
#barcelona	14,167	#job	25,486
#incdgt	13,367	#tnc	24,897
#dgt	12,978	#trndnl	24,319
#meteocat	12,260	#areacode	23,084
#endomondo	9689	#hiring	22,587
#meteo	8736	#tides	20,236
#spain	8649	#ktt	19,237
#blanco	8644	#weather	12,024
#endorphins	8559	#careerarc	11,841
#324meteo	8468	#essex	11,016
#meteo	3127	#broadbandcompareuk	10,881
#retención	8189	#bestbroadband	10,641
#precaución	7798	#photo	10,492
#elcatllar	6844	#ukweather	9978
#obra	5398	#endomondo	9770
#arameteo	4910	#jobs	9375
#sevilla	4890	#endorphins	8988
#amarillo	4030	#stalbans	7498

It is imperative to acknowledge that these mentions are derived from geolocated tweets, representing only a fraction of all tweets. Consequently, these mentions may exhibit bias, and the most popular accounts may differ from those enumerated herein.

Table 6. The most frequent users mentioned in Argentina, Mexico, Spain, and the United Kingdom in geolocalized tweets from the year 2014. The number listed is the frequency of the mentions, and the type column explains the class account. Existing personal accounts are anonymized.

Argentina			Mexico		
Account	**Frequency**	**Type**	**Account**	**Frequency**	**Type**
@clubsolotu_arg	724	Fans Club	@aicm3	3965	Airport
@vale975	649	Radio Channel	@cinemex	2629	Cinema
Personal account	543	Weather Technician	@cinepolis	2352	Cinema
@rkartista	432	Artist	@germanmontero5	2296	Singer
@aa2000oficial	417	Argentina Airports	@grupointocable	2122	Musical Group
@mauriciomacri	368	President of Argentina	@sonadoraeterno	2108	Fan Club
@abrahammateomus	339	Artist	@mariobautista_	1878	Musical Artist
@lucianopereyra	325	Singer	@smartfit_mex	1755	Gym
@todonoticias	263	News Channel	@walmartmexico	1274	Supermarket
@c5n	243	News Agency	@galeriasmx	1010	Shopping Center
@infobae	179	News Agency	@chilismexico	789	Restaurant
@radiomitre	169	Radio Station	@aeropuertodemty	729	Airport
@rialjorge	163	Journalist	@aeropuertosgap	695	Airport
@cfkargentina	156	VP of Argentina	@solosanborns	673	Store
@marialuizateodo	153	Deleted Account	@banamex	654	Bank
@starbucksar	153	Coffee Shop	@lacasadetono	644	Restaurant
@niallofficial	152	Singer	@oasis_coyoacan	625	Shopping Center
@sole_pastorutti	152	Artist	@auditoriomx	594	Entertainment Center
@brigitte2300	144	Singer	@perisur	566	Shopping Center
@lanacion	144	Newspaper	@tuado	559	Bus Company
Spain			**United Kingdom**		
Account	**Frequency**	**Type**	**Account**	**Frequency**	**Type**
@canalfiesta	1124	Radio Channel	@nationalrailenq	4797	National Rail
@aena	1068	Airports Management	@heathrowairport	2088	Airport
@aenaaeropuertos	842	Airport	@starbucksuk	1480	Coffee Shop
@dominguezja	830	Media Personality	@luvthenorth444	1401	Suspended Account
@oficialmaki	671	Compositor	@brewdog	1030	Brew Restaurant
@oficiallamorena	570	Singer	@simmons2k	928	Politician
@adif_es	462	Railway Management	Personal account	903	Chelsea Fan
@willylevy29	395	Actor	@costacoffee	902	Coffee Shop
@realmadrid	321	Football Club	Personal account	863	Chelsea Fan
@pinedademar	297	City Hall	@babs200475	819	Suspended Account
Personal account	288	William Levy's Fans	@shelleym1974	774	Suspended Account
@pablo_iglesias_	278	Politician	@luvyorkshire444	693	Suspended Account
@sanchezcastejon	273	Political leader	Personal account	659	Chelsea Fan
@psoe	265	Political Party	Personal account	634	Woman
@fcbarcelona	242	Football Team	@harrods	625	Department Store
@barcelona_cat	234	Barcelona News	Personal account	576	Chelsea Fan
@ahorapodemos	213	Political Party	Personal account	560	Chelsea Fan
@renfe	202	Railway Transportation	@selfridges	560	Department Store
@el_pais	199	Newspaper	@visitlondon	481	Travel Guide
Personal account	190	Fan Club Manager	@skynews	466	News Channel

Rank diversity curves pertaining to emojis, hashtags, and user mentions can be effectively approximated by sigmoid curves as seen in Figure 10, which is similar to many other phenomena [34]. Across all cases, user mentions emerge as the most diverse feature per country, while emojis exhibit the lowest diversity.

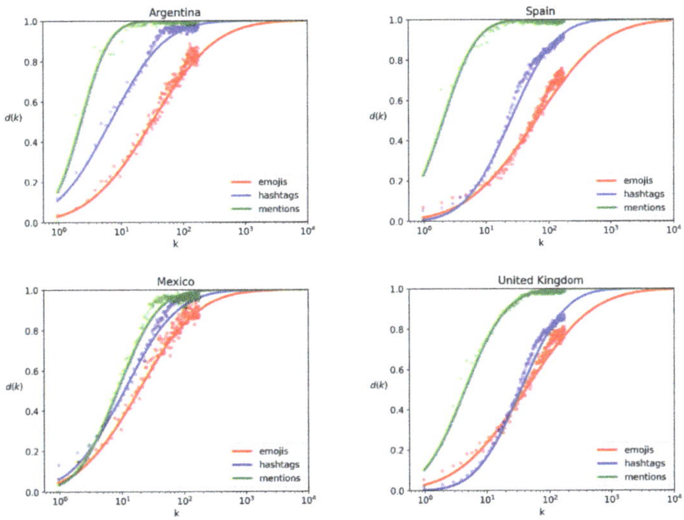

Figure 10. Rank diversity $d(k)$ of emojis, hashtags and mentions in Argentina, Mexico, Spain and United Kingdom in a time interval Δt of 24 h. Dots represent the rank diversity, and solid curves represent the sigmoid approximation.

5. Conclusions and Future Work

In conclusion, our analysis of geolocated Twitter data has provided valuable insights into language usage across different spatial, temporal, and grammatical scales.

Our findings indicate a clear relationship between higher scales and elevated values across the studied dimensions. For instance, monograms represent the lowest scale in the grammatical hierarchy, while 5-grams signify the highest. Similarly, spatial scales ranging from 3 to 3000 km, and temporal scales from 3 to 96 h, reflect the continuum of spatial and temporal granularity.

Interestingly, while higher scales generally correlate with greater rank diversity, we observed a nuanced trend in the temporal dimension. Specifically, temporal rank diversity exhibited a concave pattern with both shorter and longer time intervals demonstrating higher diversity compared to intermediate intervals.

Assessing the relative importance of each scale based on rank diversity dispersion, we identified the grammatical scale as the most influential among the three. Moreover, we observed similar levels of importance for temporal and spatial scales in Spanish-speaking countries, whereas the spatial scale emerged as less critical in English-speaking countries within the scope of the statistical measures employed.

Looking ahead, future research could delve deeper into understanding the complex interactions between these scales and their impact on language dynamics, potentially exploring additional linguistic features and employing more sophisticated analytical techniques. Additionally, investigating longitudinal trends and incorporating sociocultural factors may offer further insights into the evolving nature of language use in online platforms like Twitter.

Indeed, variations in rank diversity on Twitter are likely influenced by a multitude of factors, including the topics of interest among users in each country. For instance, countries experiencing significant economic or political upheavals may witness a narrower range of topics dominating discussions, leading to lower diversity in word usage. Conversely, the occurrence of events over shorter durations or the overlapping of multiple events could contribute to higher diversity in rank. This pattern of behavior sheds light on the observed differences across geographical areas, such as tweets originating from urban centers versus those from rural areas. Furthermore, it could also explain the differences in the usage of hashtags and mentions.

This underscores the complexity of the relationship between the considered scales and rank diversity. While the grammatical scale emerges as strongly related to changes in the speed of rank diversity increment, a comprehensive understanding of this relationship requires the consideration of multiple scales simultaneously. Therefore, future research should explore the intricate interplay between spatial, temporal, and grammatical scales to unravel the nuanced dynamics of language usage on Twitter comprehensively.

The consistent fit of the sigmoid curve to rank diversity curves across various spatial and temporal scales is intriguing. It suggests that the underlying mechanisms shaping rank diversity remain unaffected by changes in language or scales. This finding not only extends our previous research, which focused on temporal scales of years, but also underscores the robustness of the sigmoid curve in capturing language dynamics across different temporal and geographical contexts. Notably, while the diversity of monograms remains unaffected by spatial scale, higher grammatical scales exhibit varying rank diversity patterns across different spatial scales, implying greater language use variability at higher grammatical scales.

Our examination of the most frequent emojis across different languages and countries reveals intriguing cultural differences among Twitter users. Emojis, as nonverbal symbols, not only reflect cultural nuances but also potentially convey collective sentiment and biases within each country. Similarly, hashtags, as embedded metadata, play a significant role in fostering communities, evoking emotions, and highlighting relevant topics and events within a community. Understanding the dynamics of hashtags can offer valuable insights into societal trends and preferences. Furthermore, the analysis of mentions in Twitter unveils the changing relevance of various entities over time and space, reflecting evolving trends and interests.

The COVID-19 pandemic has brought to the forefront the issue of misinformation dissemination through social media platforms, prompting academic research to scrutinize its impact. Studies, such as the one by Pennycook et al. [22], have shed light on the disconnect between individuals' assessments of news accuracy and their intentions to share it, suggesting that misinformation may proliferate even when not explicitly endorsed. This underscores the importance of prioritizing accuracy to mitigate the spread of misinformation online.

As highlighted by the Pew Research Center [43], a significant portion of the population, including approximately 72 percent of Americans, relies on social media platforms for news consumption. However, the inherent tendency of false information to spread more rapidly than factual news on platforms like Twitter underscores the urgency of investigating statistical linguistics within the specific context of misinformation dissemination.

Therefore, a natural extension of current research efforts would involve delving deeper into the linguistic patterns and dynamics associated with the propagation of misinformation on social networks. By analyzing linguistic features such as vocabulary usage, grammatical structures, and temporal patterns, researchers can gain valuable insights into the mechanisms driving the dissemination of false information on social media platforms. Such insights can inform the development of strategies and interventions aimed at curbing the spread of misinformation and promoting the dissemination of accurate and reliable information during public health crises and beyond.

Recently, large language models (LLMs) have applied available data to various artificial intelligence (A.I.) [44] applications. Studying the range diversity in human-generated and LLM-generated text could help distinguish between them.

Author Contributions: All authors conceived, designed, and wrote the study, and they have all read and agreed to the published version of the manuscript.

Funding: This work was supported by UNAM-PAPIIT IN107919, IG101421, IG101324 and IV100120. We are also grateful for CONACyT 285754 grant and UNAM Posdoctoral Program (POSDOC).

Institutional Review Board Statement: Not applicable.

Data Availability Statement: The data used for simulations in this study are available on request from the corresponding author.

Conflicts of Interest: The authors declare no conflicts of interest.

References

1. Zipf, G.K. *Selective Studies and the Principle of Relative Frequency in Language*; Harvard University Press: Cambridge, MA, USA, 1932.
2. Booth, A.D. A "Law" of occurrences for words of low frequency. *Inf. Control.* **1967**, *10*, 386–393. [CrossRef]
3. Montemurro, M.A. Beyond the Zipf–Mandelbrot law in quantitative linguistics. *Phys. A Stat. Mech. Its Appl.* **2001**, *300*, 567–578. [CrossRef]
4. Newman, M.E. Power laws, Pareto distributions and Zipf's law. *Contemp. Phys.* **2005**, *46*, 323–351. [CrossRef]
5. Baek, S.K.; Bernhardsson, S.; Minnhagen, P. Zipf's law unzipped. *New J. Phys.* **2011**, *13*, 043004. [CrossRef]
6. Corominas-Murtra, B.; Fortuny, J.; Solé, R.V. Emergence of Zipf's law in the evolution of communication. *Phys. Rev. E* **2011**, *83*, 036115. [CrossRef]
7. Ferrer i Cancho, R.; Solé, R.V. Zipf's Law and Random Texts. *Adv. Complex Syst.* **2002**, *5*, 1–6. [CrossRef]
8. Gell-Mann, M.; Ruhlen, M. The origin and evolution of word order. *Proc. Natl. Acad. Sci. USA* **2011**, *108*, 17290–17295. Available online: http://www.pnas.org/content/108/42/17290.full.pdf+html (accessed on 18 August 2024). [CrossRef]
9. Kirby, S.; Dowman, M.; Griffiths, T.L. Innateness and culture in the evolution of language. *Proc. Natl. Acad. Sci. USA* **2007**, *104*, 5241–5245. Available online: http://www.pnas.org/content/104/12/5241.full.pdf+html (accessed on 18 August 2024). [CrossRef]
10. Steels, L. *Experiments in Cultural Language Evolution*; Advances in Interaction Studies; John Benjamins Publishing Company: Amsterdam, The Netherlands, 2012.
11. Baronchelli, A.; Loreto, V.; Tria, F. Language dynamics. *Adv. Complex Syst.* **2012**, *15*, 1203002. [CrossRef]
12. Perc, M. Evolution of the most common English words and phrases over the centuries. *J. R. Soc. Interface* **2012**, *9*, 3323–3328. Available online: http://rsif.royalsocietypublishing.org/content/9/77/3323.full.pdf+html (accessed on 18 August 2024). [CrossRef] [PubMed]
13. Gerlach, M.; Altmann, E.G. Stochastic Model for the Vocabulary Growth in Natural Languages. *Phys. Rev. X* **2013**, *3*, 021006. [CrossRef]
14. Alshaabi, T.; Adams, J.L.; Arnold, M.V.; Minot, J.R.; Dewhurst, D.R.; Reagan, A.J.; Danforth, C.M.; Dodds, P.S. Storywrangler: A massive exploratorium for sociolinguistic, cultural, socioeconomic, and political timelines using Twitter. *Sci. Adv.* **2021**, *7*, eabe6534. Available online: https://advances.sciencemag.org/content/7/29/eabe6534.full.pdf (accessed on 18 August 2024). [CrossRef]
15. Almodaresi, F.; Ungar, L.; Kulkarni, V.; Zakeri, M.; Giorgi, S.; Schwartz, H.A. On the Distribution of Lexical Features at Multiple Levels of Analysis. In *Proceedings of the 55th Annual Meeting of the Association for Computational Linguistics (Volume 2: Short Papers)*; Association for Computational Linguistics: Vancouver, BC, Canada, 2017; pp. 79–84. [CrossRef]
16. Michel, J.B.; Shen, Y.K.; Aiden, A.P.; Veres, A.; Gray, M.K.; Team, T.G.B.; Pickett, J.P.; Hoiberg, D.; Clancy, D.; Norvig, P.; et al. Quantitative Analysis of Culture Using Millions of Digitized Books. *Science* **2011**, *331*, 176–182. [CrossRef]
17. Rau, M.D. Language Identification by Statistical Analysis. 1974. Available online: https://apps.dtic.mil/sti/tr/pdf/ADA003518.pdf (accessed on 18 August 2024).
18. Bollen, J.; Pepe, A.; Mao, H. Modeling public mood and emotion: Twitter sentiment and socio-economic phenomena. In Proceedings of the ICWSM11, Barcelona, Spain, 17–21 July 2011; pp. 450–453.
19. Dodds, P.S.; Harris, K.D.; Kloumann, I.M.; Bliss, C.A.; Danforth, C.M. Temporal Patterns of Happiness and Information in a Global Social Network: Hedonometrics and Twitter. *PLoS ONE* **2011**, *6*, e26752. [CrossRef]
20. Morales, A.J.; Borondo, J.; Losada, J.C.; Benito, R.M. Efficiency of human activity on information spreading on Twitter. *Soc. Netw.* **2014**, *39*, 1–11. [CrossRef]
21. Morales, A.J.; Vavilala, V.; Benito, R.M.; Bar-Yam, Y. Global patterns of synchronization in human communications. *J. R. Soc. Interface* **2017**, *14*, 20161048. Available online: https://royalsocietypublishing.org/doi/pdf/10.1098/rsif.2016.1048 (accessed on 18 August 2024). [CrossRef] [PubMed]
22. Pennycook, G.; Epstein, Z.; Mosleh, M.; Arechar, A.A.; Eckles, D.; Rand, D.G. Shifting attention to accuracy can reduce misinformation online. *Nature* **2021**, *592*, 590–595. [CrossRef]
23. Morales, A.J.; Borondo, J.; Losada, J.C.; Benito, R.M. Measuring political polarization: Twitter shows the two sides of Venezuela. *Chaos Interdiscip. J. Nonlinear Sci.* **2015**, *25*, 033114. [CrossRef]
24. Hong, L.; Convertino, G.; Chi, E.H. Language Matters in Twitter: A Large Scale Study. In Proceedings of the fifth International AAAI Conference on Weblogs and Social Media, Barcelona, Spain, 17–21 July 2011.
25. Weerkamp, W.; Carter, S.; Tsagkias, M. *How People Use Twitter in Different Languages*; ACM: New York, NY, USA, 2011.
26. Morales, A.J.; Dong, X.; Bar-Yam, Y.; Pentland, A. Segregation and polarization in urban areas. *R. Soc. Open Sci.* **2019**, *6*, 190573. [CrossRef]

27. Cui, H.; Kertész, J. Competition for popularity and interventions on a Chinese microblogging site. *PLoS ONE* **2023**, *18*, e0286093. [CrossRef]
28. Cocho, G.; Flores, J.; Gershenson, C.; Pineda, C.; Sánchez, S. Rank Diversity of Languages: Generic Behavior in Computational Linguistics. *PLoS ONE* **2015**, *10*, e0121898. [CrossRef]
29. Morales, J.A.; Colman, E.; Sánchez, S.; Sánchez-Puig, F.; Pineda, C.; Iñiguez, G.; Cocho, G.; Flores, J.; Gershenson, C. Rank Dynamics of Word Usage at Multiple Scales. *Front. Phys.* **2018**, *6*, 45. [CrossRef]
30. Cocho, G.; Rodríguez, R.F.; Sánchez, S.; Flores, J.; Pineda, C.; Gershenson, C. Rank-frequency distribution of natural languages: A difference of probabilities approach. *Phys. A Stat. Mech. Its Appl.* **2019**, *532*, 121795. [CrossRef]
31. Stanisz, T.; Drożdż, S.; Kwapień, J. Complex systems approach to natural language. *Phys. Rep.* **2024**, *1053*, 1–84. [CrossRef]
32. Song, F.; Croft, W.B. A general language model for information retrieval. In *Proceedings of the Eighth International Conference on Information and Knowledge Management*; Association for Computing Machinery: New York, NY, USA, 1999; CIKM'99, pp. 316–321. [CrossRef]
33. Morales, J.A.; Sánchez, S.; Flores, J.; Pineda, C.; Gershenson, C.; Cocho, G.; Zizumbo, J.; Rodríguez, R.F.; Iñiguez, G. Generic temporal features of performance rankings in sports and games. *EPJ Data Sci.* **2016**, *5*, 33. [CrossRef]
34. Iñiguez, G.; Pineda, C.; Gershenson, C.; Barabási, A.L. Dynamics of ranking. *Nat. Commun.* **2022**, *13*, 1646. [CrossRef]
35. Ferrer i Cancho, R. The variation of Zipf's law in human language. *Eur. Phys. J. B* **2005**, *44*, 249–257. [CrossRef]
36. Evans, D.R.; Larsen-Freeman, D. Bifurcations and the Emergence of L2 Syntactic Structures in a Complex Dynamic System. *Front. Psychol.* **2020**, *11*, 2823. [CrossRef]
37. Rubin, E.J.; Gess, R. *Theoretical and Experimental Approaches to Romance Linguistics*; Current Issues in Linguistic Theory; John Benjamins Publishing Company: Amsterdam, The Netherlands, 2005.
38. Ljubešić, N.; Fišer, D. A global analysis of emoji usage. In Proceedings of the 10th Web as Corpus Workshop, Berlin, Germany, 7–12 August 2016; pp. 82–89.
39. Seargeant, P. *The Emoji Revolution: How Technology Is Shaping the Future of Communication*; Cambridge University Press: Cambridge, UK, 2019.
40. Zappavigna, M. Searchable talk: The linguistic functions of hashtags. *Soc. Semiot.* **2015**, *25*, 274–291. [CrossRef]
41. Shuai, X.; Pepe, A.; Bollen, J. How the Scientific Community Reacts to Newly Submitted Preprints: Article Downloads, Twitter Mentions, and Citations. *PLoS ONE* **2012**, *7*, e47523. [CrossRef]
42. Hemsley, J.; Stromer-Galley, J.; Semaan, B.; Tanupabrungsun, S. Tweeting to the target: Candidates' use of strategic messages and @mentions on Twitter. *J. Inf. Technol. Politics* **2018**, *15*, 3–18. [CrossRef]
43. Auxier, B.; Anderson, M. Social Media Use in 2021. *Pew Res. Center* **2021**. Available online: https://pewrsr.ch/3cYWjHA (accessed on 22 August 2024).
44. Wei, J.; Tay, Y.; Bommasani, R.; Raffel, C.; Zoph, B.; Borgeaud, S.; Yogatama, D.; Bosma, M.; Zhou, D.; Metzler, D.; et al. Emergent abilities of large language models. *arXiv* **2022**, arXiv:2206.07682.

entropy

MDPI

Article

Still No Evidence for an Effect of the Proportion of Non-Native Speakers on Natural Language Complexity

Alexander Koplenig

Department of Lexical Studies, Leibniz Institute for the German Language (IDS), 68161 Mannheim, Germany; koplenig@ids-mannheim.de

Abstract: In a recent study, I demonstrated that large numbers of L2 (second language) speakers do not appear to influence the morphological or information-theoretic complexity of natural languages. This paper has three primary aims: First, I address recent criticisms of my analyses, showing that the points raised by my critics were already explicitly considered and analysed in my original work. Furthermore, I show that the proposed alternative analyses fail to withstand detailed examination. Second, I introduce new data on the information-theoretic complexity of natural languages, with the estimates derived from various language models—ranging from simple statistical models to advanced neural networks—based on a database of 40 multilingual text collections that represent a wide range of text types. Third, I re-analyse the information-theoretic and morphological complexity data using novel methods that better account for model uncertainty in parameter estimation, as well as the genealogical relatedness and geographic proximity of languages. In line with my earlier findings, the results show no evidence that large numbers of L2 speakers have an effect on natural language complexity.

Keywords: language complexity; language models; linguistic niche hypothesis; language typology; non-native speakers; quantitative linguistics

1. Introduction

The linguistic niche hypothesis proposes that the social niche a language occupies in a community affects its structural properties. Specifically, according to the linguistic niche hypothesis, languages with large numbers of speakers tend to simplify their grammar and have a reduced structural complexity [1,2].

The linguistic niche hypothesis assumes that languages that are spoken by more people over greater geographic areas will, on average, also be learned by a greater proportion of L2 learners, i.e., often adults. Since complex morphology appears to be difficult to learn for adults, the linguistic niche hypothesis conjectures that there should be a negative selection over time against such hard-to-learn morphological paradigms for languages with a larger number of L2 speakers compared to languages that are mainly learned during childhood as L1 (first language). This, in turn, it is argued, explains the observed negative statistical association between speaker population size and morphological complexity [2,3]. In a recent study published in *Royal Society Open Science* [4], I have pointed out that since the argument outlined above is inductive by nature, its validity cannot be simply taken (more or less implicitly) for granted. Crucially, Lupyan and Dale [1] use the estimated speaker population size as a proxy for the proportion of L2 speakers [5]. In my paper [4], I tested this conjecture empirically for more than 2000 languages and showed that the results question the idea of the impact of non-native speakers on the grammatical and statistical structure of languages.

The main obstacle in this context is the fact that, as [6] points out, estimations regarding a breakdown of L1 versus L2 populations are very limited. In general, most information regarding speaker population sizes/compositions is based on *Ethnologue* [7], the most comprehensive and most widely consulted catalogue of languages that provides information and statistics for the languages of the world [8,9]. *Ethnologue* categorises each language in regard to how endangered it is using the Expanded Graded Intergenerational Disruption Scale (EGIDS) [10]. In this context, a language is categorised as vehicular if it is used as an L2 in addition to being used as an L1.

This information can be used to indirectly gain information about the proportion of L2 users: "A language at EGIDS 4 or lower is, by definition, a local language and L2 users are not expected. However, languages at EGIDS 3 and higher are vehicular and, by definition, they should have a significant number of L2 users" [7]. The great advantage here is that information at the EGIDS level is available for all languages that are listed in the Ethnologue. Table 1 provides an overview of each EGIDS level together with the mapping to vehicularity, a corresponding description taken from [10] and example languages taken from my dataset. In my paper, I used vehicularity as an indicator of whether a language is used by a large proportion of L2 speakers or not in order to test the assumed relationship between the L2 proportion and (morphological/information-theoretic) complexity. Through a series of statistical analyses, I tried to show that large L2 proportions do not seem to affect the (grammatical or information-theoretic) complexity of a language.

Table 1. Ethnologue's EGIDS scale: column 1: EGIDS value and label; column 2: type (vehicular/non-vehicular); column 3: description taken from [10]; and column 4: three example languages taken from my dataset.

EGIDS Value	Type	Description	Examples
0—International	vehicular	"The language is used internationally for a broad range of functions."	English, Mandarin Chinese, Spanish
1—National		"The language is used in education, work, mass media, government at the nationwide level."	Bengali, Standard German, Ukrainian
2—Regional		"The language is used for local and regional mass media and governmental services."	Gujarati, Igbo, Uyghur
3—Trade		"The language is used for local and regional work by both insiders and outsiders."	Hakha Chin, Tagalog, Yue Chinese
4—Educational	non-vehicular	"Literacy in the language is being transmitted through a system of public education."	Bhojpuri, Javanese, Santhali
5—Written		"The language is used orally by all generations and is effectively used in written form in parts of the community."	Hakka Chinese, Madura, Sunda
6a—Vigorous		"The language is used orally by all generations and is being learned by children as their first language."	Central Pashto, Central Atlas Tamazight, San'ani Arabic
6b—Threatened		"The language is used orally by all generations but only some of the child-bearing generation are transmitting it to their children."	Aceh, Occitan, Peripheral Mongolian
7—Shifting		"The child-bearing generation knows the language well enough to use it among themselves but none are transmitting it to their children."	Breton, Central Lalo, Tu
8a—Moribund		"The only remaining active speakers of the language are members of the grandparent generation."	Baniwa, Soqotri, Yaqui
8b—Nearly Extinct		"The only remaining speakers of the language are members of the grandparent generation or older who have little opportunity to use the language."	Qimant, Ratahan, Soo
9—Dormant		"The language serves as a reminder of heritage identity for an ethnic community. No one has more than symbolic proficiency."	Clallam, Mogholi, Quileute
10—Extinct		"No one retains a sense of ethnic identity associated with the language, even for symbolic purposes."	Madngele, Warrgamay *

* Note that for EGIDS value = 10, there are only two available languages in the used data.

In a recent comment published in the *Journal of Language Evolution*, Kauhanen, Einhaus and Walkden ([11]; KEW) challenge my findings. KEW criticised both the use of vehicularity as a (binary) indicator of whether a language has a significant number of L2 users and the idea of imputing a zero proportion of L2 speakers to non-vehicular languages whenever a direct estimate of that proportion is unavailable. While I recognise the importance of post-publication commentary on published research, I will show in this paper that both points of criticism are explicitly mentioned and analysed in my paper. In addition, I will also comment on other points raised by KEW and demonstrate that both alternative analyses offered by KEW do not stand up to closer scrutiny.

However, I agree with KEW that a further study using both better data and better methods would certainly be desirable since testing for a link between language and social structure is turning out to be more complex than I once thought, as recently summarised in an excellent review on the subject [12].

In my original paper, information-theoretic complexity was estimated with a comparatively simple non-parametric statistical entropy estimation method [13] and was solely based on a very peculiar text type, i.e., parallel translations of the Bible, and there are several important challenges that the use of the Bible as a parallel text source pose [14–16]. Additionally, complexity was measured only on the level of characters, which is problematic due to cross-linguistic differences in the mapping between phonemes and graphemes [17,18]. This paper addresses these points by adding new data on the information-theoretic complexity of languages, with estimates derived at the levels of characters, words and sub-word units using an ensemble of different language models, ranging from simple statistical models to advanced neural networks. Estimates are based on a database of 40 multilingual text collections comprising a wide variety of text types, including nearly 1.5 billion words across more than 3700 documents in over 1100 languages [19,20].

With respect to improved methods, I re-analysed the novel information-theoretic complexity data using a frequentist multi-level multi-model averaging approach [20,21], which accounts for model uncertainty related to language- and document-specific characteristics when estimating model parameters. Additionally, to better control for the genealogical and geographic relatedness of languages, I re-analysed the morphological complexity data using an approach that combines spatial autoregressive modelling [22] with frequentist multi-model averaging. Consistent with my earlier findings [4], the results indicate that the presence of large numbers of L2 speakers has no effect on complexity across languages when controlling for the estimated speaker population size.

2. Materials and Methods

Some material in this section is recycled from my prior publications [4,19,20] in accordance with the guidelines provided by the Text Recycling Research Project [23].

2.1. Original Data

Population estimates and language information. Basic information on different languages and genealogical classifications is taken from [24]. Information on geographical language areas is taken from [25]. Speaker population size and geographical range size estimates are taken from [26], who report the total number of L1 speakers based on information from *Ethnologue* [27] and calculate range sizes in km^2 based on information from Global Mapping International [28]. Aggregated information on vehicularity and L2 proportions are taken from *Ethnologue* [27].

Languages with an EGIDS value of 0, 1, 2 or 3 are categorised as vehicular, while languages with an EGIDS value of 4 to 10 are categorised as non-vehicular. Aggregated *L2* speaker proportions are taken from [27] and from [29]. The different sources are merged via the three-letter language-specific ISO 639-3 code. For a critical yet balanced assessment of *Ethnologue's* strengths and limitations, see Hammarström [9].

Morphological complexity. To construct an index of morphological complexity, ref. [30] extracted information on 28 relevant features of morphology from the *World Atlas of Languages Structures* [31] (WALS). For example, the WALS's chapter 30A, "Number of Genders," gives a range of 5 values from "None" to "Five or more". These values are then mapped to the values 1 to 5, where higher values are indicative of higher complexity. The values of each feature are normalised to the interval [0,1]. The morphological complexity score C is then calculated by summing the normalised features divided by the number of available features. Let f_i be the normalised value of feature i, and N_F be the number of features that are available in the corresponding languages, and then C can be written as:

$$C = \frac{1}{N_F} \times \sum\nolimits_{i=1}^{N_F} f_i \quad (1)$$

Greater values are indicative of higher morphological complexity; for more details and a list of all used WALS features, cf. [30]. In total, there are 1713 languages with at least one available feature. It is important to note that the amount of available WALS information varies greatly for different languages [1,5], e.g., there are only 10 languages for which information on all 28 features is available [30], but there are 393 understudied languages with only 1 or 2 available features. To account for this data sparseness, separate analyses are conducted: (i) full: a full version that incorporates all languages that have information for at least one available WALS feature; and (ii) subset: a version for a subset of languages with available information for at least six features (50% of all languages have information on at least six features).

Information-theoretic complexity. The average per-symbol information content or entropy rate of a text can be interpreted as a measure of complexity [32,33]: the harder it is, on average, to predict upcoming text, the higher the entropy rate and the greater the complexity of the text as a whole [34–37]. I used estimates for the Gospel of Mark in more than 1000 different languages based on the Parallel Bible Corpus [24], which are taken from [38]. For languages with more than one available translation, entropy estimates are averaged.

Entropy rates are estimated on the basis of the non-parametric method of [13,36] that builds on the key idea of the Lempel–Ziv compression algorithm [39]. This method does not require any prior training, produces robust estimates without the need for very long strings as input and is able to take into account the very long-range correlations typical of literary texts [40,41] that are not captured by direct parametric Markovian or "plug-in" estimators [36]. If we represent a text t as a symbolic sequence of N characters, i.e., $t = \{c_1, c_2, \ldots, c_{N-1}, c_N\}$ where c_i represents any character (including white spaces and punctuation marks) in the text at position i, the entropy rate can be estimated as [36]; cf. Equation (1):

$$H_t = \left[\frac{1}{N} \sum_{i=2}^{N} \frac{l_i}{\log_2(i)} \right]^{-1} \quad (2)$$

Here, the key quantity of interest is the match-length l_i. In order to determine the redundancy at position i, we examined the whole portion of the text up to (but not including) i and monitored how many of the initial characters of the text portion starting at i have already occurred in the same order somewhere in the preceding text, and recorded the length of the longest continuous substring. Our key quantity of interest l_i was obtained by adding 1 to the longest match-length. More details of this approach can be found in [38].

In total, the dataset includes information for 2143 different languages, with 1088 data points for entropy rates and 1581 for morphological complexity. Of these languages, 1902 are categorised as non-vehicular, while the remaining 241 are vehicular. The median estimated speaker population size across languages is 15,000. These languages represent a total of 126 language families, with significant representation from families such as Niger-Congo (16.99%), Austronesian (14.61%), Trans-New Guinea (7.93%), Sino-Tibetan (5.23%), Afro-Asiatic (4.95%) and Indo-European (4.48%), among others. The data can be downloaded from https://dx.doi.org/10.6084/m9.figshare.c.4400675 (accessed on 8 November 2024).

2.2. Additional Data on Information-Theoretic Complexity

With respect to the information-theoretic complexity estimates used in my original paper [4], there are three potential issues: (i) complexity was estimated solely at the character level; (ii) complexity was calculated using a rather simple non-parametric statistical entropy estimation method; and (iii) language-specific estimates were based only on parallel translations of the Gospel of Mark of the Bible (see Section 2.1). To address these issues in what follows, I used information-theoretic complexity estimates derived from a multilingual database that I compiled [19,20].

Regarding (i), the estimation at the character level is problematic due to cross-linguistic differences in the mapping between phonemes and graphemes [17,18]. For example, languages with deep orthographies, like English, have inconsistent mappings (e.g., "ough" in "thought" vs. "through" vs. "dough"), while languages with shallow orthographies, like Spanish, have more consistent phoneme-to-grapheme correspondences (e.g., "a" in "casa" is always pronounced the same). To mitigate this problem, information-theoretic complexity in [20] is measured at multiple levels: characters, words and the supra-character but sub-word level, by applying byte-pair encoding (BPE) [42,43]. BPE is a sub-word segmentation technique that iteratively merges the most frequent pairs of characters or character sequences, creating sub-word units that capture meaningful linguistic patterns. It plays a crucial role in modern language modelling by effectively handling morphological variations and rare words, thereby enhancing model performance across diverse languages. Moreover, BPE's ability to reveal language-specific sub-word patterns makes it particularly valuable in cross-linguistic investigations, as it highlights structural differences and typological features that are unique to each language, as recently discussed in [44].

Regarding (ii), the simplicity of the entropy estimator used in my initial paper [4], as described above, may have influenced the results. To address this, ref. [20] trained an ensemble of seven different types of language models on each document in the multilingual database, ranging from simple statistical n-gram models to state-of-the-art neural networks and transformer models. In this paper, I used the estimates from the best-performing language model for each document. For more details on language modelling and estimation, see [20].

Regarding (iii), the use of the Bible as a parallel text source poses significant challenges; for details, see [14–16]. Of particular importance is the fact that many Bible translations for minority languages, especially those produced by missionaries after World War II, were created with a specific missionary purpose in mind. These translations aim to convey the message in a manner that is easily understandable to readers or listeners, often including a considerable amount of redundant explicative content. This can complicate cross-linguistic analyses of the potential impact of social structure on language structure [16]. The compiled database helps to overcome this issue, as it includes 40 multilingual text collections encompassing a wide variety of parallel texts, such as religious texts, legal documents, movie subtitles and machine translations. Additionally, the database includes comparable corpora, which are not parallel but come from similar sources, such as newspaper articles, web crawls, Wikipedia entries, and system message translations from the Ubuntu operating system.

Overall, the database includes nearly 1.5 billion words across 3705 documents in 1104 languages. Of these languages, 882 are classified as non-vehicular, with the remaining 222 designated as vehicular. The median estimated speaker population size for these languages is 69,796. The dataset spans 99 language families, with Niger-Congo (19.38%), Austronesian (16.49%), Indo-European (9.06%), Trans-New Guinea (7.70%) and Otomanguean (4.35%) among the most represented. Further details on database compilation, including data pre-processing, document preparation, language modelling and complexity estimation methods, are available in [19,20]. The data can be downloaded from https://osf.io/xdwjc/ (accessed on 8 November 2024).

2.3. Additional Methods

To test if vehicularity significantly predicts information-theoretic complexity, I ran separate models with the entropy rate h as a measure of the information-theoretic complexity as the outcome on all three levels (words/characters/BPE). For $N = 3705$ individual documents, I fitted parametric linear mixed multi-level models (LMMs) of the form [45]:

$$\mathbf{y} = \mathbf{X}\boldsymbol{\beta} + \mathbf{Z}\mathbf{u} + \boldsymbol{\epsilon} \tag{3}$$

where $\mathbf{y}$ is the $N \times 1$ vector of estimated values of h; $\mathbf{X}$ is the $N \times p$ design matrix of p covariates, including a $N \times 1$ vector of ones for the intercept; $\boldsymbol{\beta}$ is the corresponding $p \times 1$ vector of coefficients; $\boldsymbol{\epsilon}$ is the $N \times 1$ vector of residuals, $\mathbf{Z}$ is a matrix of random predictors; and $\mathbf{u}$ is a vector of random effects that are assumed to follow a normal distribution, with mean 0 and variance-covariance matrix $\mathbf{G}$. The residual errors $\boldsymbol{\epsilon}$ are assumed to follow a normal distribution, with mean 0 and variance matrix $\sigma^2\mathbf{I}$; $\mathbf{u}\perp\boldsymbol{\epsilon}$. As fixed effects, I considered fixed effects for vehicularity, the speaker population size and their interaction. In addition, the following control variables were considered: corpus type (parallel/comparable), text length in symbols and the number of countries in which each language is spoken.

All quantitative variables (including the outcomes) are log-transformed. Text length is measured in words for words as symbols and in characters on both the character and BPE levels. The rationale behind considering the number of countries as a potential covariate is to account for the varying degrees of pluricentrism [46]. For instance, languages such as English or Spanish are spoken in several countries and may, therefore, have different codified standard forms.

Random intercepts for the following groups were included: the writing script, corpus, macro-area, macro-family, country, sub-family and language. I crossed corpus, macro-area, macro-family and writing script, and explicitly nested countries within macro-areas and language within sub-family within macro-family. To account for variations within language families and geographical units [12,47], I included random slopes for population size, i.e., the effect of population size is allowed to vary across the following groups: macro-area, country, macro-family and sub-family. In models without a fixed effect for speaker population size, potential random slopes were also excluded since excluding the fixed effect for speaker population size while including random slopes would constrain β_x to be zero and thus force the random slopes to be evenly distributed around a slope of zero.

Given the absence of clear theoretical or empirical reasons to determine which variables to include as controls, I adopted a multi-model inference approach [21] by sub-setting each full model, i.e., per symbolic level (words, characters, BPE), I generated a set of R = 17,600 candidate models with all possible variable subsets, which were then fitted to the data [20]. All models were fitted with gradient-based maximisation (maximal number of 20 iterations) and via maximum likelihood (ML). Per type and for each fixed effect, I then computed a frequentist model averaging (FMA) estimator over all R candidate models [21,48]:

$$\tilde{\beta}_x = \sum_{j=1}^{R} \omega_j \beta_{x,j} \tag{4}$$

where $\beta_{x,j}$ denotes the estimated fixed effect of variable x for model j, and ω_j is a weight computed as:

$$\omega_j = \frac{e^{\left(-\frac{1}{2}\Delta_j\right)}}{\Omega} \tag{5}$$

where $\Omega = \sum_{r=1}^{R} e^{\left(-\frac{1}{2}\Delta_r\right)}$ represents the sum of weights for all R models. To compute Δ_j, I used AIC, $\Delta_j = \text{AIC}_j - \text{AIC}_{min}$, where AIC_j denotes the AIC value computed for model j and AIC_{min} represents the minimum AIC value over all R models. Note that in models where x does not appear, $\beta_{x,j} \equiv 0$. On this basis, I computed an FMA estimator of the standard error (SE) as [21]:

$$\text{SE}(\tilde{\beta}_x) = \sum_{j=1}^{R} \omega_j \sqrt{\text{SE}\left(\beta_{x,j}\right)^2 + \left(\beta_{x,j} - \tilde{\beta}_x\right)^2} \tag{6}$$

where $\text{SE}\left(\beta_{x,j}\right)$ denotes the estimated standard error of $\beta_{x,j}$ for model j. In models where x does not appear, I set $\text{SE}\left(\beta_{x,j}\right) \equiv 0$. To assess the statistical significance, I computed a corresponding two-tailed p-value as $p = 2 \cdot \left(1 - \Phi\left(\left|\frac{\tilde{\beta}_x}{\mathrm{SE}\left(\tilde{\beta}_x\right)}\right|\right)\right)$, where $\Phi()$ denotes the cumulative standard normal distribution. Since the Akaike weights ω_j can be viewed as approximate probabilities, indicating the likelihood of each model being the best one given the data, ω_j can be used to estimate the relative importance of variable x, computed as [21]:

$$\sigma_x = \sum_{j=1}^{R} \omega_j c_{x,j} \tag{7}$$

where $c_{x,j}$ is a binary indicator that is equal to 1 if x is explicitly in model j and 0 otherwise [21]. The larger σ_x, the more important x. I computed σ_x for both the fixed effects and the random effects/slopes.

While the LMM multi-model structure specified above offers a potential solution that better accounts for both model uncertainty and the genealogical and geographic relatedness of languages compared to the approach applied in my original paper, this approach may still be incomplete if there are patterns of relatedness within random groups for which no random slopes are included—such as those below the level of sub-families or countries [12]. For this reason, an approach where relatedness is explicitly modelled as a function of spatial or phylogenetic distance is potentially preferable [12]. For the analysis of information-theoretic complexity, however, such an approach is neither feasible nor desirable due to the unbalanced nature of my multilingual database (for details, see [19]). In this context, an LMM-based method has the significant advantage of providing a principled way to account for variations between group levels while simultaneously avoiding overfitting within groups [49]. On the other hand, the data available for morphological complexity are balanced in this respect, making it possible to estimate a spatial autoregressive (SAR) model, which can be written as [22]:

$$\mathbf{y} = \mathbf{X}\boldsymbol{\beta} + (\mathbf{I} - \varphi\mathbf{W})^{-1}\boldsymbol{\epsilon} \tag{8}$$

where, in addition to the above (Equation (3)), φ is the autoregressive parameter; $\mathbf{W}$ is an $N \times N$ spatial weight matrix, representing the spatial structure between observations; and $\mathbf{I}$ represents the identity matrix. I constructed two different inverse distance matrices, (i) to control for geographic proximity, a matrix $\mathbf{W_{geo}}$ based on the geographical distances between languages was constructed, and (ii) to control for phylogenetic relatedness, a matrix $\mathbf{W_{phylo}}$ based on a phylogenetic similarity matrix provided by [50] was used. In both cases, the matrix elements are equal to the reciprocal of distance that are then normalised using spectral normalisation. In total, information on both morphological complexity, $\mathbf{W_{geo}}$ and $\mathbf{W_{phylo}}$, is available for N = 1443 languages. As above and in my original paper, I present separate analyses for the full and subset versions.

As covariates, I considered vehicularity and three quantitative variables: (i) language speaker population size, (ii) language range size and (iii) number of countries. Additionally, I included separate first-order interactions between vehicularity and each quantitative variable. Once again, I adopted an FMA inference approach by sub-setting each full model. For both the full dataset (N = 1443) and the subset (N = 804), I generated a model space of R = 35 SAR models with all possible variable subsets, fitting them to the data while modelling the spatial dependence based on either $\mathbf{W_{geo}}$ or $\mathbf{W_{phylo}}$. I then proceeded, as described above, to compute averaged estimates, statistical significance and the relative importance of each considered variable. For the FMA approach, I used an estimator that maximises the likelihood $\hat{L}$. Additional analyses were carried out using a generalised spatial two-stage least-squares estimator (GS2SLS) [51] that allows for the inclusion of more than one spatially lagged error term but does not maximise $\hat{L}$.

One of the advantages of an FMA-based approach is its ability to explore large model spaces comprehensively. In this context, note that, as written above, vehicular languages are defined as those with an EGIDS value of 0, 1, 2 or 3 (see Table 1). According to

Ethnologue's classification system, all of these EGIDS categories are expected to have a significant number of L2 speakers, with a plausible assumption that the lower the value, the higher the L2 proportion.

This thus provides the possibility of a more in-depth test of the linguistic niche hypothesis. To achieve this, I constructed a second model space consisting of $R = 8695$ candidate models, where I considered binary indicators for '0—International', '1—National', '2—Regional' and '3—Trade'. As before, I included three quantitative variables—(i) language speaker population size, (ii) language range size, and (iii) number of countries—along with first-order interactions between the quantitative variables and the binary EGIDS indicators.

All statistical analyses were carried out using Stata 18/MP. Commented code to reproduce all results is available at https://osf.io/fypx5/.

3. Results

3.1. A Rebuttal of the Critique of Kauhanen, Einhaus and Walkden (2023)

KEW criticised the use of vehicularity as a proxy of whether a language is likely to have significant numbers of L2 speakers. They stated (p. 3): "In Koplenig's analysis, languages with an EGIDS score of 3 or lower are defined to be vehicular, the rest being non-vehicular". I believe it is important to point out that this is not my definition or mapping but how languages are categorised by the *Ethnologue* [10]. KEW (p. 3) rightfully pointed out that a "considerable number of non-vehicular languages are reported by Ethnologue to be used as an L2 even though no numerical estimate of L2 users is given." I fully agree that this inconsistency is problematic and that it is thus important to ask if vehicularity is a good proxy for whether a language is used as an L2. I explicitly discussed this in the concluding section of my original paper. Here, I quote the editors of the *Ethnologue*: "Based on the use of the phrase 'vehicular language' by some as a synonym for lingua franca, we use the term vehicular to refer to the extent to which a language is used to facilitate communication among those who speak different first languages. If a language is characterised here as being Vehicular, it is used by others as an L2 in addition to being used by the community of L1 speakers." ([10]; see also Figure 1 therein). Based on this assessment, I believe that it is appropriate to use vehicularity in order to test the linguistic niche hypothesis: a language that is defined as vehicular should—according to the *Ethnologue*—be a language that is "used for communication between strangers" (KEW, p. 1; also see [52]) and "should have a significant number of L2 users" [7], p. 20. Thus, if the linguistic niche hypothesis holds, we should expect that there is a statistical association between vehicularity and complexity. In my paper [4], I demonstrated that this is not the case for either morphological or information-theoretic complexity when controlling for speaker population size.

Importantly, the problem of there being non-vehicular languages for which *Ethnologue* reports a proportion of L2 users greater than 0 is not concealed by me, but explicitly mentioned in Section 2.2 of my paper [4] and—as also mentioned therein—additional analyses are presented and discussed in Section 7 of the accompanying supplementary material, where languages categorised as non-vehicular but with L2 proportions greater than zero are excluded. The reported results generally support the results presented in the main part of the paper. KEW do not mention or take these analyses into account.

In a set of further statistical analyses, I used vehicularity to impute missing values: In correspondence with the categorisation scheme of the *Ethnologue* [7], non-vehicular languages with no available information on L2 users are assigned an L2 proportion of 0. KEW are right to point out that this step is worth discussing since this zero-imputation strategy affects almost all non-vehicular languages. Importantly, however, imputed values are only used for the non-parametric Spearman correlation analyses. Here, I tested whether there is a significant (determined by non-parametric permutation tests) monotonic relationship between (morphological or information-theoretic) complexity and the L2 proportion after removing the effect of speaker population size and vice versa (correlating complexity and speaker population size while controlling for the L2 proportion). Since, as mentioned in the paper [4] (Section 2.6), Spearman correlation coefficients and part Spearman correlation coefficients can be computed as a Pearson's correlation coefficient on the ranks of the two variables, where the zero-imputation strategy implies that all non-vehicular languages are assigned the lowest rank in each analysis—an assumption that I believe is reasonable but worthy of discussion. KEW (Section 5.1) presented three so-called complete case analyses, where all cases with missing information are removed (no imputation). Using parametric LMMs (morphological or information-theoretic) complexity is predicted by the fixed effects of the (log of) population size and the L2 proportion and a random intercept for the language family. Additionally, the two models with morphological complexity as the outcome also include a random intercept for the linguistic area. Based on their results (Tables 2 and 3), KEW argued that "population size and the proportion of L2 speakers have a declining effect on morphological complexity, and both predictors are statistically significant." (p. 6). For information-theoretic complexity as the outcome (Table 4), KEW found no evidence "for an effect of either the proportion of L2 speakers or population size" (p. 6). There are two major methodological problems with KEW's models: (i) KEW did not include any random slopes in their models due to convergence issues with the software package they used to fit LMMs; and (ii) the estimates in KEW's models are derived by restricted maximum likelihood (REML). This is highly problematic because they use Akaike's information criterion for model selection, which is defined as $\text{AIC} = -2\log(\hat{L}) + 2k$, where lower values indicate a better model and where $\log(\hat{L})$ represents the log-likelihood of the model, and k represents the number of estimated parameters [21]. For example, on p. 6, KEW argued that they "do not include interactions between the covariates in any of our models, as doing so always leads to a worse model when quantified on AIC". However, when different sets of fixed effects are considered, estimates must not be derived by REML, but by ML [53–55] since REML does not provide the full likelihood $\hat{L}$ of the model, using AIC with REML is inappropriate (REML only provides a restricted likelihood that accounts for random effects but not for thefixed effects). To solve both (i) and (ii), I generated a set of 72 candidate models consisting of all possible combinations of fixed effects for speaker population size, the L2 proportion and their interaction, as well as crossed random intercepts for language family and linguistic area and random slopes for both speaker population size and the L2 proportion for both random effects (for 16 out of 171 languages, information on the linguistic area were missing. In these cases, I manually imputed the data using the original definitions provided by the Autotyp project [25]. Further noting that here and in what follows, models that include interactions are only considered if the corresponding main effects are also included in the model). I used Stata/MP 18 for the LLMs, estimates were derived by ML, and models were fitted with gradient-based maximisation. Of a total of 216 models, 214 (or 99.07%) converged to an optimal solution, thus pointing towards problems in KEW's analyses. Table 2 summarises the results. I first checked for all three outcome versions of whether the model with the lowest AIC includes the L2 proportion as a fixed effect; row 2 of Table 2 shows that this is only the case for the versions where morphological complexity is the outcome. This means that including the speaker L2 proportion does not improve the model fit in the case of information-theoretic complexity. I then extracted the best models that include a fixed effect for the L2 proportion. Rows 4—6 list the model structure per outcome. Row 6 shows that there is only evidence for a significant effect of the L2 proportion for morphological complexity as the outcome for the subset version, i.e., when only languages with available information for at least six WALS features are considered, while for both other outcome versions, there is no evidence for a significant effect of the L2 proportion on complexity at any standard level of significance. The results obtained are not altered when an alternative version of Aikaike's information criterion, AICc [56], is used instead of AIC. AICc accounts for the sample size by including an additional bias correction term; see also Burnham and Anderson [21].

Table 2. Overview of the results of the complete case linear mixed multi-level model analyses for each outcome. β_{logPop}—estimated coefficient for the log of speaker population size; β_{L2prop}—estimated coefficient for the L2 proportion; $\beta_{\text{interaction}}$—estimated coefficient for the interaction between the L2 proportion and the log of speaker population size.

Row		Outcome		
		MC		H
1		Version: Full	Version: Subset	
2	Does the best model include a fixed effect for the L2 proportion?	Yes	Yes	No
3	Fixed effects	Population, L2 proportion and their interaction	Population and L2 proportion	Population and L2 proportion
4	Random intercepts	Family and Area	Area	Area
5	Random slopes	L2 proportion for Family	-	-
6	Estimated effects for the best model that includes a fixed effect for the L2 proportion, shown are the estimated beta coefficients for each fixed effect (parametric p-values in parentheses)	$\beta_{\text{logPop}} = -0.010$ ($p = 0.138$) $\beta_{\text{L2prop}} = -0.065$ ($p = 0.738$) $\beta_{\text{interaction}} = -0.026$ ($p = 0.134$)	$\beta_{\text{logPop}} = -0.013$ ($p < 0.01$) $\beta_{\text{L2prop}} = -0.217$ ($p < 0.01$)	$\beta_{\text{logPop}} = 0.032$ ($p < 0.01$) $\beta_{\text{L2prop}} = -0.117$ ($p = 0.212$)
7	Number of languages in the sample	148	101	94
8	How many languages in the sample have an L2 proportion > 0?	97.30%	96.04%	100.00%
9	Median L2 proportion	0.16	0.16	0.19
10	Median speaker population size	786,500	1,480,000	2,942,020
11	Spearman correlation between speaker population size and the L2 proportion (non-parametric permutation p-value in parentheses)	−0.074 ($p = 0.352$)	−0.025 ($p = 0.352$)	−0.151 ($p = 0.140$)

Table 3. Linear mixed multi-level frequentist model averaging results for information-theoretic complexity as the outcome computed on three different symbolic levels (words/characters/BPE). Per symbolic level, quantities are computed based on R = 17,600 candidate models. σ_x-values (Equation (7), rounded to two decimal places) reflect the relative importance of each variable (both fixed and random), with higher values indicating a more important variable. Values above 0.90 are shown in bold. $\tilde{\beta}_x$-values (Equation (4), rounded to four decimal places) are computed for the fixed effects only. $\tilde{\beta}_x$-values that pass a two-tailed parametric significance test at $p < 0.05$ are shown in bold. Sig. *** (**) [*] indicates that $p < 0.001$ ($p < 0.01$) [$p < 0.05$].

Type	Variable	Words		Characters		BPE	
		σ_x	$\tilde{\beta}_x$	σ_x	$\tilde{\beta}_x$	σ_x	$\tilde{\beta}_x$
Fixed effects	Vehicularity	0.39	−0.009	0.66	−0.001	0.69	−0.003
	Population size	**1.00**	**0.009** ***	**1.00**	**0.011** ***	**1.00**	**0.015** ***
	Population size x Vehicularity	0.16	0.001	0.22	−0.000	0.23	−0.001
	No. of countries	0.36	−0.002	0.85	−0.011	**0.93**	**−0.016** *
	Text length	**1.00**	**−0.424** ***	**1.00**	**−0.248** ***	**1.00**	**−0.317** ***
	Parallel (yes/no)	0.42	−0.166	0.62	−0.171	0.75	−0.255

Table 3. *Cont.*

Type	Variable	Words		Characters		BPE	
		σ_x	$\tilde{\beta}_x$	σ_x	$\tilde{\beta}_x$	σ_x	$\tilde{\beta}_x$
Random effects	Writing script	0.27		**1.00**		**1.00**	
	Corpus	**1.00**		**1.00**		**1.00**	
	Macro-area	0.00		**0.97**		0.15	
	Country	**1.00**		0.02		0.84	
	Macro-family	**1.00**		**1.00**		**1.00**	
	Sub-family	**1.00**		**1.00**		**1.00**	
	Language	0.39		**1.00**		**1.00**	
Random slopes	Macro-area	0.00		0.47		0.04	
	Country	0.28		0.01		0.31	
	Macro-family	0.32		0.29		0.58	
	Sub-family	0.81		**0.95**		**0.91**	

Table 4. Spatial autoregressive frequentist model averaging results for morphological complexity as the outcome computed both for the full ($N = 1443$) and the subset version ($N = 804$). Separate models are computed for two different types of weight matrices: $\mathbf{W}_{\mathbf{geo}}$, which is based on the geographical distances and $\mathbf{W}_{\mathbf{phylo}}$, which is based on a phylogenetic similarity matrix. Per combination of WALS feature availability and weight matrix type, quantities are computed based on $R = 35$ candidate models. σ_x-values (Equation (7), rounded to two decimal places) reflect the relative importance of each covariate, with higher values indicating a more important variable. Values above 0.90 are shown in bold. $\tilde{\beta}_x$-values (Equation (4), rounded to four decimal places) that pass a two-tailed parametric significance test at $p < 0.05$ are shown in bold. Sig. *** (**) [*] indicates that $p < 0.001$ ($p < 0.01$) [$p < 0.05$].

Version	Full				Subset			
Type of Weight Matrix	$\mathbf{W}_{\mathbf{geo}}$		$\mathbf{W}_{\mathbf{phylo}}$		$\mathbf{W}_{\mathbf{geo}}$		$\mathbf{W}_{\mathbf{phylo}}$	
Variable	σ_x	$\tilde{\beta}_x$	σ_x	$\tilde{\beta}_x$	σ_x	$\tilde{\beta}_x$	σ_x	$\tilde{\beta}_x$
Vehicularity	**0.98**	0.1106	**1.00**	−0.0428	**1.00**	−0.1626	**1.00**	−0.1853
Population size	**1.00**	**−0.0138** ***	**0.97**	**−0.0102** ***	**1.00**	**−0.0146** ***	**1.00**	**−0.0156** ***
Range size	**1.00**	**0.0155** ***	0.00	0.0000	**1.00**	**0.0169** ***	**1.00**	**0.0210** ***
No. of countries	0.35	0.0022	**1.00**	−0.0144	0.35	−0.0019	0.61	0.0177
Population size x Vehicularity	0.35	0.0030	**0.97**	−0.0004	0.39	0.0030	0.41	0.0039
Range size x Vehicularity	0.28	0.0000	0.00	0.0000	0.48	0.0059	0.42	0.0050
No. of countries x Vehicularity	0.10	−0.0018	0.01	−0.0005	0.10	0.0006	0.19	−0.0034

However, as pointed out in my original paper [4] and discussed further in Section 2 of the supplementary material for a similar sample, it is not clear whether any of the three samples are unbiased: (i) row 8 of Table 2 shows that almost all languages in all three samples have an L2 proportion that is greater than zero with a median estimate (row 9) of more than 15%, which seems rather high given the assumption of the linguistic niche hypothesis that most languages have almost no L2 speakers.

(ii) Compared to the median estimated speaker population size of 7000 for all languages listed by the *Ethnologue* [1], the median estimated speaker population size in the samples ranges between 786,500 and 2,942,020 (row 10). To place this into perspective, I drew 1,000,000 random samples of all 6880 living languages listed by *Ethnologue* [7], with each sample consisting of 100 languages. For each sample, I computed the median population size. The median estimate for the median population size is 8000, and the sample with the highest median population size is 51,000. Thus, the three samples used for KEW's complete case analyses have a median population size that is several orders of magnitude larger than what we would expect from a random sample, and the probability of randomly drawing a sample like the one used in the complete case analyses above is less than one in a million. This calls into question the appropriateness of standard parametric frequentist approaches [16,57], which is why, in my original paper [4], I used non-parametric tests that do not make any assumptions regarding the stochastic mechanism that generated the data [58].

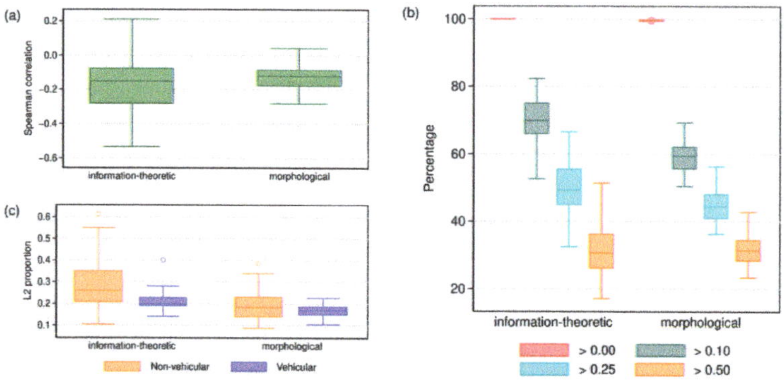

Figure 1. Descriptive results of KEW's multiple imputation approach. Per type of complexity (information-theoretic or morphological), computations are based on the 100 completed samples from KEW's multiple imputation analysis. (**a**) Spearman correlation between the imputed L2 proportion and speaker population size and (**b**) percentage of languages that have an L2 proportion of (i) more than 0, (ii) more than 0.10, (iii) more than 0.25 and (iv) more than 0.50 per type of complexity (information-theoretic or morphological). (**c**) Median L2 proportion for non-vehicular and vehicular languages per type of complexity.

(iii) There is no positive Spearman correlation between the estimated size of the speaker population and the estimated proportion of L2 speakers in any of the three samples (row 11). As I have written above and in my original paper [4], this is actually a key assumption of the linguistic niche hypothesis

Against this background, I would like to reiterate that KEW are right to point out that the imputation step is worth discussing; I do so myself in Section 2.2 of my original paper [4] and present complete case analyses without any imputation (similar to that presented by KEW, but without making parametric assumptions) in Section 10 of the supplementary material of my original paper [4]. Again, KEW neither mention nor seem to take this into account.

In their second set of analyses, KEW utilised a technique called multiple imputation to fill in the missing data [59–61]. As I will show in what follows, while the idea of imputing data in a scenario with a missingness rate of approximately 92% (KEW, p. 10) may seem highly desirable at first glance, a closer examination reveals that KEW's approach suffers from severe and systematic biases that render it an unreliable method for accurately filling in missing data in this context. Multiple imputation offers a flexible, simulation-based technique to handle missing data consisting of three steps [62]: (i) setting up an imputation model and generating m imputations (completed datasets), in KEW's analyses, $m = 100$; (ii) the completed datasets are separately analysed with standard statistical techniques; (iii) the results obtained in (ii) are pooled to provide estimates of the parameters of interest, accounting for the uncertainty due to missing data. KEW rightfully pointed out that the technique is based on the assumption that the data are either missing completely at random (MCAR) (i.e., missingness for a variable x is unrelated to the observed values of both other variables and the unobserved values of x), or missing at random (MAR) (i.e., missingness on x is uncorrelated with the unobserved value of x after other variables in the dataset have been used to predict missingness on x). Or, explained differently, after controlling for the observed variables, the probability of missingness is independent of the true value of x. If this is not true, the data are said to be missing not at random (MNAR), if the value of x itself predicts missingness [63,64]. In general, multiple imputation methods assume that data are MCAR or MAR and not MNAR. KEW (p. 11) argued that "In the current state of understanding, we feel it would be premature to conclude one way or the other, but we point out that any argument to the effect that L2 speaker proportions are MNAR would need to specify a mechanism whereby such missingness arises." Unfortunately, there is no formal test to answer this question since, as written above, the data that would be needed to determine this are, themselves, missing. However, there is a different way of representing MAR [65]: MAR implies that the distribution of x, given our imputation model, is the same whether or not x is observed. If the data are MNAR, however, the chance of observing a value of x depends on x, even after conditioning on our model. In this case, based on our imputation model, the observed data do not provide information on how the missing values differ from the observed ones [65]. As an imputation model ($\mathcal{M}_{\text{KEW}}$) to impute L2 proportions, KEW used an LMM with the (logit-transformed) L2 proportion as outcome, random intercepts for (either morphological or information-theoretic) complexity, the log of population size and the log of language range size. Now, the question is whether the chance of observing a value of the L2 proportion really does not depend on this value itself after conditioning on $\mathcal{M}_{\text{KEW}}$. I would argue that it does, or do we really believe that $\mathcal{M}_{\text{KEW}}$ tells us *all* there is to know regarding the question of how the missing values of the L2 proportions are different from the corresponding observed values? KEW (p. 11) sketched one such mechanism: a "greater proportion of L2 speakers in a speech community increases, in general, the access that outsiders have to that community, and hence also increases the likelihood of the demographic variable of L2 speaker proportion being recorded by field typologists." This seems like a reasonable assumption: the smaller the value of the L2 proportion, the bigger the chance that this value is missing. I would even say that missing the L2 proportion could be used as a standard example of an MNAR type since we almost exclusively observe higher values of L2, as written by KEW (p. 3); only in four cases does the *Ethnologue* provide "an actual numerical zero proportion estimate." But, since there is no test to formally determine this, different researchers can have different opinions regarding the type of missingness. Nevertheless, we can test the efficiency of $\mathcal{M}_{\text{KEW}}$ in order to find out if the multiple imputation will produce unbiased estimates, even in the presence of large proportions of missing data. KEW (p. 11) cited Madley-Dowd et al. [66], who demonstrated that missingness up to "90% is tolerated by the method as long as the imputation model includes all necessary predictors". Madley-Dowd et al. [66] showed that this strongly depends on the strength of the imputation model. In their simulation study, strength is determined using the coefficient of determination R^2, which measures the proportion of variance in the outcome that is predictable by the imputation model. To provide unbiased estimates in the presence of high rates of missingness, Madley-Dowd et al. [66] showed that the imputation model needs to be almost perfect with an R^2 as high as 92%. To test this for $\mathcal{M}_{\text{KEW}}$, I re-ran the specified models and computed an R^2 of ~44% for morphological complexity as a covariate in $\mathcal{M}_{\text{KEW}}$ and corresponding R^2 of ~32% for information-theoretic complexity as a covariate in $\mathcal{M}_{\text{KEW}}$.

For such an LMM, I computed the so-called conditional R^2, i.e., the variance explained by both the fixed effects and the random intercept [67]. Note that in the fully observed data, there is information for morphological complexity as a covariate in the imputation model for only 28 language families. However, as KEW wrote (p. 8), their model imputes information for 122 families. This means that ~77% of all language families are systematically missing. Similar quantities are obtained for information-theoretic complexity, where information is systematically missing for ~84% of all language families. To place this into perspective, I computed an LMM with the log of population size as the outcome and a random intercept for language family for all available data points (N = 2143). The variance of the random intercept is 6.63. I then re-ran the model but restricted the computation to data points where the L2 proportion is non-missing (N = 171). Here, the variance of the random intercept is about 3.5 times higher, with a value of 24.10. Jolani [68], who developed the imputation method used by KEW, discusses potential biases in the estimated random effects parameters that can arise from systematic missingness. Jolani presented simulation results for systematically missing rates of up to 30%; it does not seem unlikely that problems could be more pronounced for missingness rates of more than 75%.

If we compare KEW's results with the results of Madley-Dowd et al. (Table 2 in [66]), we find out that with a missingness rate of 90%, the reduction in standard error (compared to a complete cases analysis model) is arguably very unimpressive, ranging between ~0% and ~9%, to be exact for R^2 = 52%, the error reduction is 8.86%, for R^2 = 40%, the error reduction is 2.18% and for R^2 = 36%, the error reduction is 0.11%. As a guide to test for efficiency gains, Madley-Dowd et al. [66] showed that the fraction of missing information (FMI) is a valuable quantity (ranging between 0 and 1) for determining the potential efficiency gains from multiple imputation: The FMI is a measure specific to each parameter that quantifies the information loss caused by missing data, while also considering the amount of information preserved by other variables in the dataset [66]. Its interpretation is similar to an R^2, so an FMI of, say, 0.2 means that 20% of the total sampling variance can be attributed to missing data. A high value indicates a problematic variable [64]. KEW reported FMI values for both their imputations models but did not interpret them: for information-theoretic complexity as a outcome (Table 6 of KEW and so on), the FMI value is ~62%, and for morphological complexity as a outcome (Table 5), the FMI value is ~89%. This alone shows that $\mathcal{M}_{\text{KEW}}$ does not provide much information about the missing values, especially in the case of morphological complexity, where KEW reported a negative significant effect of the L2 proportion on complexity. For information-theoretic complexity, there is no indication of an effect at any standard level of statistical significance. To further investigate this, I extracted all 100 imputed completed samples based on $\mathcal{M}_{\text{KEW}}$ from the code provided by KEW. First, I computed the Spearman correlation between the imputed L2 proportion and speaker population size for each completed dataset and for both types of complexity. Plot (a) of Figure 1 presents the results: for both types of complexity, the Spearman correlation is negative in most samples, and in 75% of all samples, the Spearman correlation is lower than $-0.08/-0.09$ for information-theoretic/morphological complexity. This seems rather implausible and—as written above and in my original paper [4]—contradicts a basic assumption of the linguistic niche hypothesis.

Table 5. Spatial autoregressive frequentist model averaging results for morphological complexity as the outcome computed both for the full (N = 1443) and the subset versions (N = 804). Separate models are computed for two different types of weight matrices: $\mathbf{W}_{\textbf{geo}}$, which is based on the geographical distances and $\mathbf{W}_{\textbf{phylo}}$, which is based on a phylogenetic similarity matrix. Per combination of WALS feature availability and weight matrix type, quantities are computed based on R = 8695 candidate models. σ_x-values (Equation (7), rounded to two decimal places) reflect the relative importance of each covariate, with higher values indicating a more important variable. Values above 0.90 are shown in bold. $\tilde{\beta}_x$-values (Equation (4), rounded to four decimal places) that pass a two-tailed parametric significance test at $p < 0.05$ are shown in bold. Sig. *** (**) [*] indicates that $p < 0.001$ ($p < 0.01$) [$p < 0.05$].

Version	Full				Subset			
Type of Weight Matrix	$\mathbf{W}_{\textbf{geo}}$		$\mathbf{W}_{\textbf{phylo}}$		$\mathbf{W}_{\textbf{geo}}$		$\mathbf{W}_{\textbf{phylo}}$	
Variable	σ_x	$\tilde{\beta}_x$	σ_x	$\tilde{\beta}_x$	σ_x	$\tilde{\beta}_x$	σ_x	$\tilde{\beta}_x$
0—International	0.75	0.0814	0.75	−0.0953	0.75	0.0760	**0.93**	0.1489
1—National	0.87	−0.0225	0.77	0.0624	**0.91**	−0.0929	**0.98**	−0.0987
2—Regional	0.87	−0.3396	0.76	−0.1728	**0.96**	−0.3483	0.81	−0.2544
3—Trade	0.83	−0.1009	0.51	−0.0167	0.81	−0.3084	**0.99**	**−0.5257** *

Table 5. *Cont.*

Version	Full				Subset			
Type of Weight Matrix	W_{geo}		W_{phylo}		W_{geo}		W_{phylo}	
Variable	σ_x	$\tilde{\beta}_x$	σ_x	$\tilde{\beta}_x$	σ_x	$\tilde{\beta}_x$	σ_x	$\tilde{\beta}_x$
Population size	**1.00**	**−0.0144 *****	**1.00**	**−0.0115 *****	**1.00**	**−0.0147 *****	**1.00**	**−0.0166 *****
Range size	**1.00**	**0.0155 *****	0.00	−0.0000	**1.00**	**0.0170 *****	**1.00**	**0.0200 *****
No. of countries	0.55	0.0043	**1.00**	−0.0120	0.52	0.0010	**0.91**	0.0333
Population size x International	0.21	−0.0095	0.19	−0.0007	0.21	−0.0054	0.27	−0.0130
Range size x International	0.20	−0.0043	0.00	0.0000	0.21	−0.0064	0.26	−0.0088
No. of countries x International	0.12	0.0033	0.17	0.0042	0.11	0.0012	0.23	0.0006
Population size x National	0.25	0.0008	0.38	−0.0063	0.23	−0.0011	0.28	−0.0022
Range size x National	0.32	−0.0058	0.00	−0.0000	0.27	0.0027	0.28	0.0013
No. of countries x National	0.14	0.0024	0.35	−0.0148	0.16	0.0054	0.28	0.0082
Population size x Regional	0.59	0.0173	0.50	0.0093	0.38	0.0047	0.33	0.0053
Range size x Regional	0.30	0.0048	0.00	0.0000	0.65	0.0215	0.46	0.0147
No. of countries x Regional	0.14	0.0029	0.18	0.0017	0.14	−0.0013	0.20	−0.0001
Population size x Trade	0.24	0.0008	0.06	0.0002	0.58	0.0172	0.77	0.0279
Range size x Trade	0.28	0.0036	0.00	0.0000	0.30	0.0048	0.36	0.0060
No. of countries x Trade	0.13	0.0004	0.12	−0.0043	0.17	0.0074	0.29	0.0119

For each sample, I then computed the percentage of languages that have an L2 proportion of (i) more than 0, (ii) more than 0.10, (iii) more than 0.25 and (iv) more than 0.50. Plot (b) of Figure 1 visualises the results. For information-theoretic complexity, all languages have an L2 proportion > 0; for morphological complexity, the median across the samples is 99.74%. Both results do not seem plausible with respect to the linguistic niche hypothesis. For the remaining quantities, the results seem to be equally implausible: for information-theoretic complexity, the median percentages are 69.89% for the L2 proportion > 0.10, 49.44% for the L2 proportion > 0.25 and 30.74% for the L2 proportion > 0.50; for morphological complexity, the median percentages are 59.48% for the L2 proportion > 0.10, 44.47% for the L2 proportion > 0.25 and 31.57% for the L2 proportion > 0.50. In my view, it is very hard to argue that a model that assumes that almost a third of all languages have an L2 proportion of over 50% reflects the linguistic reality.

Finally, I computed the median estimated L2 proportion in each sample for both non-vehicular, i.e., $\mathcal{L}_{\text{non-vehic}}$, and vehicular languages, i.e., $\mathcal{L}_{\text{vehic}}$. Plot (c) of Figure 1 presents the results: for both types of complexity, the completed datasets based on $\mathcal{M}_{KEW}$ show a lower L2 proportion for the vehicular languages compared to non-vehicular languages in the majority of cases (80 out of 100 for information-theoretic complexity and 68 out of 100 for morphological complexity). Specifically, the median $\mathcal{L}_{\text{vehic}}$ is 0.20 for information-theoretic complexity and 0.17 for morphological complexity, whereas the median $\mathcal{L}_{\text{non-vehic}}$ is 0.26 for information-theoretic complexity and 0.18 for morphological complexity. As written above and in my original paper [4], this completely contradicts the categorisation scheme of the *Ethnologue* and basic typological intuitions.

3.2. Updated Results

Table 3 summarises the results for information-theoretic complexity as the outcome. Consistent with the findings from my original paper [4], while there is a stable positive effect of population size on information-theoretic complexity, no evidence is found for an effect of vehicularity at any of the three symbolic levels (words, characters and BPE). Additionally, the σ_x-values (Equation (7)) reflecting relative variable importance indicate that vehicularity is not an important factor in explaining information-theoretic complexity (all σ_x-values < 0.70). In contrast, all σ_x-values for speaker population reach the maximum value of one, underscoring its importance in predicting information-theoretic complexity. This suggests that my initial results generalise across different language models, symbolic levels and text types.

Table 4 summarises the results for the SAR model, averaging with $R = 35$ candidate models, applied to both (i) the full dataset and the subset that includes only languages with at least six available WALS features and (ii) two different weight matrices, $\mathbf{W_{geo}}$ and $\mathbf{W_{phylo}}$. Consistent with my original paper [4], there is strong evidence of a statistically significant negative effect of population size on morphological complexity (at $p < 0.001$) in all four investigated scenarios. Three out of four corresponding σ_x-values are maximal, with the remaining value also indicating high variable importance ($\sigma_x = 0.97$). The only other variable that reaches statistical significance is language range size, which shows a significant effect on morphological complexity in three out of four cases. Interestingly, however, all three corresponding $\tilde{\beta}_x$-values are positive. Based on the linguistic niche hypothesis, one would expect the opposite [1].

Regarding vehicularity, the $\tilde{\beta}_x$-value is negative in all four scenarios, which, at first glance, aligns with the linguistic niche hypothesis. However, in none of these cases is the corresponding coefficient significantly different from zero (all p-values > 0.05). Interestingly, all corresponding σ_x-values indicate high importance for vehicularity.

To understand this seemingly counter-intuitive result, I computed the σ_x-value of all $R = 13$ candidate models that do not include population size as a covariate. In all four scenarios, the resulting σ_x-value is very low, with three out of four values below 0.001 and the remaining value below 0.05. This suggests that the apparent importance of vehicularity is actually driven by the strong effect of population size. To further explore this, I ran a SAR model of morphological complexity using population size as the predictor, with two spatially lagged error terms specified by each weighting matrix, using the GS2SLS estimator. From this model, I calculated uncorrelated residuals, which serve as estimates of the uncorrelated error term. I then computed the Pearson correlation between these residuals and vehicularity. For both the full and the subset versions, the resulting negative correlation coefficient was insignificant ($p = 0.071$ for the full and $p = 0.338$ for the subset).

Next, I reversed the analysis: I ran the SAR model with morphological complexity as the outcome and vehicularity as the predictor, including the two error terms. Again, I calculated uncorrelated residuals and correlated them with population size. In this case, for both the full and the subset versions, the negative correlation coefficients were significant ($p < 0.01$ for the full version and $p < 0.05$ for the subset version). This further indicates that the observed effect of vehicularity is actually driven by differences in population size.

Table 5 presents the findings from the larger candidate space ($R = 8695$). The results from this FMA align well with those based on the smaller model space (Table 4): the only consistent evidence for any effect on morphological complexity is the negative $\tilde{\beta}_x$-value of population size across all four scenarios (all p-values < 0.001). Similarly, language range size shows a significant positive effect on morphological complexity in three out of four cases, with the exception being for the full version when $\mathbf{W_{phylo}}$ is the autoregressive structure. With respect to the four binary EGIDS indicators, Table 5 shows that there is statistically significant evidence for an effect only in one out of sixteen cases: for the subset version with $\mathbf{W_{phylo}}$ as the weight matrix, the $\tilde{\beta}_x$-value of -0.5257 for the binary indicator for '3—Trade' is significant at $p < 0.05$. All other $\tilde{\beta}_x$-values do not reach statistical significance. In addition, none of the first-order interactions passes the significance test. The fact that we find almost no significant negative effects thus constitutes strong evidence against the linguistic niche hypothesis.

Regarding relative variable importance, only the σ_x-values for population size are consistently above 0.90. For the EGIDS indicator variables, only five σ_x-values are above 0.90. As for the small candidate space, I computed the σ_x-value of all $R = 803$ candidate models that do not include population size as a covariate. In all four scenarios, the resulting σ_x-value is below 0.001. Again, this indicates that the results are driven by the strong effect of population size.

4. Discussion

In sum, I would like to thank KEW for providing me the opportunity to revisit the relationship between language complexity and the proportion of non-native speakers. However, as pointed out above, I had already addressed their two main points of critique in my original paper [4]. In this context, I would like to thank one of the reviewers of my original paper, to whom I owe the consideration of these two points of criticism—during peer review, the reviewer pointed out that both "the relationship between the two variables and the fact that not all languages with a vehicularity of 0 have 0 L2 speakers needs to be dealt with openly" and that there needs to be an analysis that "removes the 78 languages with a vehicularity index of 0 and a proportion of L2 speakers > 0". Here, I refer the interested reader to the review reports, which I have deliberately chosen to make freely available online.

While KEW acknowledged and apologised for their oversight in personal communication, they did not retract their paper or issue any form of amendment or clarification. I, therefore, used the first part of the Results section (Section 3.1) to address their criticism. I hope to have convincingly demonstrated that the alternative analyses offered by KEW do not stand up to closer scrutiny: (i) Only one in three linear mixed-effects model analyses based on complete cases supports the linguistic niche hypothesis at all, and there are good reasons to doubt the appropriateness of the samples used to test for an effect of L2 proportions on complexity. (ii) The multiple imputation analyses suffer from similar biases, and it is clear from the interpretation of the FMI values reported by KEW that the imputation model does not provide much information about the missing values.

Nevertheless, KEW and I might agree that neither non-imputation nor imputation of L2 proportions is an ideal strategy. This brings us back to the use of vehicularity as an indicator of high L2 vs. low L2 languages. To drive home my point, let me provide an illustrative example that shows why it is possible to use such proxy variables to test claims between continuous variables: In a study examining the relationship between occupational exposure to a certain chemical and the risk of developing a specific health outcome, researchers may want to accurately measure the level of exposure to the chemical for each participant. However, it may be challenging to obtain accurate measurements of exposure, especially if the exposure occurred in the past or if the exposure was intermittent. Instead, they may rely on the assessment of medical experts, such as occupational health physicians, to classify each participant as having either high or low exposure based on their job history, work practices and other relevant factors. While this strategy does not seem to be perfect, and information certainly becomes lost when turning a continuous variable into a categorical one, it seems justified to statistically compare the incidence of the specific health outcome between the high- and low-exposure groups to assess whether there is a significant association between exposure to the chemical and the risk of developing the health outcome.

In our scenario, exposure to the chemical element is the proportion of L2 speakers, the outcome is language complexity, and the occupational health physicians are the field linguists of the *Ethnologue* that classify languages into high L2 language, i.e., vehicular languages, and for low L2 language, i.e., non-vehicular languages. Based on this logic, I aimed to demonstrate in the second part of the Results section (Section 3.2), using additional data on information-theoretic complexity and quantitative methods that better account for phylogenetic relatedness and geographic proximity, that when comparing a high L2 language with a low L2 language, both are statistically indistinguishable in terms of their morphological or information-theoretic complexity, provided that both languages have a comparable speaker population size. Furthermore, I conducted a more granular analysis using binary indicators for several EGIDS values that, according to the *Ethnologue*, should correspond to a substantial number of L2 speakers. This analysis also fails to support the linguistic niche hypothesis.

In the Appendix, I present the results of two additional analysis approaches, both based on a semi-parametric spatial filtering technique, which is outlined in Appendix A. The focus is on two key points:

(i) This paper primarily uses FMA, which is a method that integrates information from multiple plausible models while accounting for resulting uncertainty in the estimation process [21,48]. However, there are influential critiques cautioning against its naïve application. In particular, critics make the case against using model-averaged coefficients based on AIC weights [69–73]. While [21] claimed that AIC weights can be interpreted as model probabilities, those weights are only approximate, and as such, it can be problematic to interpret them as estimates of the predictor variable's importance [70,71]. As an alternative to FMA, I thus considered Bayesian model averaging in Appendix B, as it provides a principled and unified way to estimate the model weights as posterior model probabilities that are readily interpretable [74,75].

(ii) Statistical significance was assessed using parametric frequentist approaches, whose appropriateness I called into question in my original paper [4]. In Appendix C, I present the results of a non-parametric test that does not make any assumptions regarding the stochastic mechanism that generated the data [58].

For results, see Tables A1–A4.

The results of both approaches are closely aligned with the findings presented in the main paper.

In a recent paper, Shcherbakova et al. [76] followed my suggestion to use vehicularity as an indicator of the proportion of L2 speakers. Using Grambank, a novel and extensive database of grammatical features [77], and employing statistical methods that differ from mine to account for the effects of genealogical and geographic non-independence of languages [78], they found that the only effects of vehicularity on grammatical complexity were weakly or moderately positive. Consistent with the results presented in both my original paper [4] and this paper, their findings provide no support for the "specific claim of the 'linguistic niche hypothesis' that grammatical complexity should reduce with an increased number of non-native speakers" [76]. The title of their paper neatly sums this up nicely: "Societies of strangers do not speak grammatically simpler languages".

Now, critiques like KEW could argue that the categorisation of languages into high L2/vehicular and low L2/non-vehicular by the *Ethnologue* is incorrect. However, then this begs the question: If we do not trust the *Ethnologue* regarding this categorisation, why should we trust them regarding the—arguably more challenging—assessment of both the number of L1 and L2 speakers? As the saying goes—you cannot have your cake and eat it.

All in all, I thus do not have the impression that KEW's critique weakens the argumentation laid out in my original paper in any way. Moreover, the updated data and analyses provided here indicate that there is still no evidence for an effect of the proportion of non-native speakers on language complexity.

Funding: This research received no external funding.

Institutional Review Board Statement: Not applicable.

Data Availability Statement: Commented Stata 18 code to reproduce all results is available at https://osf.io/fypx5/ (accessed on 8 November 2024). The original data can be downloaded from https://dx.doi.org/10.6084/m9.figshare.c.4400675 or https://github.com/erc-starfish/koplenig-reply. The updated data on information-theoretic complexity can be downloaded from https://osf.io/xdwjc/ (accessed on 8 November 2024). The data used to generate the phylogenetic and geographic weight matrices are available at https://osf.io/cufv7/ and https://cdstar.eva.mpg.de//bitstreams/EAEA0-B701-6328-C3E3-0/languages_and_dialects_geo.csv (accessed on 8 November 2024). For the 1,000,000 random samples, each consisting of 100 languages selected from the 6880 living languages listed in the Ethnologue (Section 2), I used the Ethnologue Global Dataset (20th edition), a licenced product with restricted terms of use under a personal research licence.

Acknowledgments: I very much thank Vera Kempe, the thorough reviewer of my original paper. I also thank Sascha Wolfer and Peter Meyer for their input and feedback and Henri Kauhanen, Sarah Einhaus and George Walkden for their honest response to the notification of my intention to submit this paper. GitHub Copilot was used to accelerate the code writing. ChatGPT was used to improve readability and language and to assist in finding the example presented in the concluding remarks, which illustrates the use of proxy binary variables to test relationships between continuous variables.

Conflicts of Interest: I declare no conflicts of interest.

Appendix A Eigenvector Spatial Filtering

I first consider linear models of the form:

$$\mathbf{y} = \mathbf{F}\gamma + \epsilon \tag{A1}$$

where $\mathbf{y}$ is the $n \times 1$ vector morphological complexity values, n denotes the number of languages, $\mathbf{F}$ is a $n \times q$ matrix of q eigenvectors, γ is a $n \times 1$ vector of parameters and ϵ is the $n \times 1$ vector of residuals. Initially, $\mathbf{F}$ only consists of a $n \times 1$ vector of ones for the intercept. The other vectors of $\mathbf{F}$ are computed based on a transformed version of $\mathbf{W}$ defined as [79]:

$$\mathbf{M} \equiv \left(\mathbf{I} - \frac{11^{\mathrm{T}}}{n}\right)\mathbf{W}\left(\mathbf{I} - \frac{11^{\mathrm{T}}}{n}\right) \tag{A2}$$

where $\mathbf{I}$ is the $n \times n$ identity matrix, $\mathbf{1}$ is an $n \times 1$ vector of ones and T denotes the matrix transpose. The eigensystem decomposition of $\mathbf{M}$ generates n eigenvalues and n corresponding eigenvectors. The eigenvalues are then sorted in descending order, denoted as $\lambda = (\lambda_1, \lambda_2, \lambda_3, \ldots \lambda_n)$, so that the largest eigenvalue receives the subscript 1, the second largest eigenvalue receives the subscript 2 and so on. As in the paper, for $\mathbf{W}$, I considered $\mathbf{W_{geo}}$ and $\mathbf{W_{phylo}}$. The corresponding set of eigenvectors can then be denoted as $\mathbf{E_{geo}} = \left(\mathbf{E}_{\text{geo},1}, \mathbf{E}_{\text{geo},2}, \mathbf{E}_{\text{geo},3}, \ldots, \mathbf{E}_{\text{geo},n}\right)$, and $\mathbf{E_{phylo}} = \left(\mathbf{E}_{\text{phylo},1}, \mathbf{E}_{\text{phylo},2}, \mathbf{E}_{\text{phylo},3}, \ldots, \mathbf{E}_{\text{phylo},n}\right)$.

To test for potential autocorrelation among the residuals, I tested the following null hypothesis:

$$H_0 : E\left[\epsilon\epsilon^{\mathrm{T}}\right] = \sigma^2\mathbf{I} \tag{A3}$$

where $\mathbf{I}$ is the $n \times n$ identity matrix, and T denotes the matrix transpose. To test H_0, I computed the modified version of Moran's I [80], suggested by Kelejian and Prucha [81], based on $\mathbf{W_{geo}}$ and $\mathbf{W_{phylo}}$. Since $I^2 \sim X^2(2)$ [81], I tested H_0 via a standard X^2-test with two degrees of freedom. As suggested by [82], for $\mathbf{W_{geo}}$ I defined a search set of relevant eigenvectors, $\mathbf{E}^*_{\mathbf{geo}}$, using the following search algorithm, which includes all x eigenvectors with a corresponding eigenvalue $\lambda_x \geq (0.1\lambda_{\text{geo},1})$; likewise, for $\mathbf{W_{phylo}}$. The algorithm [82] then iterates over the eigenvectors in the search set $\mathbf{E}^*$, consisting of $\mathbf{E}^*_{\mathbf{geo}}$ and $\mathbf{E}^*_{\mathbf{phylo}}$. At each step, Equation (A1) is fitted to the data, and the eigenvector $\mathbf{E}$ that reduces the X^2-value the most is selected into $\mathbf{F}$ and removed from $\mathbf{E}^*$. This iterative procedure is repeated $X^2(2) < 1$, corresponding to a p-value of approximately 0.61.

The great advantage is that it is easy to extend Equation (A1):

$$\mathbf{y} = \mathbf{F}\gamma + \mathbf{X}\beta + \epsilon \tag{A4}$$

where, in addition to the above, $\mathbf{X}$ is a $n \times p$ design matrix of p covariates. Here, standard linear least square regression, which is computationally very efficient, provides consistent estimates for β [83] because the identified set $\mathbf{F}$ of the control variables isolates the spatial and phylogenetic relationships between languages, enabling the construction of models as though the languages were in fact independent [84]. Table A1 summarises the results of the spatial filtering for both the full and the subset versions. Before filtering, there is strong evidence of autocorrelation in the residuals for both versions (both X^2-values > 95 at $p < 0.0001$, row 3). R^2 unfiltered (row 4) represents the coefficient of determination before filtering. For the full version, the algorithm identifies a total of 15 eigenvectors, with 12 from $\mathbf{E}^*_{\mathbf{geo}}$ and 3 from $\mathbf{E}^*_{\mathbf{phylo}}$ (row 5). For the subset version, 19 eigenvectors are chosen, 16 from $\mathbf{E}^*_{\mathbf{geo}}$ and 3 from $\mathbf{E}^*_{\mathbf{phylo}}$. After filtering, there is no indication of autocorrelation among the residuals (both X^2-values < 1 at $p > 0.66$, row 6). In both cases, the coefficient of determination increases significantly (row 7). In what follows, I used the identified sets for both versions, i.e., $\mathbf{F_{full}}$ and $\mathbf{F_{subset}}$, as input for Equation (A4).

Table A1. Spatial filtering results for both the full and the subset versions.

Version	Full	Subset
N	1443	804
X^2 unfiltered	95.88 ($p < 0.0001$)	269.95 ($p < 0.0001$)
R^2 unfiltered	5.45%	9.79%
Cardinality of F	$N_{geo} = 12$, $N_{phylo} = 3$	$N_{geo} = 16$, $N_{phylo} = 3$
X^2 filtered	0.82 ($p = 0.663$)	0.61 ($p = 0.737$)
R^2 filtered	18.11%	39.28%

Appendix B Bayesian Model Averaging (BMA)

In a BMA, model averaging is conducted in a Bayesian framework [74,75]. For any model M_j of the R candidates, we can compute its analytical posterior model probability (PMP) via Bayes theorem [85]:

$$\mathrm{P}(M_j|D) = \frac{\mathrm{P}(D|M_j)\mathrm{P}(M_j)}{\sum_{r=1}^{R} \mathrm{P}(D|M_r)\mathrm{P}(M_r)} \tag{A5}$$

where D is the given data and $\mathrm{P}(M_j)$ is an assigned prior. In what follows, I allowed the data to speak for itself by assuming a non-informative prior, i.e., I assumed equal probability for all R models [75]. $\mathrm{P}(M_j|D)$ can be used to compute a BMA estimator as:

$$\tilde{\beta}_x = \sum_{j=1}^{R} \mathrm{P}(M_j|D)\beta_{x,j} \tag{A6}$$

where $\beta_{x,j}$ denotes the estimated effect of variable x for model j. The posterior inclusion probability (PIP) for predictor variable x can be defined as the sum of PMPs that include x [75]:

$$\mathrm{PIP}_x = \sum_{j=1}^{R} \mathrm{P}(M_j|D)c_{x,j} \tag{A7}$$

where $c_{x,j}$ is a binary indicator that is equal to 1 if x is explicitly in model j and 0 otherwise [21]. The higher PIP_x the higher its importance. As predictors, I first considered the vehicularity, speaker population size, range size, the number of countries and all higher-order interactions, yielding a total of $R = 2129$ candidates.

As in the main part of the paper, models that include interactions are only considered if the corresponding main effects and all corresponding lower-order interactions are also included in the model. Additionally, note that all quantitative variables are log-transformed and that the results presented in this section and in Appendix C do not change qualitatively if the quantitative predictors are additionally standardised to a mean of 0 and a standard deviation of 2 prior to estimation, as suggested by Gelman (https://statmodeling.stat.columbia.edu/2009/07/11/when_to_standar/, accessed on 8 November 2024).

The identified set of control variables, $\mathrm{F_{full}}$ or $\mathrm{F_{subset}}$, are always included. (The use of spatial filtering rather than fitting spatial autoregressive models is due to the high computational cost of fitting such non-linear models across a large model space [83]). Table A2 presents the results. For the full version, the only variable with a moderately high PIP of 66.06% is the population size, while all other predictors have PIPs below 25%. Correspondingly, the BMA variable-inclusion summary shows that a model with only population size (and $\mathbf{F}_{\text{full}}$) achieves a PMP of 51.90%. All other models only play a minor role. The results for the subset version are highly comparable, with only population size as an important predictor, with a PIP of 88.52%. The low PIPs for vehicularity again question the linguistic niche hypothesis [1].

I continued by considering a larger model space. As in the main part of the paper, I considered binary indicators for the first four levels of EGIDS (0, 1, 2, 3). Out of curiosity, I have added a fifth EGIDS indicator for EGIDS = 4 ("Literacy in the language is being transmitted through a system of public education"; see Table 1 in the paper). Again, I consider the following three quantitative variables: population size, range size and number of countries. To keep the model space manageable, I only additionally considered possible higher-order correlations between the EGIDS indicators and the first two quantitative variables (population size and range size). In total, the model space consists of R = 237,232 models. Table A3 demonstrates that the results are in close agreement with the smaller candidate space: only speaker population size is an important predictor of morphological complexity.

Table A2. Bayesian model averaging results for both the full and the subset versions for the small candidate set, with R = 2129 considered candidate models. Note that each model in both versions always includes the corresponding set of control variables, F_{full} or F_{subset}, identified above (Appendix A). Per version, the first part of the table presents the BMA estimates for each predictor, $\widetilde{\beta}_x$, ranked by their PIP. Predictors with a PIP of less than 1% are omitted. The second part of the table presents results for the top five models ranked by their PMP. 'x' in the variable-inclusion summary indicates that the corresponding predictor is included in the model.

		Version: Full			
	$\widetilde{\beta}_x$	**PIP$_x$(%)**			
Population size	−0.00505	66.06%			
Vehicularity	−0.01369	23.34%			
N countries	−0.00669	21.18%			
Range size	−0.00049	8.35%			
		BMA variable-inclusion summary			
	Rank 1 (PMP = 51.90%)	**Rank 2 (PMP = 12.83%)**	**Rank 3 (PMP = 11.26%)**	**Rank 4 (PMP = 5.79%)**	**Rank 5 (PMP = 4.84%)**
Population size	x			x	x
Vehicularity		x		x	
N countries			x		x
		Version: subset			
	$\widetilde{\beta}_x$	**PIP$_x$(%)**			
Population size	−0.00630	88.52%			
Range size	0.00139	17.30%			
Vehicularity	−0.00495	12.58%			
N countries	−0.00012	5.52%			
Pop x Range	−0.00002	1.54%			
		BMA variable-inclusion summary			
	Rank 1 (PMP = 65.50%)	**Rank 2 (PMP = 12.88%)**	**Rank 3 (PMP = 6.72%)**	**Rank 4 (PMP = 3.51%)**	**Rank 5 (PMP = 2.56%)**
Population size	x	x		x	
Range size		x			
Vehicularity			x	x	

Table A3. Bayesian model averaging results for both the full and the subset versions for the small candidate set, with R = 237,232 considered candidate models. Note that each model in both versions always includes the corresponding set of control variables, F_{full} or F_{subset}, identified above (see Appendix A). Per version, the first part of the table presents the BMA estimates for each predictor, $\tilde{\beta}_x$, ranked by their PIP. Predictors with a PIP of less than 1% are omitted. The second part of the table presents results for the top five models ranked by their PMP. 'x' in the variable-inclusion summary indicates that the corresponding predictor is included in the model.

Version: Full					
	$\tilde{\beta}_x$	**PIP$_x$(%)**			
Population size	−0.00570	73.76%			
N countries	−0.00842	21.62%			
EGIDS = 1	−0.00708	12.15%			
EGIDS = 4	−0.00322	8.86%			
Range size	−0.00055	8.74%			
EGIDS = 0	−0.00861	5.87%			
EGIDS = 2	−0.00323	4.46%			
EGIDS = 3	−0.00033	2.75%			
BMA variable-inclusion summary					
	Rank 1 (PMP = 49.12%)	**Rank 2 (PMP = 10.66%)**	**Rank 3 (PMP = 4.46%)**	**Rank 4 (PMP = 4.32%)**	**Rank 5 (PMP = 3.61%)**
Population size	x		x	x	x
EGIDS = 1				x	
EGIDS = 4					x
N countries		x	x		
Version: subset					
	$\tilde{\beta}_x$	**PIP$_x$(%)**			
Population size	−0.00574	82.43%			
EGIDS = 1	−0.01421	23.17%			
Range size	0.00111	14.13%			
EGIDS = 0	−0.01684	11.36%			
EGIDS = 4	−0.00353	7.63%			
EGIDS = 3	−0.00060	4.24%			
EGIDS = 2	−0.00153	3.99%			
N countries	−0.00057	3.97%			
BMA variable-inclusion summary					
	Rank 1 (PMP = 47.90%)	**Rank 2 (PMP = 7.46%)**	**Rank 3 (PMP = 6.25%)**	**Rank 4 (PMP = 5.83%)**	**Rank 5 (PMP = 3.65%)**
Population size	x	x		x	x
Range size		x			
EGIDS = 0					x
EGIDS = 1			x	x	

Appendix C Non-Parametric Permutation Testing

To test in a non-parametric setting, we write:

$$\mathbf{y} = \mathbf{Z}\gamma + \mathbf{X}\beta + \epsilon \tag{A8}$$

where, in addition to the above, **Z** is a matrix of nuisance/control variables. The predictor of interest, represented by **X**, is either vehicularity or population size. In what follows, I used a generic variant of the *Freedman-Lane* permutation testing procedure that does not make any assumptions about the mechanism that generated the data [58,86]

1. Let y denote the morphological complexity variable. Per version (full/subset), y is regressed onto the predictor of interest and the control variables. This model is used to calculate the test statistic T_0, which is the t-statistic of the estimated parameter for vehicularity or population size.
2. Further, y is regressed only onto the control variables in a reduced model, and fitted values $\hat{y}$ and residuals $\hat{\varepsilon}$ are obtained. This means that $\hat{y}$ contains the sample mean plus contributions based on the control variables.
3. The residuals $\hat{\varepsilon}$ are randomly permuted, called the resulting variable $\hat{\varepsilon}^*$, and a new variable is computed that is defined as the sum of the fitted values and the randomly permuted residuals, i.e., $y^* = \hat{y} + \hat{\varepsilon}^*$.
4. y^* is regressed onto the predictor of interest and the control variables in order to calculate the test statistic of interest and call this statistic T_j^*.
5. Steps c and d are repeated 10,000 times to calculate the reference distribution of T^*.
6. Count the number of times where $|T_j^*| \geq |T_0|$ and divide that number by 10,000. The result is the p-value.

The intuitive idea behind this permutation test is that if the null hypothesis holds—meaning the predictor of interest is unrelated to morphological complexity—then the derived datasets, i.e., those with randomly permuted residuals, should be indistinguishable from the original dataset. As Freedman and Lane [58] described it: "a small reported significance level indicates an unusual data set". Table A4 presents the results. Vehicularity is only a significant predictor of morphological complexity when speaker population size is not included as a control. Conversely, population size remains a significant predictor of morphological complexity regardless of the set of control variables.

Table A4. Results of the permutation test. For each version (full/subset), morphological complexity is regressed onto the predictor of interest and the control variables. F_{full} or F_{subset} represents the set of control variables, which are identified above (Appendix A). The fourth column shows the estimated effect and the t-value in brackets. The last column shows the permutation-based p-value. Sig. *** (**) [*] indicates that $p < 0.001$ ($p < 0.01$) [$p < 0.05$].

Version	Predictor of Interest	Control Variables	$\hat{\beta}_x(t)$	p_{perm}
full	Vehicularity	$\mathbf{F_{full}}$	−0.071 (−3.415) ***	0.0000
		Population size + $\mathbf{F_{full}}$	−0.041 (−1.691)	0.1000
subset		$\mathbf{F_{subset}}$	−0.052 (−2.971) **	0.0022
		Population size + $\mathbf{F_{subset}}$	−0.023 (−1.082)	0.2786
full	Population size	$\mathbf{F_{full}}$	−0.008 (−3.801) ***	0.0000
		Vehicularity + $\mathbf{F_{full}}$	−0.006 (−2.373) *	0.0192
subset		$\mathbf{F_{subset}}$	−0.007 (−3.652) ***	0.0004
		Vehicularity + $\mathbf{F_{subset}}$	−0.005 (−2.373) *	0.0178

References

1. Lupyan, G.; Dale, R. Language Structure Is Partly Determined by Social Structure. *PLoS ONE* **2010**, *5*, e8559. [CrossRef] [PubMed]
2. Dale, R.; Lupyan, G. Understanding the Origins of Morphological Diversity: The Linguistic Niche Hypothesis. *Adv. Complex Syst.* **2012**, *15*, 1150017. [CrossRef]
3. Trudgill, P. Contact and Simplification: Historical Baggage and Directionality in Linguistic Change. *Linguist. Typology* **2001**, *5*, 371–374.
4. Koplenig, A. Language Structure Is Influenced by the Number of Speakers but Seemingly Not by the Proportion of Non-Native Speakers. *R. Soc. Open Sci.* **2019**, *6*, 181274. [CrossRef] [PubMed]
5. Nettle, D. Social Scale and Structural Complexity in Human Languages. *Philos. Trans. R. Soc. B Biol. Sci.* **2012**, *367*, 1829–1836. [CrossRef] [PubMed]
6. Bentz, C.; Winter, B. Languages with More Second Language Learners Tend to Lose Nominal Case. *Lang. Dyn. Change* **2013**, *3*, 1–27. [CrossRef]
7. Simons, G.F.; Fennig Charles, D. *Ethnologue: Languages of the World*, 20th ed.; SIL International: Dallas, TX, USA, 2017.
8. Brunn, S.D.; Kehrein, R. (Eds.) *Handbook of the Changing World Language Map*; Springer Reference; Springe: Cham, Switzerland, 2020; ISBN 978-3-030-02438-3.
9. Hammarström, H. Ethnologue 16/17/18th Editions: A Comprehensive Review. *Language* **2015**, *91*, 723–737. [CrossRef]
10. Lewis, M.P.; Simons, G.F. Assessing Endangerment: Expanding Fishman's GIDS. *Rev. Roum. De Linguist.* **2010**, *55*, 103–120.
11. Kauhanen, H.; Einhaus, S.; Walkden, G. Language Structure Is Influenced by the Proportion of Non-Native Speakers: A Reply to Koplenig (2019). *J. Lang. Evol.* **2023**, *8*, 90–101. [CrossRef]
12. Bromham, L. Chapter 4. Solving Galton's Problem: Practical Solutions for Analysing Language Diversity and Evolution. In *Current Issues in Linguistic*; Evans, B., Gallego, M.K., Miceli, L., Eds.; Benjamins Publishing Company: Amsterdam, The Netherlands, 2024; Volume 367, pp. 74–108, ISBN 978-90-272-1820-9/978-90-272-4631-8. [CrossRef]
13. Kontoyiannis, I.; Algoet, P.H.; Suhov, Y.M.; Wyner, A.J. Nonparametric Entropy Estimation for Stationary Processes and Random Fields, with Applications to English Text. *IEEE Trans. Inf. Theory* **1998**, *44*, 1319–1327. [CrossRef]
14. Kroneman, D. The Lord Is My Shepherd: An Exploration Into the Theory and Practice of Translating Biblical Metaphor. Doctoral Dissertation, Vrije Universiteit, Amsterdam, The Netherlands, 2004.
15. de Vries, L.J. Some Remarks on the Use of Bible Translations as Parallel Texts in Linguistic Research. *Sprachtypol. Universalienforschung* **2007**, *60*, 148–157. [CrossRef]
16. Koplenig, A. Quantifying the Efficiency of Written Language. *Linguist. Vanguard* **2021**, *7*, 20190057. [CrossRef]
17. Cotterell, R.; Mielke, S.J.; Eisner, J.; Roark, B. Are All Languages Equally Hard to Language-Model? In Proceedings of the 2018 Conference of the North American Chapter of the Association for Computational Linguistics: Human Language Technologies, Volume 2 (Short Papers), New Orleans, LA, USA, 1–6 June 2018; Association for Computational Linguistics: New Orleans, LA, USA, 2018; pp. 536–541.
18. Moran, S.; Cysouw, M. *The Unicode Cookbook For Linguists: Managing Writing Systems Using Orthography Profiles*; Translation and Multilingual Natural Language Processing; Language Science Press: Berlin, Germany, 2018.
19. Koplenig, A.; Wolfer, S.; Meyer, P. A Large Quantitative Analysis of Written Language Challenges the Idea That All Languages Are Equally Complex. *Sci. Rep.* **2023**, *13*, 15351. [CrossRef] [PubMed]
20. Koplenig, A.; Wolfer, S.; Rüdiger, J.O.; Meyer, P. Human Languages Trade off Complexity against Efficiency. **2024**. [CrossRef]
21. Burnham, K.P.; Anderson, D.R. (Eds.) *Model Selection and Multimodel Inference*; Springer: New York, NY, USA, 2004; ISBN 978-0-387-95364-9.
22. Anselin, L. Spatial Econometrics. In *A Companion to Theoretical Econometrics*; Baltagi, B.H., Ed.; Wiley: Hoboken, NJ, USA, 2003; pp. 310–330. ISBN 978-0-631-21254-6.
23. Hall, S.; Moskovitz, C.; Pemberton, M. Understanding Text Recycling: A Guide for Researchers V.1. Text Recycling Research Project. 2021. Available online: https://digitalcommons.georgiasouthern.edu/writing-linguistics-facpubs/257/ (accessed on 8 November 2024).
24. Mayer, T.; Cysouw, M. Creating a Massively Parallel Bible Corpus. In *Proceedings of the Ninth International Conference on Language Resources and Evaluation (LREC'14)*; Reykjavik, Iceland, 26–31 May 2014, Choukri, K., Declerck, T., Loftsson, H., Maegaard, B., Mariani, J., Moreno, A., Odijk, J., Piperidis, S., Eds.; European Language Resources Association (ELRA): Reykjavik, Iceland, 2014.
25. Nichols, J.; Witzlack-Makarevich, A.; Bickel, B. *The AUTOTYP Genealogy and Geography Database: 2013 Release*; University of Zurich: Zurich, Switzerland, 2013.
26. Amano, T.; Sandel, B.; Eager, H.; Bulteau, E.; Svenning, J.-C.; Dalsgaard, B.; Rahbek, C.; Davies, R.G.; Sutherland, W.J. Global Distribution and Drivers of Language Extinction Risk. *Proc. R. Soc. B Biol. Sci.* **2014**, *281*, 20141574. [CrossRef]
27. Eberhard, D.M.; Simons, G.F.; Fenning, C.D. (Eds.) *Ethnologue: Languages of the World*, 22nd ed.; SIL International: Dallas, TX, USA, 2019.
28. Global Mapping International. 2010 World Language Mapping System Version 16.0. Colorado Springs, CO: Global Mapping International. Available online: http://www.worldgeodatasets.com/language/ (accessed on 8 November 2024).
29. Bentz, C.; Verkerk, A.; Kiela, D.; Hill, F.; Buttery, P. Adaptive Communication: Languages with More Non-Native Speakers Tend to Have Fewer Word Forms. *PLoS ONE* **2015**, *10*, e0128254. [CrossRef]

30. Bentz, C.; Ruzsics, T.; Koplenig, A.; Samardzic, T. A Comparison Between Morphological Complexity Measures: Typological Data vs. Language Corpora. In Proceedings of the Workshop on Computational Linguistics for Linguistic Complexity (CL4LC), Osaka, Japan, 11 December 2016; The COLING 2016 Organizing Committee: Osaka, Japan, 2016.
31. Dryer, M.S.; Haspelmath, M. (Eds.) *WALS Online*; Max Planck Institute for Evolutionary Anthropology: Leipzig, Germany, 2013.
32. Ren, G.; Takahashi, S.; Tanaka-Ishii, K. Entropy Rate Estimation for English via a Large Cognitive Experiment Using Mechanical Turk. *Entropy* **2019**, *21*, 1201. [CrossRef]
33. Meister, C.; Pimentel, T.; Wiher, G.; Cotterell, R. Locally Typical Sampling. *arXiv* **2022**, arXiv:2202.00666. [CrossRef]
34. Chater, N.; Vitányi, P. 'Ideal Learning' of Natural Language: Positive Results about Learning from Positive Evidence. *J. Math. Psychol.* **2007**, *51*, 135–163. [CrossRef]
35. Kolmogorov, A.N. Three Approaches to the Quantitative Definition of Information. *Int. J. Comput. Math.* **1968**, *2*, 157–168. [CrossRef]
36. Kontoyiannis, I. *The Complexity and Entropy of Literary Styles*; NSF Technical Report; Stanford University: Stanford, CA, USA, 1996; Volume 97.
37. Cover, T.M. Kolmogorov Complexity, Data Compression, and Inference. In *The Impact of Processing Techniques on Communications*; Skwirzynski, J.K., Ed.; Springer: Dordrecht, The Netherlands, 1985; pp. 23–33. ISBN 978-94-010-8760-5.
38. Koplenig, A.; Meyer, P.; Wolfer, S.; Müller-Spitzer, C. The Statistical Trade-off between Word Order and Word Structure–Large-Scale Evidence for the Principle of Least Effort. *PLoS ONE* **2017**, *12*, e0173614. [CrossRef] [PubMed]
39. Wyner, A.D.; Ziv, J. Some Asymptotic Properties of the Entropy of a Stationary Ergodic Data Source with Applications to Data Compression. *IEEE Trans. Inf. Theor.* **1989**, *35*, 1250–1258. [CrossRef]
40. Ebeling, W.; Neiman, A. Long-Range Correlations between Letters and Sentences in Texts. *Phys. A Stat. Mech. Its Appl.* **1995**, *215*, 233–241. [CrossRef]
41. Montemurro, M.A.; Pury, P.A. Long-Range Fractal Correlations in Literary Corpora. *Fractals* **2002**, *10*, 451–461. [CrossRef]
42. Mielke, S.J.; Cotterell, R.; Gorman, K.; Roark, B.; Eisner, J. What Kind of Language Is Hard to Language-Model? In Proceedings of the 57th Annual Meeting of the Association for Computational Linguistics, Florence, Italy, 28 July–2 August 2019; Association for Computational Linguistics: Florence, Italy, 2019; pp. 4975–4989.
43. Sennrich, R.; Haddow, B.; Birch, A. Neural Machine Translation of Rare Words with Subword Units. In Proceedings of the 54th Annual Meeting of the Association for Computational Linguistics (Volume 1: Long Papers), Berlin, Germany, 7–12 August 2016; Association for Computational Linguistics: Berlin, Germany, 2016; pp. 1715–1725.
44. Gutierrez-Vasques, X.; Bentz, C.; Samardžić, T. Languages Through the Looking Glass of BPE Compression. *Comput. Linguist.* **2023**, *49*, 943–1001. [CrossRef]
45. Baayen, R.H.; Davidson, D.J.; Bates, D.M. Mixed-Effects Modeling with Crossed Random Effects for Subjects and Items. *J. Mem. Lang.* **2008**, *59*, 390–412. [CrossRef]
46. Stewart, W.A. A Sociolinguistic Typology for Describing National Multilingualism. In *Readings in the Sociology of Language*; Fishman, J.A., Ed.; De Gruyter: Berlin, Germany, 1968; pp. 531–545. ISBN 978-90-279-1528-3.
47. Bromham, L.; Yaxley, K.J. Neighbours and Relatives: Accounting for Spatial Distribution When Testing Causal Hypotheses in Cultural Evolution. *Evolut. Hum. Sci.* **2023**, *5*, e27. [CrossRef]
48. Buckland, S.T.; Burnham, K.P.; Augustin, N.H. Model Selection: An Integral Part of Inference. *Biometrics* **1997**, *53*, 603–618. [CrossRef]
49. Gelman, A.; Hill, J. *Data Analysis Using Regression and Multilevel/Hierarchical Models*; Analytical methods for social research; Cambridge University Press: New York, NY, USA, 2007.
50. Jäger, G. Global-Scale Phylogenetic Linguistic Inference from Lexical Resources. *Sci. Data* **2018**, *5*, 180189. [CrossRef]
51. Kelejian, H.H.; Prucha, I.R. Specification and Estimation of Spatial Autoregressive Models with Autoregressive and Heteroskedastic Disturbances. *J. Econom.* **2010**, *157*, 53–67. [CrossRef] [PubMed]
52. Wray, A.; Grace, G.W. The Consequences of Talking to Strangers: Evolutionary Corollaries of Socio-Cultural Influences on Linguistic Form. *Lingua* **2007**, *117*, 543–578. [CrossRef]
53. Verbeke, G.; Molenberghs, G. *Linear Mixed Models for Longitudinal Data*; Springer series in statistics; Corr. 2. print. Springer: New York, NY, USA; Berlin/Heidelberg, Germany, 2001; ISBN 978-0-387-95027-3.
54. Zuur, A.F.; Ieno, E.N.; Walker, N.; Saveliev, A.A.; Smith, G.M. *Mixed Effects Models and Extensions in Ecology with R*; Statistics for Biology and Health; Springer: New York, NY, USA, 2009; ISBN 978-0-387-87457-9.
55. Faraway, J.J. *Extending the Linear Model with R: Generalized Linear, Mixed Effects and Nonparametric Regression Models*, 2nd ed.; Chapman and Hall/CRC: Boca Raton, FL, USA, 2016; ISBN 978-1-315-38272-2.
56. Hurvich, C.M.; Tsai, C.-L. Regression and Time Series Model Selection in Small Samples. *Biometrika* **1989**, *76*, 297–307. [CrossRef]
57. Koplenig, A. Against Statistical Significance Testing in Corpus Linguistics. *Corpus Linguist. Linguist. Theory* **2019**, *15*, 321–346. [CrossRef]
58. Freedman, D.A.; Lane, D. A Nonstochastic Interpretation of Reported Significance Levels. *J. Bus. Econ. Stat.* **1983**, *1*, 292–298. [CrossRef]
59. Rubin, D.B. Inference and Missing Data. *Biometrika* **1976**, *63*, 581–592. [CrossRef]
60. Rubin, D.B. (Ed.) *Multiple Imputation for Nonresponse in Surveys*; Wiley Series in Probability and Statistics; John Wiley & Sons, Inc.: Hoboken, NJ, USA, 1987; ISBN 978-0-470-31669-6.

61. van Buuren, S. *Flexible Imputation of Missing Data*, 2nd ed.; CRC Press: Boca Raton, FL, USA, 2018.
62. StataCorp. *Multiple-Imputation Reference Manual*; Stata Press: College Station, TX, USA, 2021; ISBN 978-1-59718-341-3.
63. Medeiros, R. Handling Missing Data in Stata: Imputation and Likelihood-Based Approaches. In Proceedings of the 2016 Swiss Stata Users Group Meeting, Düsseldorfon, Germany, 17 November 2016.
64. UCLA: Statistical Consulting Group Multiple Imputation in Stata. Advanced Research Computing: Statistical Methods and Data Analytics 2023. Available online: https://stats.oarc.ucla.edu/stata/seminars/mi_in_stata_pt1_new/ (accessed on 8 November 2024).
65. Bartlett, J. Handling Missing Data in Stata-a Whirlwind Tour; Stata Users Group: 2012. Available online: https://ideas.repec.org/p/boc/isug12/01.html (accessed on 8 November 2024).
66. Madley-Dowd, P.; Hughes, R.; Tilling, K.; Heron, J. The Proportion of Missing Data Should Not Be Used to Guide Decisions on Multiple Imputation. *J. Clin. Epidemiol.* **2019**, *110*, 63–73. [CrossRef]
67. Nakagawa, S.; Schielzeth, H. A General and Simple Method for Obtaining R2 from Generalized Linear Mixed-Effects Models. *Methods Ecol. Evol.* **2013**, *4*, 133–142. [CrossRef]
68. Jolani, S. Hierarchical Imputation of Systematically and Sporadically Missing Data: An Approximate Bayesian Approach Using Chained Equations. *Biom. J.* **2018**, *60*, 333–351. [CrossRef]
69. Cade, B.S. Model Averaging and Muddled Multimodel Inferences. *Ecology* **2015**, *96*, 2370–2382. [CrossRef]
70. Galipaud, M.; Gillingham, M.A.F.; David, M.; Dechaume-Moncharmont, F. Ecologists Overestimate the Importance of Predictor Variables in Model Averaging: A Plea for Cautious Interpretations. *Methods Ecol. Evol.* **2014**, *5*, 983–991. [CrossRef]
71. Galipaud, M.; Gillingham, M.A.F.; Dechaume-Moncharmont, F. A Farewell to the Sum of Akaike Weights: The Benefits of Alternative Metrics for Variable Importance Estimations in Model Selection. *Methods Ecol. Evol.* **2017**, *8*, 1668–1678. [CrossRef]
72. Banner, K.M.; Higgs, M.D. Considerations for Assessing Model Averaging of Regression Coefficients. *Ecol. Appl.* **2017**, *27*, 78–93. [CrossRef] [PubMed]
73. Dormann, C.F.; Calabrese, J.M.; Guillera-Arroita, G.; Matechou, E.; Bahn, V.; Bartoń, K.; Beale, C.M.; Ciuti, S.; Elith, J.; Gerstner, K.; et al. Model Averaging in Ecology: A Review of Bayesian, Information-theoretic, and Tactical Approaches for Predictive Inference. *Ecol. Monogr.* **2018**, *88*, 485–504. [CrossRef]
74. Steel, M.F.J. Model Averaging and Its Use in Economics. *J. Econ. Lit.* **2020**, *58*, 644–719. [CrossRef]
75. StataCorp. *Stata 18 Bayesian Model Averaging Reference Manual*; Stata Press: College Station, TX, USA, 2023.
76. Shcherbakova, O.; Michaelis, S.M.; Haynie, H.J.; Passmore, S.; Gast, V.; Gray, R.D.; Greenhill, S.J.; Blasi, D.E.; Skirgård, H. Societies of Strangers Do Not Speak Less Complex Languages. *Sci. Adv.* **2023**, *9*, eadf7704. [CrossRef]
77. Skirgård, H.; Haynie, H.J.; Blasi, D.E.; Hammarström, H.; Collins, J.; Latarche, J.J.; Lesage, J.; Weber, T.; Witzlack-Makarevich, A.; Passmore, S.; et al. Grambank Reveals the Importance of Genealogical Constraints on Linguistic Diversity and Highlights the Impact of Language Loss. *Sci. Adv.* **2023**, *9*, eadg6175. [CrossRef]
78. Dinnage, R.; Skeels, A.; Cardillo, M. Spatiophylogenetic Modelling of Extinction Risk Reveals Evolutionary Distinctiveness and Brief Flowering Period as Threats in a Hotspot Plant Genus. *Proc. R. Soc. B.* **2020**, *287*, 20192817. [CrossRef]
79. Griffith, D.A. A Spatial Filtering Specification for the Autologistic Model. *Environ. Plan. A* **2004**, *36*, 1791–1811. [CrossRef]
80. Moran, P.A.P. Notes on Continuous Stochastic Phenomena. *Biometrika* **1950**, *37*, 17–23. [CrossRef]
81. Kelejian, H.H.; Prucha, I.R. On the Asymptotic Distribution of the Moran I Test Statistic with Applications. *J. Econom.* **2001**, *104*, 219–257. [CrossRef]
82. Tiefelsdorf, M.; Griffith, D.A. Semiparametric Filtering of Spatial Autocorrelation: The Eigenvector Approach. *Environ. Plan. A* **2007**, *39*, 1193–1221. [CrossRef]
83. Oberdabernig, D.A.; Humer, S.; Crespo Cuaresma, J. Democracy, Geography and Model Uncertainty. *Scott. J. Political Econ.* **2018**, *65*, 154–185. [CrossRef]
84. Griffith, D.; Chun, Y. Spatial Autocorrelation and Spatial Filtering. In *Handbook of Regional Science*; Fischer, M.M., Nijkamp, P., Eds.; Springer: Berlin/Heidelberg, Germany, 2014; pp. 1477–1507. ISBN 978-3-642-23429-3.
85. Hoeting, J.A.; Madigan, D.; Raftery, A.E.; Volinsky, C.T. Bayesian Model Averaging: A Tutorial (with Comments by M. Clyde, David Draper and E. I. George, and a Rejoinder by the Authors. *Statist. Sci.* **1999**, *14*, 382–417. [CrossRef]
86. Winkler, A.M.; Ridgway, G.R.; Webster, M.A.; Smith, S.M.; Nichols, T.E. Permutation Inference for the General Linear Model. *NeuroImage* **2014**, *92*, 381–397. [CrossRef]

entropy

MDPI

Article

Is Word Order Responsive to Morphology? Disentangling Cause and Effect in Morphosyntactic Change in Five Western European Languages

Julie Nijs *, Freek Van de Velde and Hubert Cuyckens

Department of Linguistics, KU Leuven, 3000 Leuven, Belgium; freek.vandevelde@kuleuven.be (F.V.d.V.); hubert.cuyckens@kuleuven.be (H.C.)

* Correspondence: julie.nijs@kuleuven.be

Abstract: This study examines the relationship between morphological complexity and word order rigidity, addressing a gap in the literature regarding causality in linguistic changes. While prior research suggests that the loss of inflectional morphology correlates with the adoption of fixed word order, this study shifts the focus from correlation to causation. By employing Kolmogorov complexity as a measure of linguistic complexity alongside Granger Causality to examine causal relationships, we analyzed data from Germanic and Romance languages over time. Our findings indicate that changes in morphological complexity are statistically more likely to cause shifts in word order rigidity than vice versa. The causal asymmetry is robustly borne out in Dutch and German, though waveringly in English, as well as in French and Italian. Nowhere, however, is the asymmetry reversed. Together, these results can be interpreted as supporting the idea that a decline in morphological complexity causally precedes a rise in syntactic complexity, though further investigation into the underlying factors contributing to the differing trends across languages is needed.

Keywords: Kolmogorov complexity; Granger causality; morphological complexity; word order rigidity; diachronic linguistics

1. Introduction

Most Indo-European languages, especially the Germanic and the Romance languages, have transitioned from more 'synthetic' structures, where grammatical relationships are expressed through inflections, such as case endings on nouns or verbal affixation, to more 'analytic' structures, where grammatical relationships are conveyed primarily through word order and function words, such as articles, prepositions, and auxiliaries [1–6]. The loss of bound morphology, as a strategy for the expression of morphosyntactic relationships, can be expected to be compensated by word order. Specifically, the existence of a case system tends to inversely correlate with fixed word order, and the deterioration of the verbal morphology can be compensated by auxiliaries and, again, word order. In West Germanic languages, the various verbal positions, viz. the difference between V1, V2, and V-final, signal distinctions in the domain of mood and illocution. This trade-off between morphology and word order is well-known [7,8] and has been established on the basis of corpus data [9–12], as well as in experimental studies [13,14].

It is a moot point among linguists to determine how the various changes are chronologically and causally related. One perspective argues that the loss of inflectional morphology, such as case endings, necessitated the development of a fixed word order (see 'configurationality' in [15,16]) to preserve clarity and comprehension in sentences. According to this view, as languages lost their inflections, speakers relied more on word order to convey grammatical relationships [17,18]. Another perspective suggests that the rise of a fixed word order might have contributed to the loss of inflectional morphology. As certain word orders became more prevalent and entrenched, the need for explicit morphological markers diminished, leading to their eventual erosion [19]. A third viewpoint questions the existence of a direct causal relationship between the two phenomena. Instead, it posits that the decline of inflectional morphology and the emergence of fixed word order could be parallel developments influenced by other language-internal factors such as discourse strategies that aim at rhetorical effectiveness rather than grammatical efficiency [20].

While linguists have always had a strong interest in *how* languages change, they have also given due attention to the (internal) *causes* of language change [21]. Still, the field of linguistics has long been somewhat reluctant to explore causality [22]. When one assumes the knock-on effects of several internal factors, linguists insist one should adduce evidence that goes beyond a simple post hoc ergo propter hoc explanation. In this respect, the temporal co-existence of two historical processes does not necessarily imply that these processes are causally related, even when this relation is highly plausible. This is also well-known in correlational studies: correlation does not imply causation. More recently, significant efforts have been made to explore causation in linguistics, and more specifically the relation between morphology and word order, using methods from adjacent fields of diachronic studies [12,23,24]. These studies provide evidence that morphological changes are indeed the chronological and causal precursors of changes in word order. What has not yet been explored in terms of causality is the relationship between morphological complexity and word order rigidity, which is the focus of the study at hand.

Following [23,24], we used a statistical technique borrowed from economics, namely Granger Causality, to examine the (potentially causal) relationship between morphological complexity and word order rigidity. Rosemeyer and Van de Velde (2021) [24] analyzed a specific aspect of morphosyntax: the relation between word order and clefting in Brazilian Portuguese. For the dependent variable, Moscoso del Prado Martín (2014) [23] used a more 'holistic' metric, namely Shannon entropy [25], to look at inflectional diversity (i.e., inflectional entropy) and syntactic diversity (i.e., entropy of the parse tree) in Icelandic. Our study can be seen as expanding upon these two previous studies, by both broadening the scope of the study and shifting the focus from the relation between inflectional morphology and word order to morphological complexity and word order rigidity (as a proxy for syntactic complexity). We combined Granger Causality (see Section 2.2) with Kolmogorov complexity (see Sections 2.1 and 2.4), an information-theoretic notion that can be used for measuring linguistic complexity [26]. By using Kolmogorov complexity, we avoided having to zoom in on a specific domain within the morphosyntax, and we could instead assess the aggregate morphological complexity and word order rigidity. Both techniques have very specific data requirements: Granger Causality requires uninterrupted time series, while Kolmogorov complexity works best with comparable texts (see Section 2.3). Good data are hard to come by, but with some leeway in the text collections, we were able to examine six different corpora of five different large languages spoken (and written) in Western Europe: Dutch, English, German, Italian, and French, with enough historical depth in their text records to venture into diachronic terrain.

2. Data and Methods

2.1. Assessing Linguistic Complexity with Kolmogorov Complexity

The study of linguistic complexity is riddled with a variety of terminological and methodological issues. However, two principal concepts of complexity can generally be distinguished: absolute complexity and relative complexity [27–29]. **Absolute complexity** approaches complexity as independent of, and unrelated to, any specific language user. It is concerned with the inherent features or phenomena of a linguistic system (e.g., the morphological markers for plural). From an information-theoretic perspective, it relates to the length of the shortest possible description of a given linguistic feature or phenomenon [30]. As it focusses on the inherent complexity of a linguistic system, independent of any agent or user, absolute complexity can be viewed as an 'objective' measure. For instance, absolute complexity may be interested in which languages employ more or less complex strategies in specific linguistic domains. **Relative complexity**, on the other hand, adopts a 'subjective' perspective, being agent-related and dependent on the experience of language users. This approach defines complexity in terms of the cost, processing effort, or acquisition difficulty as perceived by, and relative to, a language user, such as a speaker, hearer, or learner [27,28]. Typically, the language user referred to in discussions of relative complexity is the second-language learner. Consequently, relative complexity is often equated with the difficulty experienced in second-language acquisition [31–34]. As an example, relative complexity is, among other things, interested in whether linguistic optionality (i.e., the choice between two morphosyntactic variants) causes production difficulty for the speaker.

Information theory is a branch of applied mathematics and computer science that deals with the quantification, transmission, storage, and processing of information. The main goal of information theory is to measure and analyze the amount of information contained in a message or signal. It provides a mathematical framework assessing how efficiently information can be transmitted through a communication channel and how much redundancy or noise might be present in the data. Information theory was developed by Claude Shannon in the 1940s [25] and has since become a fundamental concept in various fields, including communication systems, cryptography, data compression, and more recently, linguistics.

Kolmogorov complexity [35] measures the intrinsic complexity or information content of a data sequence in terms of the length of the shortest possible algorithm required to generate that particular sequence. In essence, Kolmogorov complexity quantifies the compressibility of data; data with low complexity can be compressed efficiently, while data with high complexity cannot be compressed much. The application of Kolmogorov complexity to linguistic complexity was initiated by Juola (1998) [36]. Subsequent investigations delving into this area have been conducted by Bane [37], Juola [38], Sadeniemi et al. [39], Ehret and Szmrecsanyi [40], Ehret [26] and Nijs, Van de Velde and Cuyckens [41].

Kolmogorov complexity calculates the complexity of texts in an unsupervised manner, by exploiting the (ir)regularity and redundancy of linguistic surface structures. In essence, it gauges the repetition and recurrence of orthographic character sequences within a given text. The complexity of a text sample is defined by the length of the shortest possible description that fully represents that sample. When comparing the complexity of two texts, the goal is to describe them as concisely as possible. The text for which the most concise description is the longest is considered the most complex, whereas the text for which the most concise description is the shortest is deemed the least complex. More reducibility means less complexity. For example, the shortest description of the sequence *abababababab* would be 'ab'*4, while the shortest description of a sequence like *abhirfemholp* would be 'abhirfemholp' (i.e., *abhirfemholp* cannot be further reduced). As the most concise description of the sequence *abababababab* is shorter than the most concise description of *abhirfemholp*, the sequence *abababababab* is less complex. A high-complexity sequence consists of random data that requires a lengthy algorithm to reproduce, while a low-complexity sequence has a clear and repetitive pattern that can be represented by a concise algorithm. In a similar way, natural language is characterized by repetition that can be exploited to arrive at the shortest possible description of a text; examples include frequently occurring N-grams such as the Dutch *ik ben geweest* (I have been) or plural suffixes *-en* and *-s* in the Dutch *plant-en* (plant-PLURAL) and Dutch *vogel-s* (bird-PLURAL). While Kolmogorov complexity is a counterintuitive notion to most scholars of language as it completely ignores semantics, it has been proven to be a workable metric of complexity in various fields of applied mathematics.

Unfortunately, Kolmogorov complexity is strictly speaking incomputable because of the Halting Problem [42]. The Halting Problem is concerned with figuring out whether a given program will eventually finish its task (halt) or will keep running forever. Alan Turing showed that there is no general algorithm (set of instructions) that can determine whether any arbitrary program will halt or run forever. The Halting Problem demonstrates the existence of undecidable problems in computation. This implies that there is no algorithm to determine the shortest possible program that generates any given string of data. In other words, there is no general algorithm that can always provide the shortest possible description for all possible inputs.

Since Kolmogorov complexity relies on finding the shortest program to describe data, the undecidability of the Halting Problem implies that it is unfeasible to measure Kolmogorov complexity for all possible inputs. Though Kolmogorov complexity fundamentally remains a theoretical ideal, there are often practical workarounds to approximate Kolmogorov complexity in specific cases, such as assessing the morphological and syntactic complexity of texts. One common procedure involves using data compression algorithms, which achieve compression by finding patterns and redundancies in data and represent it as concisely as possible. Data compression techniques, such as gzip, which was also used in our study, are effective at reducing the size of data by concisely encoding recurring patterns. The size of the compressed data can serve as an approximation of its Kolmogorov complexity.

2.2. Determining Causality with Granger Causality

In this study, we wanted to test the causal precedence between changes in morphological complexity and word order rigidity by utilizing a statistical technique known as Granger Causality [43,44], initially developed and used in the field of economics. **Granger Causality** allows us to determine the causal precedence between two time series, denoted as A and B. Specifically, the technique assesses the added value of using prior values of time series B to predict time series A above and beyond the capacity of predicting time series A by prior values of itself, i.e., pure 'autoregression'. It then tries to do the reverse, using prior values of A to predict B on top of B's own prior values. Granger Causality is established when an asymmetry arises and only one of the comparisons of the two nested models comes out as significant. The underlying idea is that if time series A causally influences time series B, the dynamics of A contain information that can help reduce the prediction errors (residuals) in B.

As an example, Granger Causality analysis can help assess whether changes in interest rates (A) causally influence housing prices (B) or vice versa. If using prior values of interest rates significantly improves the prediction of housing prices, beyond what housing prices' own past values can predict, then Granger Causality from A to B is established. On the other hand, if prior housing prices do not significantly improve the prediction of interest rates, there would be no Granger Causality from B to A. This asymmetry would suggest that fluctuations in interest rates influence housing prices, but not the other way around, highlighting the predictive power of A's dynamics in explaining B.

Granger Causality thus capitalizes on the temporal relationship between two time series, where the first series can be considered a Granger-cause of the second series if changes in the second series frequently follow corresponding changes in the first series. However, it should be noted that Granger Causality alone does not provide definitive proof of causation. Its primary function is to exploit chronological asymmetry to rule out causation in the opposite direction. It does not entirely eliminate the possibility that the two time series may be coincidentally correlated, although this risk is mitigated by incorporating p-values to assess the correlation. Another limitation of Granger Causality is that it does not exclude a scenario where trend A Granger-causes trend B because both trends are actually influenced by a deeper trend C (what is known as a 'confounder'), and B happens to exhibit a delayed reaction. One example could be the relationship between ice cream sales and drowning incidents. During summer, higher temperatures drive both trends: people buy more ice cream and engage in water activities, increasing drowning risks. Granger Causality might falsely suggest that ice cream sales predict drowning incidents because there is some delay between the two events, but the true cause is the confounding factor—temperature.

The presence of a so-called 'unit root' in a time series increases the risk of Type I errors (i.e., the risk of a false positive) [45]. A **unit root** is the term for an overall upward or downward trend over time. To mitigate this effect, the time series can be detrended by differencing, i.e., by calculating the difference between consecutive data points [23]. Essentially, detrending means that the underlying trend is removed from a time series, leaving behind the fluctuations or variations that are not explained by the trend. The goal of detrending is to focus on the remaining patterns or residuals, which may contain important information about short-term behavior or irregularities in the data. However, this detrending process may lead to information loss, particularly in data-sparse time series, which are often encountered in historical linguistics. We tried to determine whether there was a unit root in the time series using the Augmented Dickey–Fuller Test [46]. The results of the test will be discussed below per corpus, together with the results of the Granger Causality test.

Another cause of Type I errors is a greater temporal lag between two time series. As said, Granger Causality essentially compares two regression models: a model with only the autoregression of time series A, where the outcome y at time step t is predicted by its past values, together with an error term ε (1); and a model where y is predicted by the autoregression of time series A and the past values of time series B (the term $\sum_{i=1}^{k} \beta_i x_{t-i}$) (2). A Chi square test or an F-test can be used on the residual deviance to see whether the two regression models differ significantly [47,48]. Under the null hypothesis, all the β terms approach 0, which means that the models are not significantly different. The null hypothesis can be rejected when model (2) outperforms model (1), which means that time series A is Granger-caused by time series B. The result of the F-test depends on the number of time lags k (i.e., the number of time steps the model uses to look back) used. The higher the number of lags, the more β factors there are that can spuriously differ from 0. The higher the lag between the two time series, the greater the risk of Type I errors. To mitigate this risk, it is best to keep the number of lags low [49]. In the case at hand, we restricted ourselves to lags 1 to 10.

$$y_t = \alpha_0 + \sum_{i=1}^{k} \alpha_i y_{t-i} + \varepsilon_t \tag{1}$$

$$y_t = \alpha_0 + \sum_{i=1}^{k} \alpha_i y_{t-i} + \sum_{i=1}^{k} \beta_i x_{t-i} + \varepsilon_t \tag{2}$$

It is important to test the causation in two ways: (i) do changes in morphological complexity Granger-cause changes in word order rigidity, or (ii) do changes in word order rigidity Granger-cause changes in morphological complexity? We hypothesize that the causal relation runs from morphological complexity to word order rigidity and not the other way around, following the most common hypothesis that word order rigidity 'compensates' for the loss in inflectional morphology. Accordingly, in test (i), the difference between the two regression models has to be significant, and in test (ii), the difference between the two regression models has to be non-significant.

2.3. *Corpora*

Both Kolmogorov complexity and Granger Causality have a set of requirements that influence the selection of the data. The ideal type of data for a Kolmogorov complexity analysis is parallel data, which ensures that topic and orthography have a minimal influence on the complexity measure. However, there are very few parallel, diachronic corpora that allow us to look at diachronic change. While parallel data are ideal, it is not altogether impossible to obtain meaningful results with semi-parallel or even comparable data (see Ehret (2017: pp. 131–159) [26] for an application of Kolmogorov complexity to British English registers and English learner corpora). Since Kolmogorov complexity relies on the compression of reoccurring patterns, a second requirement is that the input data need to be large enough to ensure that such reoccurring patterns are indeed present. Additionally, when using semi-parallel or comparable data, all the texts in the input need to be the same size. Following [26,40], we define texts as having the 'same size' if they have the same number of sentences. We adhered to this principal corpus-internally; i.e., per corpus, we sampled an equal number of sentences for each year. However, this number was not kept equal across the different corpora. For each corpus, we selected as much data per year as was available in the year with the least data. Accordingly, we did not draw conclusions across the different corpora or languages but only answered the research question for each individual corpus, so that each corpus independently contributes evidence for the hypotheses. Some corpora contained a lot of data, even in the year with the least data, but sampling too many sentences was also not ideal, as this would have increased computation time. In practice, we took samples ranging from 1500 sentences (for DTA) to 5000 sentences (for C-CLAMP).

Granger Causality works with time series data. A time series is a sequence of data points collected or recorded at successive, evenly spaced time intervals. The two main requirements for this technique are that the time series is uninterrupted and has sufficient length. Interruptions or gaps in the time series data may distort or obscure the true causal relationships between variables, leading to inaccurate results. Granger Causality analysis relies on the concept of time lags, where the past values of one variable are used to predict the future values of another variable. If the time series data are interrupted, the temporal sequence required for accurate time lag analysis is disrupted as well. A sufficient length of time series data is required to establish meaningful relationships between variables.

Selecting corpora and sampling is thus a difficult balancing act. We selected diachronic corpora containing as much data as possible over a sufficiently long time span (required for Granger Causality), while aiming for homogeneity (required for Kolmogorov complexity) within each corpus. Some of the corpora did not have enough data for an analysis per year, so for these, we opted for an analysis per decade instead. We created a random sample per time unit for the analysis. Table 1 gives an overview of the corpora, the level of granularity (year or decade), the sample size, the time span, and the genre of the corpus.

Table 1. Overview of the corpora.

Corpus	Language	Year/Decade	Sample Size (Sentences)	Time Span	Genre
C-CLAMP [50]	Dutch	Year	5000	1837–1999	Cultural and literary journals
The Corpus of Historical American English (COHA) [51]	American English	Year	4000	1830–1999	Only the fiction texts are taken into account
The Corpus of Late Modern English Texts, version 3.0 (CLMET3.0) [52]	British English	Decade	2000	1710–1920	Narrative fiction, narrative non-fiction, drama, letters, and treatise
Deutsches Textarchiv (DTA) [53]	German	Year	1500	1863–1927	Not genre-consistent
L'Unità [54]	Italian	Year	5000	1948–2014	Newspapers
Frantext [55]	French	Decade	5000	1600–1940	Not genre-consistent

2.4. Measuring Morphological Complexity and Word Order Rigidity

The objective of this study was to investigate whether changes in morphological complexity can predict changes in word order rigidity, as a proxy for syntactic complexity. To achieve this, we needed to calculate the morphological complexity ratio and word order rigidity ratio. However, simply applying compression to the entire text was insufficient, since we wished to analyze morphological complexity and word order rigidity separately. Instead, we had to modify the text at the morphological and syntactic level before compression. The complexity ratios were determined by dividing the file size of the compressed and distorted files (resulting from steps 2 and 3 below) by the compressed file without distortion (resulting from step 1) [26,38]. To be more specific, the following steps were taken:

1. Compress the original file using Gzip.
2. Introduce distortion to the original file, creating the distorted file.
3. Compress the distorted file once again using Gzip.
4. Calculate the complexity ratio.

Morphological distortion is implemented by randomly removing 10% of the orthographically transcribed characters in the text. This alteration disrupts the information at the morphological level and compromises morphological regularity, affecting inflectional endings (e.g., *giving* can become *givig*) and roots and stems (e.g., *fossil* can become *fssil*). The random removal of characters essentially gives rise to new word forms generated haphazardly, which means that the number of unique tokens in the text increases. The distortion will have a smaller effect on the compression of complex texts with a high surface token diversity, because they already contain a higher number of unique tokens before distortion. The effect of the distortion on the compression of less complex texts will be comparatively larger, because these texts have a low surface token diversity before distortion, but their diversity increases remarkably after distortion.

Syntactic distortion happens in the same way as morphological distortion, but instead of characters, words are removed. This process disrupts the word order rules (e.g., a 4-gram such as *it should have been* can become *it should been*), resulting in a greater number of unique n-grams. Consequently, the compressibility of the text is adversely affected, leading to decreased compression performance. Complex texts, with strict word order, will be more affected by distortion, because they contain more structural surface redundancies. In other words, complex texts have more recurring n-grams that can be jumbled during distortion. Syntactically simple texts with free word order already have fewer structural surface redundancies and will therefore be less affected by the distortion (thus facilitating compressibility).

The distortion method employed implies that the syntactic complexity ratio reflects the rigidity of the word order, rather than providing a comprehensive representation of the overall syntactic complexity of a language. While it is true that word order is not the exclusive means by which languages convey functions or meanings through syntactic strategies, it is a strategy at the opposite end from affixal morphology on the morphology–syntax cline. Additional strategies encompass semi-morphological aspects such as agreement and cross-reference. With this approach, we followed Bakker (1998) [56], who also investigated syntactic complexity on the basis of word order rigidity or 'flexibility', where flexible languages are syntactically simple and inflexible languages are syntactically complex. From the perspective of language processing (rooted within relative complexity), rigid word order is normally defined as syntactically simple, because strict word order is easier to process than free word order. The methodology of the research at hand is embedded within the absolute view on complexity, which focusses on the number of word order rules and constraints. In this context, a language with strict word order is considered more complex, because it imposes more restrictions or constraints on the rules governing word order. This idea of complexity as constraints goes back at least to Greenberg (1960) [57], who calculated the proportion of word order links over the total number of 'nexus' (Greenberg's term for grammatical relations). To avoid confusion, we will henceforth talk about word order rigidity instead of syntactic complexity.

The morphological complexity ratio and the word order rigidity ratio were calculated as follows:

$$morphological\ complexity\ ratio = -\frac{compressed\ file\ size\ after\ morphological\ distortion}{compressed\ file\ size} \quad (3)$$

$$word\ order\ rigidity\ ratio = \frac{compressed\ file\ size\ after\ syntactic\ distortion}{compressed\ file\ size} \quad (4)$$

It is important to observe that the word order rigidity ratio is represented by a positive value, while the morphological complexity ratio is indicated by a negative value. This distinction follows from the nature of the distortion process employed. In the case of morphological distortion, we anticipated that a complex text would be less affected, resulting in a negative value. Conversely, with syntactic distortion, we expected a complex text to be more impacted, leading to a positive value. Making the morphological complexity ratio negative and the syntactic complexity ratio positive ensured easier interpretation of the results. Indeed, in both ratios, higher values (marked by a smaller difference between numerator and denominator) indicated greater complexity.

The use of random deletion means that it is in theory possible that similar types of elements are repeatedly deleted in a particular text. For example, with regard to morphological distortion, it is possible that in a text, only roots or only affixes are deleted. With regard to syntactic distortion, it is possible that the same words are deleted, while others are left untouched. In order to mitigate this aleatoric effect of the randomization in our study, for each text, the morphological complexity ratio and word order rigidity ratio were computed over 100 iterations, whose means were then calculated. In other words, for each text, the mean morphological complexity ratio and mean word order rigidity ratio were retained.

Finally, we had to turn these complexity measures into two time series: one for morphology and one for syntax (word order rigidity). Importantly, a time series requires a measurement at each time step; in other words, it is not allowed to have gaps. If we were to have created a time series with time steps of one year, such gaps would have presented a problem. Accordingly, we operated with time steps of one decade rather than one year. For each of these time steps, we calculated the mean morphological complexity ratio and the mean word order rigidity ratio. These means make up our time series.

3. Analysis and Results

For each corpus, we examined whether there is a causal relationship between morphological complexity and word order rigidity. First, a covariate-augmented Dickey–Fuller test [58] was applied to the time series to check for a unit root (see Section 2.2). Table 2 gives an overview of the results of the test. A significant result ($p < 0.05$), indicated in bold, means that the time series does not contain a unit root and detrending is not necessary. However, it is recommended that the other time series, which do contain a unit root, are detrended. It is not possible to compare a detrended time series with one that has not been detrended, because a detrended time series is one time step shorter than its original. This is why we applied Granger Causality to both the original and the detrended time series in both directions (i.e., from morphology to word order and the other way around). We examined the relationship between morphology and word order up to 10 time steps (lags) back in time. For corpora with yearly time series data, we used a lag of up to 10 years. However, time series from corpora with data measured per decade were shorter, so fewer lags were preferable to ensure accurate analysis. Thus, for CLMET and Frantext, we opted for 5 lags.

Table 2. p-values of the covariate-augmented Dickey–Fuller test; significant results at an α level of 0.05.

	CCLAMP	COHA	CLMET	DTA	L'Unità	Frantext
Morphology	0.2916	**0.001219**	0.6638	**0.0002868**	0.2934	0.8443
Word order	**0.0005695**	**0.004377**	0.5149	0.06443	0.2232	**0.04433**

Additionally, we examined the relationship between morphological complexity and word order rigidity through linear regression, to see whether there is indeed a trade-off; i.e., we tried to see if years/decades with high morphological complexity have low word order rigidity and vice versa. The results can be appreciated in Tables 3–8 and Figures 1–6. Each corpus produced the same results: there is indeed a significant negative correlation between morphological complexity and word order rigidity.

Table 3. Linear regression model of morphological complexity over word order rigidity for C-CLAMP (adjusted R-squared = 0.313).

	Estimate	Std. Error	t Value	Pr (>|t|)
Intercept	5.6206	0.7691	7.308	<0.0001
Word order rigidity	−7.1578	0.8275	−8.650	<0.0001

Table 4. Linear regression model of morphological complexity over word order rigidity for COHA (adjusted R-squared = 0.5626).

	Estimate	Std. Error	t Value	Pr (>|t|)
Intercept	1.5096	0.1720	8.774	<0.0001
Word order rigidity	−2.7166	0.1838	−14.778	<0.0001

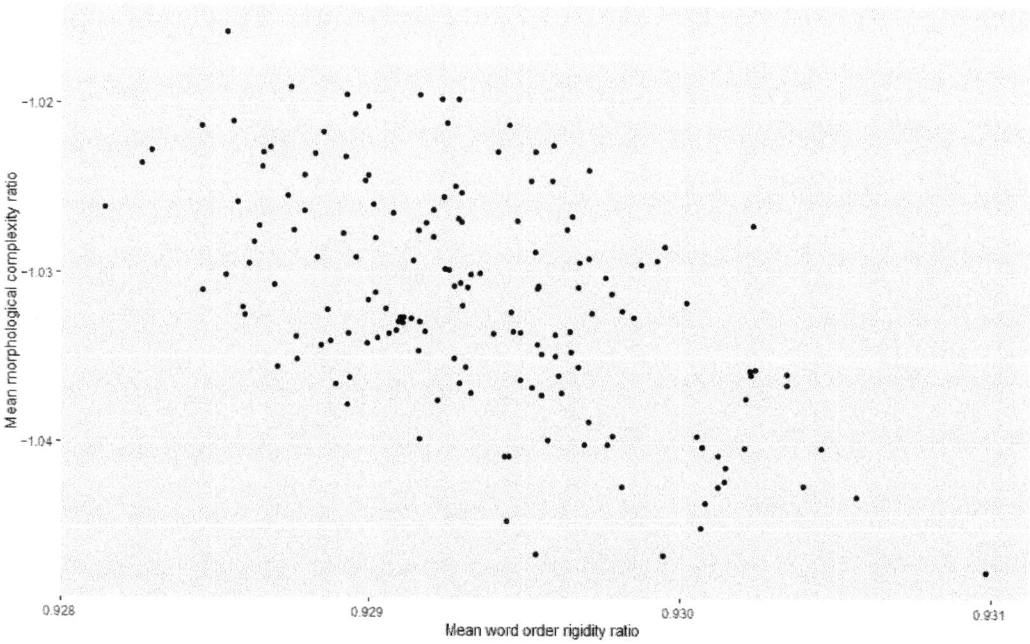

Figure 1. Visualization of the linear regression model; scatterplot of the morphological complexity and word order rigidity measures per time step in C-CLAMP.

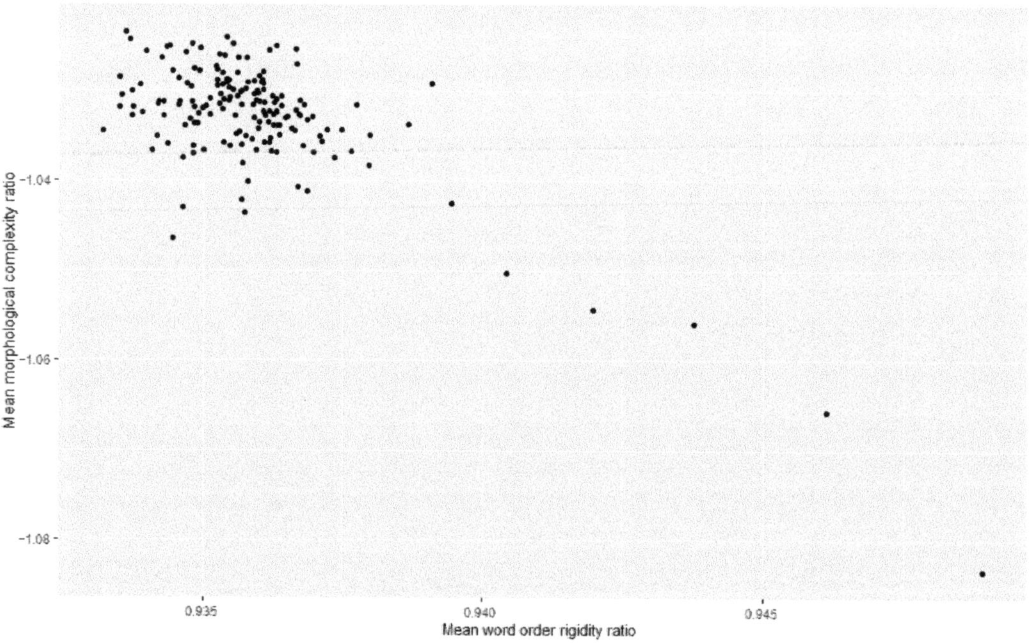

Figure 2. Visualization of the linear regression model; scatterplot of the morphological complexity and word order rigidity measures per time step in COHA.

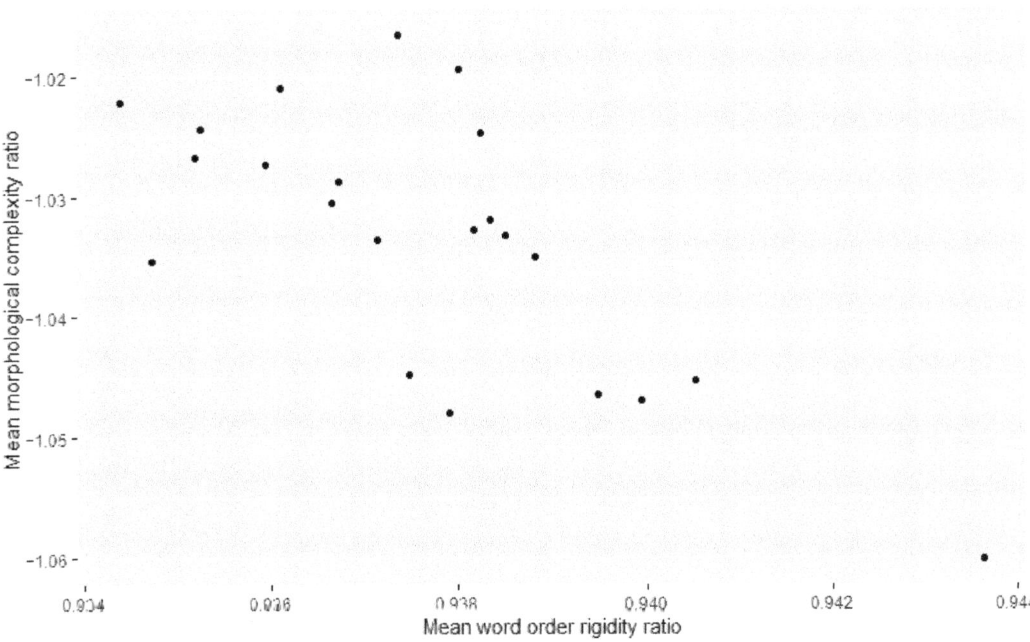

Figure 3. Visualization of the linear regression model; scatterplot of the morphological complexity and word order rigidity measures per time step in CLMET.

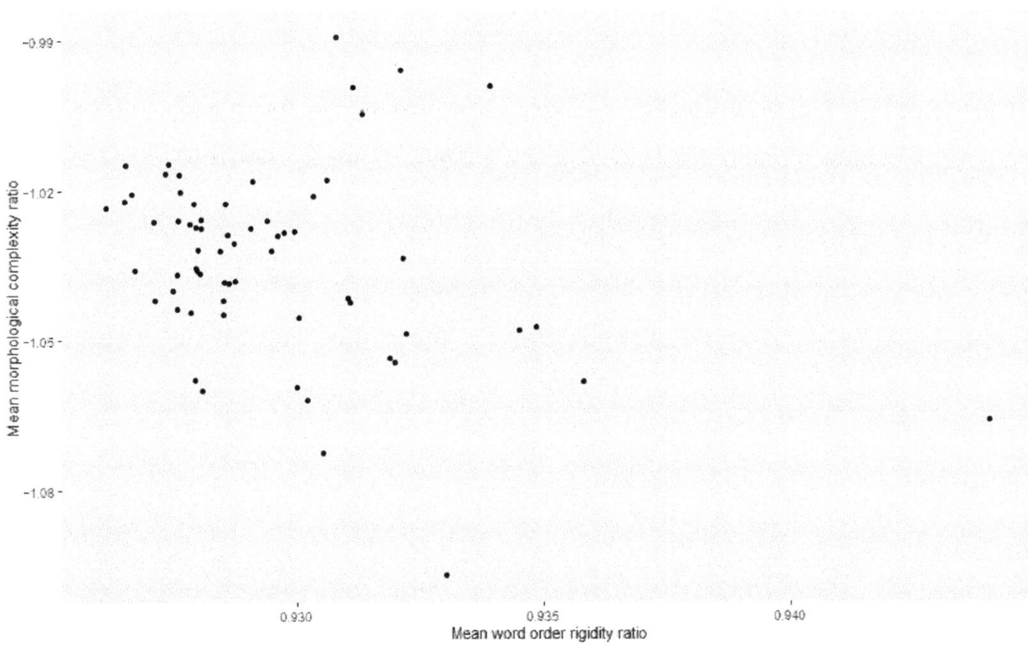

Figure 4. Visualization of the linear regression model; scatterplot of the morphological complexity and word order rigidity measures per time step in DTA.

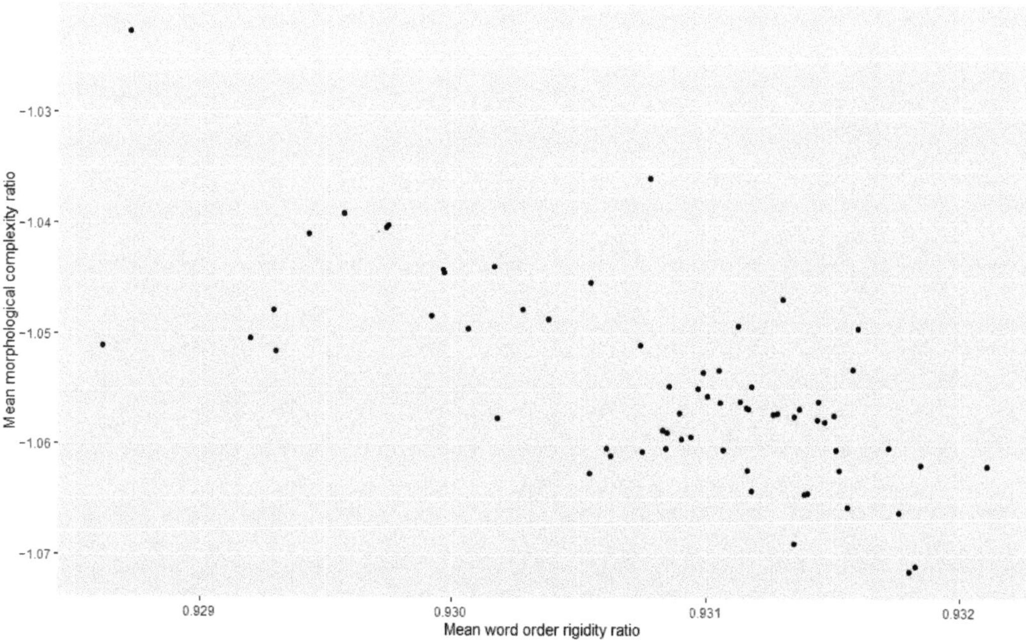

Figure 5. Visualization of the linear regression model; scatterplot of the morphological complexity and word order rigidity measures per time step in L'Unità corpus.

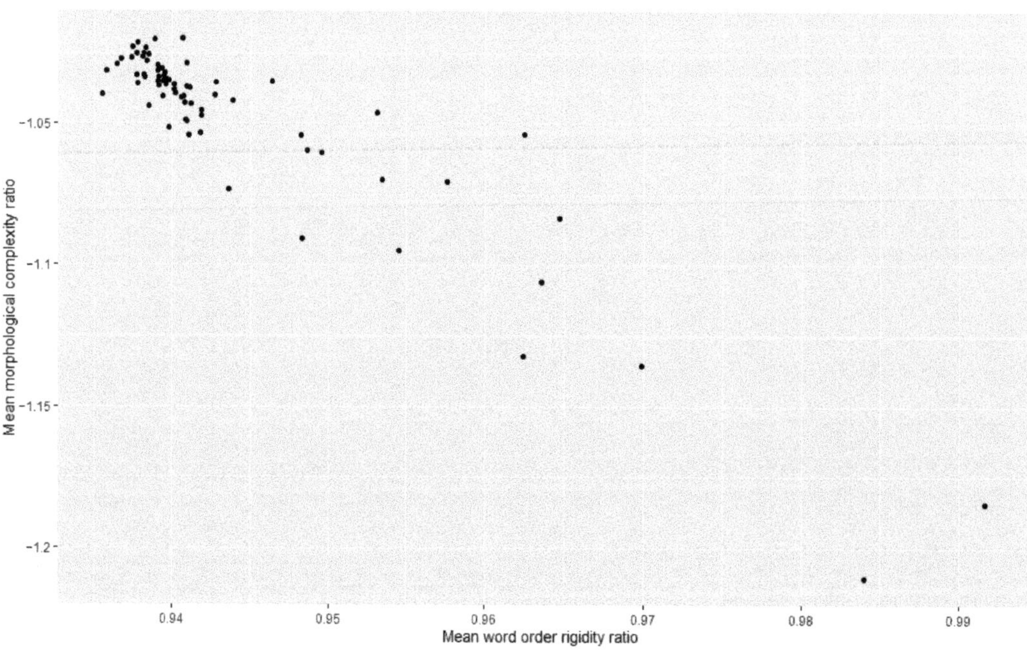

Figure 6. Visualization of the linear regression model; scatterplot of the morphological complexity and word order rigidity measures per time step in Frantext.

Table 5. Linear regression model of morphological complexity over word order rigidity for CLMET 3.0 (adjusted R-squared = 0.4795).

	Estimate	Std. Error	t Value	Pr (>∣t∣)
Intercept	2.4469	0.7716	3.171	<0.0001
Word order rigidity	−3.7117	0.8229	−4.510	<0.0001

Table 6. Linear regression model of morphological complexity over word order rigidity for DTA (adjusted R-squared = 0.08424).

	Estimate	Std. Error	t Value	Pr (>∣t∣)
Intercept	0.4010	0.3980	1.008	0.3155
Word order rigidity	−1.5496	0.4274	−3.625	<0.0001

Table 7. Linear regression model of morphological complexity over word order rigidity for L'Unità corpus (adjusted R-squared = 0.4864).

	Estimate	Std. Error	t Value	Pr (>∣t∣)
Intercept	6.1967	0.9100	6.810	<0.0001
Word order rigidity	−7.7905	0.9776	−7.969	<0.0001

Table 8. Linear regression model of morphological complexity over word order rigidity for Frantext (adjusted R-squared = 0.8519).

	Estimate	Std. Error	t Value	Pr (>∣t∣)
Intercept	1.7899	0.1423	12.58	<0.0001
Word order rigidity	−3.0067	0.1507	−19.95	<0.0001

Turning to the Granger Causality analysis, the results can be found in Tables 9–14 and Figures 7–12. The results for the Germanic languages, except for the CLMET corpus, show that changes in morphological complexity indeed Granger cause changes in word order rigidity if the time series are not detrended. The only exception is the COHA corpus, where changes in morphological complexity can still be said to Granger-cause changes in word order rigidity even when the time series have been detrended. Most of the Germanic corpora show significance at several lags, which indicates that the results are less likely to be spurious. C-CLAMP and DTA both show significance at the lower time steps, lags 1–2 and lags 1–4, respectively. While COHA shows significance at the higher lags of 4–10, CLMET is the only Germanic corpus that shows no significant results in either direction.

Table 9. *p*-values of the Granger Causality analysis per lag with and without detrending for C-CLAMP.

	Without Detrending		With Detrending	
Lag	Morphology Causes Word Order	Word Order Causes Morphology	Morphology Causes Word Order	Word Order Causes Morphology
1	**0.001402**	0.1755	0.4752	0.5094
2	**0.02456**	0.5463	0.7249	0.1887
3	0.07971	0.3151	0.8092	0.2566
4	0.06928	0.3661	0.8956	0.3564
5	0.1328	0.4958	0.9473	0.3461

Table 9. *Cont.*

Lag	Without Detrending		With Detrending	
	Morphology Causes Word Order	Word Order Causes Morphology	Morphology Causes Word Order	Word Order Causes Morphology
6	0.1856	0.4948	0.9623	0.3843
7	0.1364	0.4516	0.8261	0.4534
8	0.195	0.5532	0.8464	0.2641
9	0.2664	0.3497	0.831	0.3763
10	0.248	0.4806	0.8451	0.3036

Table 10. *p*-values of the Granger Causality analysis per lag with and without detrending for COHA.

Lag	Without Detrending		With Detrending	
	Morphology Causes Word Order	Word Order Causes Morphology	Morphology Causes Word Order	Word Order Causes Morphology
1	0.858	0.8623	0.1816	0.8783
2	0.4833	0.7976	0.1682	0.9423
3	0.3735	0.9036	**0.00112**	0.9318
4	**0.0005979**	0.8865	**0.001819**	0.9047
5	**0.001591**	0.8979	**0.005217**	0.8006
6	**0.009824**	0.852	**0.004183**	0.7589
7	**0.007966**	0.8743	**0.009972**	0.8227
8	**0.0139**	0.9062	**0.01866**	0.8045
9	**0.02486**	0.8858	**0.02058**	0.941
10	**0.03655**	0.9473	**0.01578**	0.8172

Table 11. *p*-values of the Granger Causality analysis per lag with and without detrending for CLMET 3.0.

Lag	Without Detrending		With Detrending	
	Morphology Causes Word Order	Word Order Causes Morphology	Morphology Causes Word Order	Word Order Causes Morphology
1	0.2877	0.5308	0.2774	0.2774
2	0.2544	0.6607	0.1641	0.1641
3	0.3219	0.08349	0.2621	0.2621
4	0.4308	0.3023	0.3442	0.3442
5	0.68	0.3935	0.3706	0.3706

Table 12. *p*-values of the Granger Causality analysis per lag with and without detrending for DTA.

Lag	Without Detrending		With Detrending	
	Morphology Causes Word Order	Word Order Causes Morphology	Morphology Causes Word Order	Word Order Causes Morphology
1	**0.005058**	0.35	0.2139	0.3681
2	**0.01473**	0.3295	0.05311	0.8995

Table 12. *Cont.*

Lag	Without Detrending		With Detrending	
	Morphology Causes Word Order	Word Order Causes Morphology	Morphology Causes Word Order	Word Order Causes Morphology
3	**0.03366**	0.9487	0.07508	0.9928
4	**0.01518**	0.9929	0.05801	0.71
5	0.05535	0.8028	0.2451	0.9412
6	0.1125	0.9778	0.3804	0.9212
7	0.1532	0.953	0.3523	0.7832
8	0.08149	0.8473	0.2842	0.8567
9	0.09821	0.9037	0.317	0.9328
10	0.05143	0.9121	0.2332	0.8846

Table 13. *p*-values of the Granger Causality analysis per lag with and without detrending for L'Unità corpus.

Lag	Without Detrending		With Detrending	
	Morphology Causes Word Order	Word Order Causes Morphology	Morphology Causes Word Order	Word Order Causes Morphology
1	0.734	0.7296	0.2733	0.8145
2	0.3631	0.8849	0.5469	0.8706
3	0.5807	0.9053	0.338	0.8134
4	0.475	0.8846	0.4782	0.9604
5	0.6295	0.9537	0.5042	0.9769
6	0.6002	0.975	0.4484	0.9221
7	0.5087	0.9233	0.4815	0.9574
8	0.475	0.9689	0.587	0.9481
9	0.5008	0.9369	0.6787	0.934
10	0.5579	0.8835	0.8587	0.9117

Table 14. *p*-values of the Granger Causality analysis per lag with and without detrending for Frantext.

Lag	Without Detrending		With Detrending	
	Morphology Causes Word Order	Word Order Causes Morphology	Morphology Causes Word Order	Word Order Causes Morphology
1	**0.01698**	**0.02445**	**0.01689**	**0.007708**
2	0.05865	**0.02889**	0.188	**0.006095**
3	0.1806	**0.01083**	0.6095	**0.04238**
4	0.6813	0.0687	0.6362	0.06929
5	0.7898	**0.04298**	0.6639	0.05629

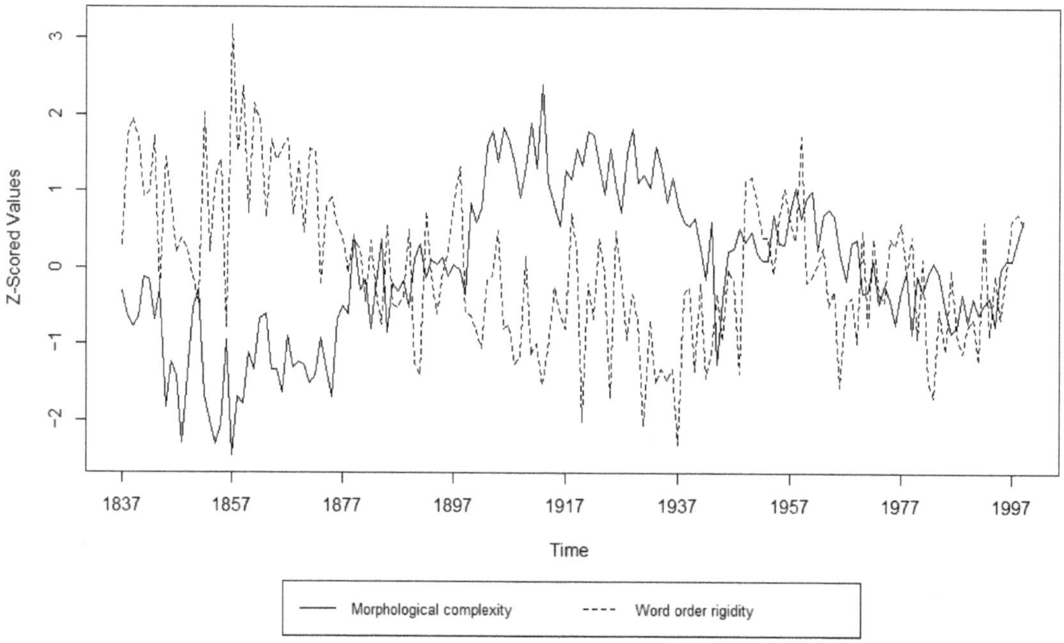

Figure 7. Visualization of the Granger Causality analysis; z-scored morphological complexity and word order rigidity measures per year in C-CLAMP.

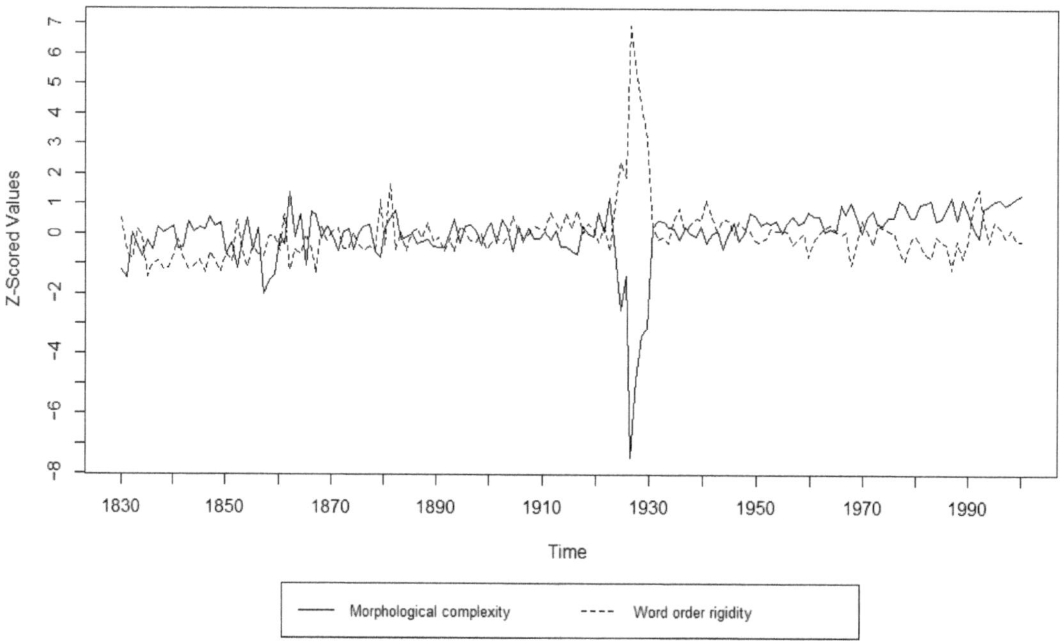

Figure 8. Visualization of the Granger Causality analysis; z-scored morphological complexity and word order rigidity measures per year in COHA.

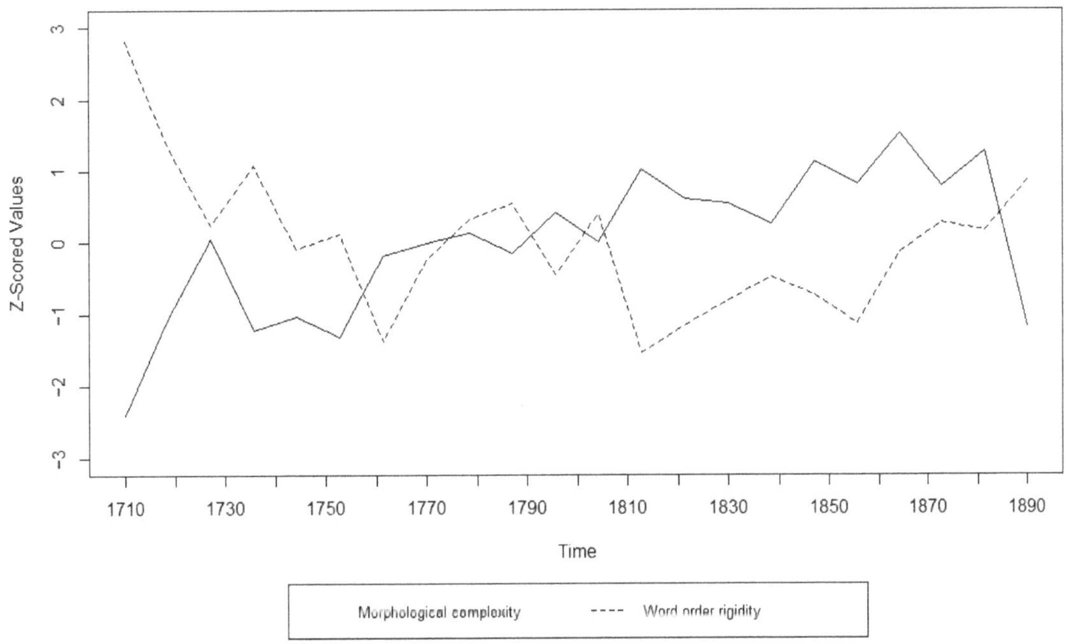

Figure 9. Visualization of the Granger Causality analysis; z-scored morphological complexity and word order rigidity measures per year in CLMET.

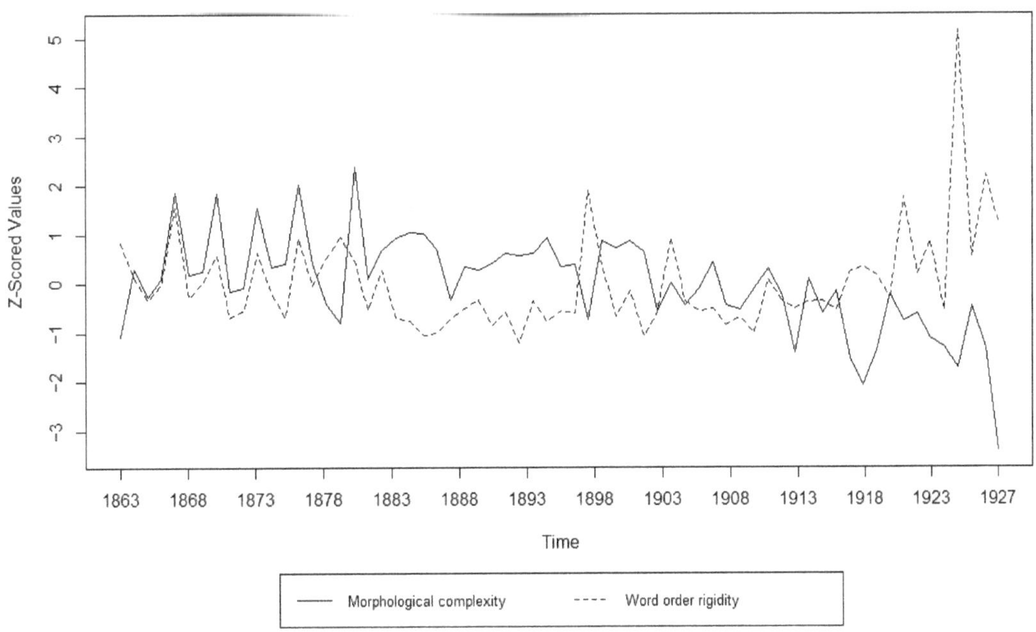

Figure 10. Visualization of the Granger Causality analysis; z-scored morphological complexity and word order rigidity measures per year in DTA.

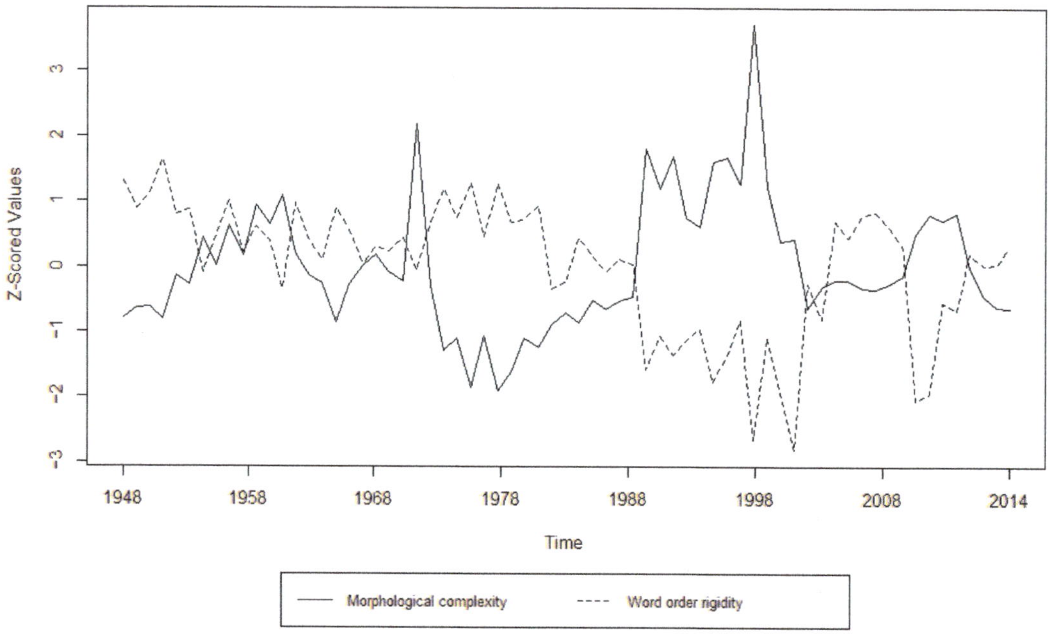

Figure 11. Visualization of the Granger Causality analysis; z-scored morphological complexity and word order rigidity measures per year in L'Unità corpus.

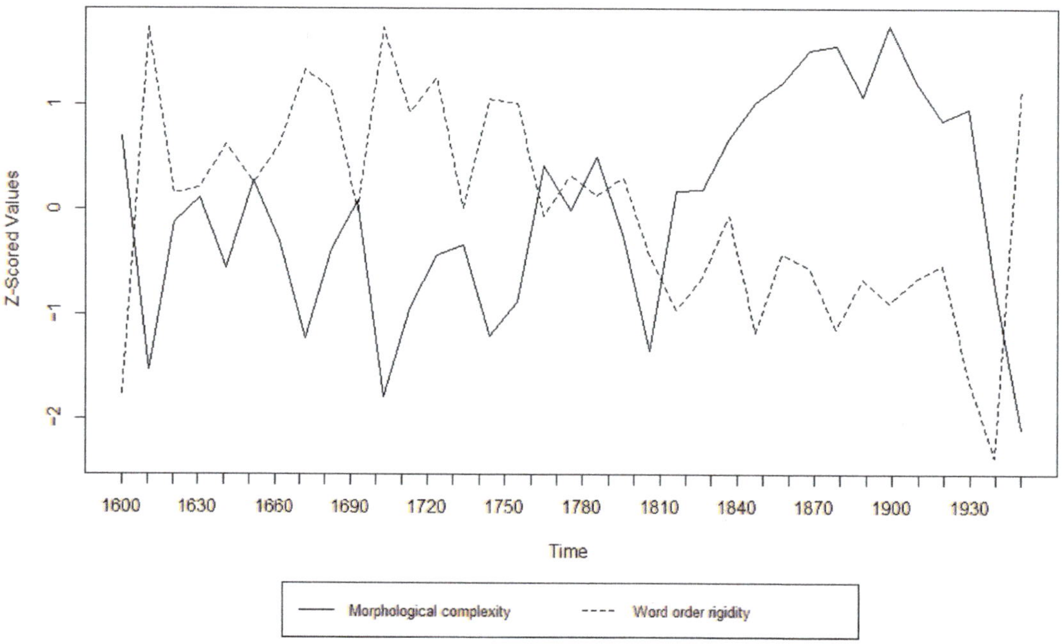

Figure 12. Visualization of the Granger Causality analysis; z-scored morphological complexity and word order rigidity measures per decade in Frantext.

The results for the Romance languages, however, do not weigh in favorably for a causal relation. The Italian corpus shows no significant effects in either direction, with or without the detrending of the time series. The French corpus shows significant effects in both directions at lag 1 with and without detrending. Since causation can only run in one direction, we have to reject this result. At higher lags (lags 2, 3, and 5 without detrending and lags 2 and 3 with detrending), we found a significant ($p < 0.05$) causal relation running from word order rigidity to morphological complexity. This means that, strictly speaking, in Frantext, changes in word order rigidity Granger-cause changes in morphological complexity at these higher lags. This result is dubious, given the bidirectional 'causation' at lag 1.

4. Discussion and Conclusions

In this study, we set out to examine the relationship between morphological complexity and word order rigidity in terms of causality. Whether changes in morphology drive changes in word order or the other way around is still a much debated moot point within linguistics [17–20]. The most commonly held hypothesis with the most substantial quantitative evidence is that morphological changes typically precede shifts in word order [12,23,24]. However, the question of whether this hypothesis holds true in relation to linguistic complexity has not been thoroughly investigated to date. We have tried to answer this question using both Kolmogorov complexity as a 'holistic' (i.e., aggregate) measure of morphological complexity and word order rigidity and Granger Causality as a means of identifying causation.

For the Germanic languages in our sample, the aforementioned hypothesis was borne out in all but one corpus. Remarkably, the corpus where the model fails to detect the expected Granger Causality is the only one with a lower-resolution measurement, at the level of decades rather than years. It would appear, then, that the method requires enough variability and enough measurement points. The anomaly could be an artefact of the data, though we do not want to overstate our claims and must retain some caution.

The low-resolution problem may also explain why the method fails to give robust results for the Romance languages in our sample: the French corpus has time measured in decades, while the Italian corpus has time measured in years but covers only half a century. This suggests that we are in urgent need of better corpora in historical linguistics.

The difference in the results displayed by the Germanic and the Romance languages could also have a linguistic explanation. One possibility is that the Germanic languages simply exhibit more outspoken differences in complexity over time. Note that overall, the timespans which we examined are not very extensive. Deflection is a process that takes several centuries. It is not unlikely, therefore, that it does not (yet) clearly manifest itself in the short time spans covered by our corpora. In this sense, the Germanic languages may very well be the odd ones out, because they do show these differences, again over quite short time spans. Indeed, Germanic languages form a branch of the Indo-European languages that is known for its rampant reduction in morphological complexity [2]—though it is generally held that German holds out better, at least in the written language, with numerous inflection intricacies that are unknown to its large West Germanic siblings and even to its Romance cousins like French and Italian.

Another explanation may be found in the system of clitic pronouns in the Romance languages. Clitic pronouns cannot be considered fully independent words and thus should not be considered completely analytical from the point of view of morphological typology. Furthermore, they exhibit highly variable behavior in the Romance languages (Pescarini 2021) [59]. It is a possibility that systems of clitic pronouns may blur the analysis.

However, Ehret (2017) [26] has conducted a similar analysis; i.e., she tested Kolmogorov complexity on a synchronic corpus including both Romance and Germanic languages, where linguistically sensible results were obtained, indicating that the analysis is robust.

Even so, the results point to an asymmetry, which can best be stated in conditional form: if there is a significant Granger Causality to be detected in a language (which may depend on its time resolution), then it is never the case that word order rigidity Granger-causes changes in morphological complexity, whereas it is common to find morphological complexity Granger-causing changes in word order rigidity. This tentatively supports the word-order-as-a-compensation strategy for the West European languages, confirming findings from earlier studies [7–12]. This study can also be compared to other works that use Kolmogorov complexity to gain insight into linguistic complexity [26,36,38,39]. The trade-offs that we observe within each language are particularly similar to those identified between languages in the studies by Ehret (2017) [26] and Ehret and Szmrecsanyi (2016) [40].

Importantly, the results work best when the time series have not been detrended. Detrending a time series involves removing trends, i.e., long-term systematic changes, from the data to focus on the more immediate or cyclical behaviors. However, in linguistics, these long-term systematic changes are exactly what is of interest in the data. Deflection is a long-term diachronic process that spans several centuries [2,5]. It could thus be argued that with linguistic phenomena, and especially given the specific research question that we are interested in, detrending may be unnecessary or even obfuscating.

Author Contributions: Conceptualization, J.N., F.V.d.V. and H.C.; data curation, J.N.; formal analysis, J.N.; funding acquisition, F.V.d.V. and H.C.; investigation, J.N.; methodology, J.N. and F.V.d.V.; project administration, F.V.d.V. and H.C.; resources, J.N.; software, J.N.; supervision, F.V.d.V. and H.C.; validation, J.N.; visualization, J.N.; writing—original draft, J.N.; writing—review and editing, J.N., F.V.d.V. and H.C. All authors have read and agreed to the published version of the manuscript.

Funding: This work was supported by the Research Foundation Flanders (FWO) under grant number G071719N.

Data Availability Statement: The data sets and scripts are available at https://zenodo.org/records/14162717 (accessed on 14 November 2024). The samples of the corpora are available at https://zenodo.org/records/14162733 (accessed on 14 November 2024).

Acknowledgments: We would like to thank Pierpaolo Basile for providing a version of L'Unita corpus split by year and Anthe Sevenants for his help with scraping the Frantext corpus.

Conflicts of Interest: The authors declare no conflict of interest.

References

1. Alkire, T.; Rosen, C.G. *Romance Languages: A Historical Introduction*; Cambridge University Press: Cambridge, UK, 2010.
2. Harbert, W. *The Germanic Languages (Cambridge Language Surveys)*; Cambridge University Press: Cambridge, UK, 2007.
3. Jespersen, O. *Progress in Language with Special Reference to English*; Swan Sonnenschein: London, UK, 1894.
4. Posner, R. *The Romance Languages (Cambridge Language Surveys)*; Cambridge University Press: Cambridge, UK, 1996.
5. Sapir, E. *Language: An Introduction to the Study of Speech*; Harcourt: New York, NY, USA, 1921.
6. Szmrecsanyi, B. Analyticity and syntheticity in the history of English. In *The Oxford Handbook of the History of English*; Nevalainen, T., Traugott, E.C., Eds.; Oxford University Press: Oxford, UK, 2012; pp. 654–665. Available online: https://research.manchester.ac.uk/en/publications/analyticity-and-syntheticity-in-the-history-of-english (accessed on 4 November 2024).
7. Haiman, J. *Natural Syntax: Iconicity and Erosion*; Cambridge University Press: Cambridge, UK, 1985.
8. Givón, T. *Syntax: An Introduction*; John Benjamins: Amsterdam, The Netherlands, 2001; Volume I. [CrossRef]
9. Fenk-Oczolon, G.; Fenk, A. Complexity trade-offs between the subsystems of language. In *Language Complexity: Typology, Contact, Change*; Miestamo, M., Sinnemäki, K., Karlsson, F., Eds.; John Benjamins: Amsterdam, The Netherlands, 2008; pp. 43–65.

10. Sinnemäki, K. Complexity trade-offs in core argument marking. In *Language Complexity: Typology, Contact, Change*; Miestamo, M., Sinnemäki, K., Karlsson, F., Eds.; John Benjamins: Amsterdam, The Netherlands, 2008; pp. 68–88.
11. Koplenig, A.; Meyer, P.; Wolfer, S.; Müller-Spitzer, C. The statistical trade-off between word order and word structure—Large-scale evidence for the principle of least effort. *PLoS ONE* **2017**, *12*, e0173614. [CrossRef]
12. Levshina, N. Cross-linguistic trade-offs and causal relationships between cues to grammatical subject and object, and the problem of efficiency-related explanations. *Front. Psychol.* **2021**, *12*, 648200. [CrossRef] [PubMed]
13. Fedzechkina, M.; Newport, E.L.; Jaeger, T.F. Balancing effort and information transmission during language acquisition: Evidence from word order and case marking. *Cogn. Sci.* **2017**, *41*, 416–446. [CrossRef] [PubMed]
14. Fedzechkina, M.; Jaeger, T.F. Production efficiency can cause grammatical change: Learners deviate from the input to better balance efficiency against robust message transmission. *Cognition* **2020**, *196*, 104115. [CrossRef] [PubMed]
15. Faarlund, J.T. From ancient Germanic to modern Germanic languages. In *Language Typology and Language Universals*; Haspelmath, M., König, E., Österreicher, W., Reible, W., Eds.; Mouton de Gruyter: Berlin, Germany, 2001; Volume 2, pp. 1706–1719. [CrossRef]
16. Van de Velde, F. Dutch. In *Oxford Research Encyclopedia of Linguistics*; Kürschner, S., Dammel, A., Eds.; Oxford University Press: Oxford, UK, 2024. [CrossRef]
17. Harris, M. *The Evolution of French Syntax: A Comparative Approach (Longman Linguistics Library 22)*; Longman: London, UK, 1978.
18. Bentz, C.; Christiansen, M. Linguistic adaptation: The trade-off between case marking and word order in Germanic and Romance languages. In *Eastward Flows the Great River: Festschrift in Honor of Prof. William S-Y. Wang on his 80th Birthday*; City University of Hong Kong Press: Hong Kong, China, 2013; pp. 46–58.
19. Bourciez, É.; Bourciez, J. *Éléments de Linguistique Romane (Tradition de L'humanisme 10)*, 5th ed.; Klincksieck: Paris, France, 1967.
20. Detges, U. How useful is case morphology? The loss of the Old French two-case system within a theory of Preferred Argument structure. In *The Role of Semantic, Pragmatic, and Discourse Factors in the Development of Case*; Barðdal, J., Chelliah, S.L., Eds.; Studies in Language Companion Series; John Benjamins: Amsterdam, The Netherlands, 2009; Volume 108, pp. 93–120. [CrossRef]
21. Campbell, L. *Historical Linguistics: An Introduction*, 3rd ed.; MIT Press: Cambridge, MA, USA, 1999.
22. Aitchison, J. *Language Change: Progress or Decay?* 4th ed.; Cambridge University Press: Cambridge, UK, 2013.
23. Moscoso del Prado Martín, F. Grammatical change begins within the word: Causal modeling of the co-evolution of Icelandic morphology and syntax. *Proc. Annu. Meet. Cogn. Sci. Soc.* **2014**, *36*, 2657–2662.
24. Rosemeyer, M.; Van de Velde, F. On cause and correlation in language change: Word order and clefting in Brazilian Portuguese. *Lang. Dyn. Change* **2021**, *11*, 130–166. [CrossRef]
25. Shannon, C. A mathematical theory of communication. *Bell Syst. Tech. J.* **1948**, *27*, 379–423. [CrossRef]
26. Ehret, K. An Information-Theoretic Approach to Language Complexity: Variation in Naturalistic Corpora. Ph.D. Thesis, Albert-Ludwig-Universität Freiburg, Freiburg, Germany, 2017.
27. Miestamo, M. On the feasibility of complexity metrics. In *FinEst Linguistics. Proceedings of the Annual Finnish and Estonian Conference of Linguistics [Publications of the Department of Estonian of Tallinn University 8]*; Kerge, K., Maria-Maren, S., Eds.; Tallinna Ülikooli Kirjastus: Tallinn, Estonia, 2006; pp. 11–26.
28. Miestamo, M. Grammatical complexity in a cross-linguistic perspective. In *Language Complexity: Typology, Contact, Change*; Miestamo, M., Sinnemäki, K., Karlsson, F., Eds.; John Benjamins: Amsterdam, The Netherlands, 2008; pp. 23–41.
29. Miestamo, M. Implicational hierarchies and grammatical complexity. In *Language Complexity as an Evolving Variable*; Sampson, G., Gil, D., Trudgill, P., Eds.; Oxford University Press: Oxford, UK, 2009; pp. 80–97.
30. Dahl, Ö. *The Growth and Maintenance of Linguistic Complexity*; John Benjamins: Amsterdam, The Netherlands, 2004.
31. Kusters, W. *Linguistic Complexity: The Influence of Social Change on Verbal Inflection*; LOT: Utrecht, The Netherlands, 2003.
32. Kusters, W. Complexity in linguistic theory, language learning and language change. In *Language Complexity: Typology, Contact, Change*; Miestamo, M., Karlsson, F., Eds.; John Benjamins: Amsterdam, The Netherlands, 2008; pp. 3–21.
33. Szmrecsanyi, B.; Kortmann, B. Between simplification and complexification: Non-standard varieties of English around the world. In *Language Complexity as an Evolving Variable*; Sampson, G., Gil, D., Trudgill, P., Eds.; Oxford University Press: Oxford, UK, 2009; pp. 64–79.
34. Trudgill, P. Contact and simplification: Historical baggage and directionality in linguistic change Linguist. *Typology* **2001**, *5*, 371–374.
35. Kolmogorov, A. Three approaches to the quantitative definition of information. *Int. J. Comput. Math.* **1968**, *2*, 157–168. [CrossRef]
36. Juola, P. Measuring linguistic complexity: The morphological tier. *J. Quant. Linguist.* **1998**, *5*, 206–213. [CrossRef]
37. Bane, M. Quantifying and Measuring Morphological Complexity. In Proceedings of the 26th West Coast Conference on Formal Linguistics; Chang, C., Haynie, H., Eds.; Cascadilla Proceedings Project: Somerville, MA, USA, 2008; pp. 69–76.
38. Juola, P. Assessing linguistic complexity. In *Language Complexity: Typology, Contact, Change*; Miestamo, M., Sinnemäki, K., Karlsson, F., Eds.; John Benjamins: Amsterdam, The Netherlands, 2008; pp. 89–108.

39. Sadeniemi, M.; Kettunen, K.; Lindh-Knuutila, T.; Honkela, T. Complexity of European Union Languages: A comparative approach. *J. Quant. Linguist.* **2008**, *15*, 185–211. [CrossRef]
40. Ehret, K.; Szmrecsanyi, B. An information-theoretic approach to assess linguistic complexity. In *Complexity, Isolation, and Variation*; Baechler, R., Seiler, G., Eds.; De Gruyter: Berlin, Germany, 2016; pp. 71–94. [CrossRef]
41. Nijs, J.; Van de Velde, F.; Cuyckens, H. An information-theoretic approach to morphosyntactic complexity in English, Dutch and German. *J. Quant. Linguist.* **2024**, *31*, 275–297. [CrossRef]
42. Turing, A.M. On computable numbers, with an application to the Entscheidungsproblem. *Proc. Lond. Math. Soc.* **1937**, *42*, 230–265. [CrossRef]
43. Granger, C.W.J. Investigating causal relations by econometric models and cross-spectral methods. *Econometrica* **1969**, *37*, 424. [CrossRef]
44. Thurman, W.N.; Fisher, M.E. Chickens, Eggs, and Causality, or Which Came First? *Am. J. Agric. Econ.* **1988**, *70*, 237–238. [CrossRef]
45. Rodriguez-Caballero, C.V.; Ventosa-Santaularia, D. Granger Causality and Unit Roots. *J. Stat. Econom. Methods* **2014**, *3*, 97–114.
46. Van de Velde, F.; Petré, P. Historical linguistics. In *The Routledge Handbook of English Language and Digital Humanities*; Routledge: London, UK, 2020.
47. Fox, J.; Weisberg, S. *An R Companion to Applied Regression*, 2nd ed.; Sage: Thousand Oaks, CA, USA, 2011.
48. Speelman, D. Logistic regression: A confirmatory technique for comparisons in corpus linguistics. In *Corpus Methods for Semantics: Quantitative Studies in Polysemy and Synonymy*; Glynn, D., Robinson, J.A., Eds.; John Benjamins: Amsterdam, The Netherlands, 2014; pp. 487–533.
49. Thornton, D.L.; Batten, D.S. Lag-Length Selection and Tests of Granger Causality Between Money and Income. *J. Money Credit. Bank.* **1985**, *17*, 164–178. [CrossRef]
50. Piersoul, J.; De Troij, R.; Van de Velde, F. 150 years of written Dutch: The construction of the Dutch Corpus of Contemporary and Late Modern Periodicals. *Ned. Taalkunde* **2021**, *26*, 339–362. [CrossRef]
51. Davies, M. Expanding horizons in historical linguistics with the 400-million word Corpus of Historical American English. *Corpora* **2012**, *7*, 121–157. [CrossRef]
52. De Smet, H. CLMET 3.0. 2011. Available online: https://perswww.kuleuven.be/~u0044428/clmet3_0.htm (accessed on 19 February 2024).
53. Berlin-Brandenburg Academy of Sciences. 2024. Deutsches Textarchiv. Available online: https://www.deutschestextarchiv.de/ (accessed on 16 February 2024).
54. Basile, P.; Caputo, A.; Caselli, T.; Cassotti, P.; Varvara, R. A Diachronic Italian Corpus based on "L'Unità": Italian Conference on Computational Linguistics 2020. In *CLiC-It 2020 Italian Conference on Computational Linguistics 2020*; CEUR Workshop Proceedings (CEUR-WS.org): Bologna, Italy, 2020; p. 2769.
55. ATILF. Frantext Text Base (Online). *ATILF-CNRS & University of Lorraine*. 1998. Available online: https://www.frantext.fr/ (accessed on 29 July 2024).
56. Bakker, D. Flexibility and consistency in word order patterns in the languages of Europe. In *Constituent Order in the Languages of Europe*; Siewierska, A., Ed.; De Gruyter Mouton: Berlin, Germany, 1998; pp. 383–420. [CrossRef]
57. Greenberg, J.H. A auantitative approach to the morphological typology of language. *Int. J. Am. Linguist.* **1960**, *26*, 178–194. [CrossRef]
58. Lupi, C. Unit Root CADF Testing with *R. J. Stat. Softw.* **2009**, *32*, 1–19. [CrossRef]
59. Pescarini, D. Pronoun Systems in the Romance Languages. In *Oxford Research Encyclopedia of Linguistics*; Aronoff, M., Ed.; Oxford University Press: Oxford, UK, 2021. [CrossRef]

entropy

MDPI

Article

Punctuation Patterns in *Finnegans Wake* by James Joyce Are Largely Translation-Invariant

Krzysztof Bartnicki [1], Stanisław Drożdż [2,3,*], Jarosław Kwapień [2] and Tomasz Stanisz [2]

1 Independent Researcher, 53-201 Wrocław, Poland
2 Complex Systems Theory Department, Institute of Nuclear Physics, Polish Academy of Sciences, 31-342 Kraków, Poland
3 Faculty of Computer Science and Telecommunications, Cracow University of Technology, 31-155 Kraków, Poland
* Correspondence: stanislaw.drozdz@ifj.edu.pl

Abstract: The complexity characteristics of texts written in natural languages are significantly related to the rules of punctuation. In particular, the distances between punctuation marks measured by the number of words quite universally follow the family of Weibull distributions known from survival analyses. However, the values of two parameters marking specific forms of these distributions distinguish specific languages. This is such a strong constraint that the punctuation distributions of texts translated from the original language into another adopt quantitative characteristics of the target language. All these changes take place within Weibull distributions such that the corresponding hazard functions are always increasing. Recent previous research shows that James Joyce's famous novel *Finnegans Wake* is subject to such an extreme distribution from the Weibull family that the corresponding hazard function is clearly decreasing. At the same time, the distances of sentence-ending punctuation marks, determining the sentence length variability, have an almost perfect multifractal organization to an extent found nowhere else in the literature thus far. In the present contribution, based on several available translations (Dutch, French, German, Polish, and Russian) of *Finnegans Wake*, it is shown that the punctuation characteristics of this work remain largely translation-invariant, contrary to the common cases. These observations may constitute further evidence that *Finnegans Wake* is a translinguistic work in this respect as well, in line with Joyce's original intention.

Keywords: complex systems; natural language; multiscaling; punctuation; sentence length variability; discrete Weibull distribution

1. Introduction

A complexity science perspective provides valuable insights into the study of natural language as multiple traits identify it as a complex system [1,2]. Methods widely employed in analyzing complex systems—rooted in concepts from information theory [3–5], time series analysis [6–10], network science [11–19], and power-law probability distributions [20–23]—have been utilized to explore the quantitative characteristics of natural language. Understanding the mechanisms underlying these characteristics can significantly enhance natural language processing (NLP) and generation techniques, a particularly pertinent goal in the era of large language models (LLMs) [24,25], which are foundational to generative AI systems such as OpenAI's ChatGPT, Microsoft's Bing Chat, and Google's Bard.

One language property recently analyzed using statistical approaches is punctuation use in written texts. Research demonstrates that the distribution of word counts between consecutive punctuation marks in literary texts generally follows a discrete Weibull distribution [26–28]. Remarkably, the two parameters of this Weibull distribution are largely language-specific, reflecting the distinctive punctuation characteristics of different languages. Moreover, translations of texts into another language in general retain the Weibull distribution but with parameter values corresponding to the target language [26,27]. Sentence-ending punctuation, such as periods, deviates from this rigidity, however. The intervals thus defined, representing sentence lengths, exhibit greater variability and are not as strictly bound by the Weibull distribution, enabling a more diverse range of patterns.

In this context, *Finnegans Wake* by James Joyce is a challenging object for quantitative scientific analyses as it constitutes one of the most complex and enigmatic works of literature in the English language. It is known for its experimental style, inventive language that convolutes English with dozens of other languages, and it thus induces intricate, layered meanings. Many words therein have multiple meanings, inviting readers to find several interpretations for each word, phrase, and sentence. *Finnegans Wake* is structured to reflect the logic of a dream rather than a conventional narrative. The book begins and ends mid-sentence, creating a circular structure where the last line connects back to the first, symbolizing an endless cycle. There is no clear plot or sequence; instead, the narrative spirals, looping back on itself with characters, symbols, and themes that morph and shift. It is supposed to explore how language mirrors human consciousness and the complexity of memory, experience, and culture. Already, from this perspective, it is natural to anticipate the appearance of various types of long-range correlations in the text of this book, which do not exist in traditional written texts.

Indeed, *Finnegans Wake*'s punctuation patterns, which reflect mutual arrangement in the organization of phrases and sentences, contain correlations incomparably more amazing than in any other work that has ever been subjected to this type of analysis. The most spectacular dimension of *Finnegans Wake* manifests in its full multifractality glory through the patterns of sentence length variability, largely paralleling model mathematical cascades. The second unusual property of this work is that, the longer the sequence of words uninterrupted by punctuation, the less likely it is that a punctuation mark will appear after the next word, which is a unique property among books. In the formal language of probabilistic survival analysis, this means that the hazard function expressing the need for a punctuation mark to appear if it has not been present in the sequence of words is decreasing. It is as if for a living organism the probability of surviving subsequent years increases with the age of that organism even at its age of maturity.

2. Materials and Methods

2.1. *Finnegans Wake*

James Joyce's *Finnegans Wake* is a unique and challenging work of modernist literature, known for its complex structure, innovative use of language, and unconventional narrative techniques. The opening sentence is a continuation of the final sentence, creating a loop that reflects the cyclical themes in the book. Influenced by Giambattista Vico's theory of historical cycles, the book explores patterns of rise, fall, and renewal across human history. The narrative mimics the structure of a dream, delving into subconscious associations and symbolic imagery. Central to the story is the Earwicker family, representing archetypes of father, mother, and children. The work draws on mythological, biblical, and literary sources, embedding archetypes like the trickster, the hero, and the everyman. The text eschews a traditional plot for a fragmented cyclical approach, mirroring the logic of dreams.

The book is divided into four parts. Part I (8 chapters) introduces the central characters and themes. It sets the stage for the dreamlike narrative, blending personal and universal histories. Part II (4 chapters) focuses on the Earwicker family, including explorations of their interpersonal dynamics and symbolic representations of societal archetypes. Part III (4 chapters) contrasts rationality and structure versus creativity and chaos of the two brothers. Finally, Part IV (1 chapter) acts as a coda, bringing together the cyclical motifs of renewal and closure as ALP's (Anna Livia Plurabelle) monologue leads back to the beginning.

Finnegans Wake is often described as one of the most difficult works of literature due to its linguistic density and abstract narrative. It has been interpreted as a "universal dream", capturing the collective unconscious and human experience across time and space. The text is rich with references to world literature, history, philosophy, and folklore. Sentences often break grammatical rules, defying conventional readability but inviting interpretive exploration. For all these reasons, *Finnegans Wake* is often described as one of the most difficult works of literature due to its linguistic density and abstract narrative. Therefore, the multiplicity of interpretations and the diversity of opinions—often contradictory even on the same issues—expressed by experts on *Finnegans Wake* in the existing extensive literature on the subject are inevitable. The reader interested in exploring the current state of views on *Finnegans Wake* is referred to the doctoral thesis of Krzysztof Bartnicki [29], co-author of the present work, who is also the author of the Polish translation of *Finnegans Wake* [30].

Translating James Joyce's *Finnegans Wake* into other languages thus presents an array of unique challenges due to its experimental nature and narrative full of obscurities, ambiguities, and polyglot infusions. The text often prioritizes sound and rhythm over conventional grammar, creating a musicality that is integral to its meaning. Translating this musical quality into another language, while maintaining the sense of the text, is a delicate balancing act. Translators face the challenge of balancing fidelity to Joyce's complexity with readability for their audience. While *Finnegans Wake* is intentionally opaque, readers in the target language may lack the cultural or linguistic tools that English-speaking readers can rely on. Joyce's narrative incorporates multiple voices, tones, and registers, often blending them seamlessly. Translating this heteroglossia is difficult, especially in languages with less flexibility in tonal or dialectal variation. These are the reasons why available complete translations of this book into other languages are not very numerous but exist for several European languages. In addition to the original, five such widely recognized translations are used in the quantitative analysis presented here. They include translations into Dutch [31], French [32], German [33], Polish [30], and Russian [34].

2.2. Discrete Weibull Distribution

Punctuation can be understood as a mechanism for interrupting the continuous flow of words, thus enhancing the clarity of the message and providing necessary pauses for the reader. These functions align well with the framework of survival analysis [35]. Prior studies have established that distances between punctuation marks follow distributions that can be modeled using the discrete Weibull distribution [26,27]. This distribution is defined by the following probability mass function [36]:

$$f(k) = (1-p)^{k^\beta} - (1-p)^{(k+1)^\beta}, \quad p \in (0,1), \quad \beta > 0 \tag{1}$$

and its cumulative distribution function:

$$\mathcal{F}(k) = 1 - (1-p)^{k^\beta}. \tag{2}$$

The cumulative distribution function represents the probability that the random variable exceeds the value k. The discrete Weibull distribution generalizes the geometric distribution, which is recovered when $\beta = 1$, resulting in a constant probability $\mathcal{F}(k)$ over time. When $\beta > 1$, the probability increases with time, while, for $\beta < 1$, it decreases. This distribution has broad applications in fields such as survival analysis, weather prediction, and textual data analysis [35,37,38]. In the context of punctuation, $f(k)$ denotes the probability that a punctuation mark will appear precisely after k words.

A complementary approach to the above generalization can be readily and transparently formulated in terms of the hazard function $\lambda(k)$ expressing the conditional probability that the kth trial will result in a success provided that no success has occurred in the preceding $k-1$ trials:

$$\lambda(k) = \frac{f(k)}{1 - \mathcal{F}(k-1)}. \quad (3)$$

In the case of the discrete Weibull distribution, it becomes [39]

$$\lambda(k) = 1 - (1-p)^{k^{\beta} - (k-1)^{\beta}}. \quad (4)$$

For data that exactly follow a Weibull distribution, $\beta > 1$ corresponds to $\lambda(k)$, which is an increasing function of k. In other words, the probability of success increases with the number of preceding unsuccessful trials. The opposite applies to $\beta < 1$. In the *memoryless* case of $\beta = 1$, the hazard function is constant. Since $p = \lambda(1)$, the parameter p reflects the probability of putting a punctuation mark right after the first word following the last punctuation mark.

2.3. Multifractal Detrended Fluctuation Analysis (MFDFA)

Self-similarity, or the absence of a characteristic scale, is a defining feature of natural complex systems. Empirically, this property often manifests as a non-trivial temporal organization in measurement outcomes, represented as a time series. Specifically, complexity is frequently linked to a cascade-like hierarchy of data points exhibiting multiscaling, underscoring the importance of practical methods for identifying such structures in complex systems research [40]. Drawing on current research, multifractal detrended fluctuation analysis (MFDFA) has proven to be a highly reliable method for analyzing multiscaling structures [41,42]. This technique builds upon the widely adopted detrended fluctuation analysis (DFA) [43], offering a multiscale approach to studying hierarchical and multiscaling patterns. Below, we provide a concise outline of the key steps involved in the MFDFA algorithm.

Consider a time series $U = \{u_i\}_{i=1}^{T}$, consisting of T consecutive measurements of an observable u. This series is divided into M_s non-overlapping windows, each of length s, starting from both ends of U, resulting in a total of $2M_s$ windows. To address potential non-stationarity in the signal, a detrending procedure is applied within each window to an integrated signal (referred to as the profile) $X = \{x_i\}_{i=1}^{s}$, with elements defined as

$$x_i = \sum_{j=1}^{i} u_j. \quad (5)$$

The detrending process involves fitting a polynomial $P^{(m)}$ of order m (with $m = 2$ used throughout this study) to the data X within each window $\nu = 0, \ldots, 2M_s - 1$. The variance of the detrended signal obtained after subtracting the fitted polynomial is then computed as follows:

$$f^2(\nu, s) = \frac{1}{s} \sum_{i=1}^{s} (x_i - P^{(m)}(i))^2. \quad (6)$$

In the next step, a family of fluctuation functions of order q is defined using the average variance computed across all windows:

$$F_q(s) = \left\{ \frac{1}{2M_s} \sum_{\nu=0}^{2M_s-1} \left[f^2(\nu, s) \right]^{q/2} \right\}^{1/q}. \tag{7}$$

Here, q is a real number. The fluctuation functions $F_q(s)$ are computed for various values of the scale s and the index q. Typically, the minimum s is selected to exceed the length of the longest sequence of constant values in U, while the maximum ss is set to $T/5$. Unlike s, there is no standard range for q. Since q is associated with the moments of the signal, extreme values should be avoided for time series with heavy tails to ensure meaningful results. If the fluctuation functions depend on s as power laws

$$F_q(s) \sim s^{h(q)} \tag{8}$$

for a number of different choices of q, it indicates that the time series under study is either monofractal (when $h(q)$ is constant in q) or multifractal otherwise. The function $h(q)$ is called the generalized Hurst exponent because, for $q = 2$, $h(q) = H$, where H is the standard Hurst exponent [44,45]. From a visual perspective, fractal $F_q(s)$ organizations result in straight lines on double logarithmic plots.

A practical representation of the multifractal characteristics of data is the singularity spectrum $f(\alpha)$, which can be derived from $h(q)$ by the Legendre transform:

$$\begin{aligned} \alpha &= h(q) + qh'(q), \\ f(\alpha) &= q[\alpha - h(q)] + 1, \end{aligned} \tag{9}$$

Here, α represents the measure of data-point singularity, equivalent to the Hölder exponent. Geometrically, $f(\alpha)$ can be interpreted as the fractal dimension of the subset of the data characterized by a specific Hölder exponent α [46]. For a monofractal time series, the pair $(\alpha, f(\alpha))$ reduces to a single point. In contrast, for a multifractal time series, it typically forms a downward-pointing parabola. The broader the singularity spectrum $f(\alpha)$, the greater the richness of multifractality in the time series, which serves as an indicator of its complexity content. In some cases, the $f(\alpha)$ parabola may appear distorted or asymmetric, suggesting that data points of varying amplitudes exhibit different scaling behaviors [47–51].

3. Results and Discussion

3.1. Inter-Punctuation Intervals (IPIs)

The results illustrating the characteristics of the distributions of distances between all the consecutive punctuation marks for *Finnegans Wake* in the original and its five above-mentioned translations are shown in Figure 1. The panels in the left column show the corresponding normalized counts for increasing values of the distance k measured by the number of words between consecutive punctuation marks. These empirical distributions are then fitted by the formula of Equation (1), and parameters p and β of the best fits are explicitly provided in the corresponding panels. For *Finnegans Wake* in the original, it is a good fit, which was already presented in Ref. [52], but, remarkably, parameter β here is less than 1, in contrast to the multiplicity of all the other literary texts studied [26], written in the major European languages, even those classified as belonging to experimental literature [52]. In general, the Weibull functional form describes the distributions of distances between consecutive punctuation marks well, but the corresponding parameters are somewhat different and on average are language-specific, so they even undergo appropriate transformations in translations. The special case of *Finnegans Wake* analyzed here, in its uniqueness, turns out to be much less susceptible to changing values of these parameters. For the translations into Dutch and French, they even remain almost exactly unchanged.

For the translation into Polish, β approaches the limit of 1 but does not exceed it, while, for typical texts in this language, $\langle\beta\rangle \approx 1.4$. For the two remaining translations, German and Russian, β exceeds the value of 1, especially clear for the Russian translation considered here, although, in both cases, it remains below the average values characteristic of these languages [26]. In this context, it should also be noted that the quality of the fit is slightly worse in these two cases, especially for small values of k.

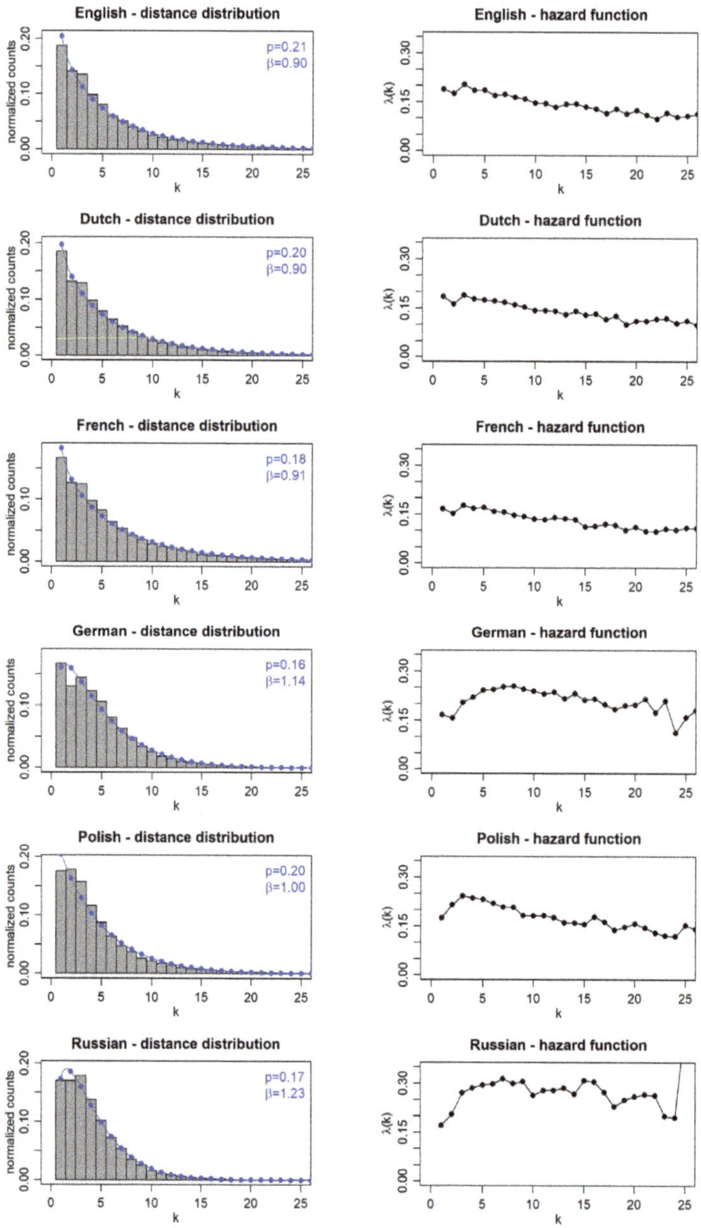

Figure 1. The distributions of the distances between consecutive punctuation marks (**left column**) and the corresponding hazard functions (**right column**) in *Finnegans Wake* and its translations.

A complementary insight into the specificity of the punctuation distribution is obtained through the hazard function $\lambda(k)$ defined by Equation (3). It provides a better perspective on the asymptotics of the distribution at larger values of k. These functions, calculated directly from empirical data, (not from the p and β fit parameters using the Weibull distribution) are presented in parallel on the right side of Figure 1. As indicated, from this perspective, all the translations demonstrate similar decreasing asymptotics, with the Dutch and French translations having almost the same course as the English original. Apart from an initial slightly larger increase, the Polish translation behaves similarly. Such initial increases in $\lambda(k)$ are more pronounced for the German and Russian translations, but, for larger k, the decreases start to prevail even in these cases.

3.2. Long-Range Correlations in IPIs

The distributions of fluctuations are some of the important characteristics of time series. With the same distributions, however, such series may differ in the arrangement of successive values relative to each other. The MFDFA formalism presented above is well suited to quantifying possible long-range correlations of this type and is therefore applied to the time series of successive IPIs analyzed here, proceeding through the entire text of the book.

The series of distances measured in word counts between all the consecutive punctuation marks for the original *Finnegans Wake* and its five translations are visualized in the left panels of Figure 2. The same scale is used on the vertical axis to make the relative values directly visible. As immediately evident, these series for the Dutch and French translations have strongly similar values and courses to those for the English original. The Polish translation is also quite consistent with these values. For the German translation, some correspondence is also noticeable, but there are clearly more small numbers, so the series is longer than in the above cases. An extreme case is the Russian translation, where the values are on average even significantly smaller, so their number in the series is even greater than for the German translation.

The nature of the long-range correlations encoded in these series is quantified within the MFDFA in terms of the fluctuation functions $F_q(s)$ expressed in Equation (7). These functions, calculated for $-4 \leq q \leq 4$ and plotted in the right-hand panels of Figure 2, respectively, show clear scaling (straight line on the log–log scale), but this is not a convincing instance of multifractal scaling because the dependence of these functions on q is weak. These dependencies are rather close to monofractal. The slope of $F_q(s)$ lines is, however, clearly larger than 0.5, in particular the Hurst exponents ($H = h(q = 2)$, red line in this Figure) determined by Equation (8), found in the range $0.65 \leq H \leq 0.75$. This indicates clear long-range correlations of a persistent nature. The Polish translation, which is at the upper end of the H-value range, is the most extreme in this respect.

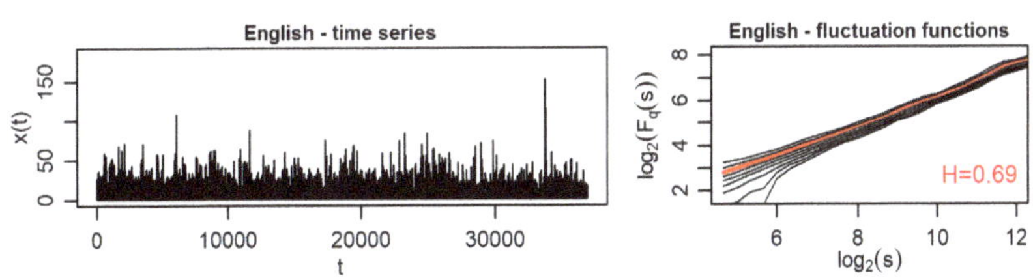

Figure 2. *Cont.*

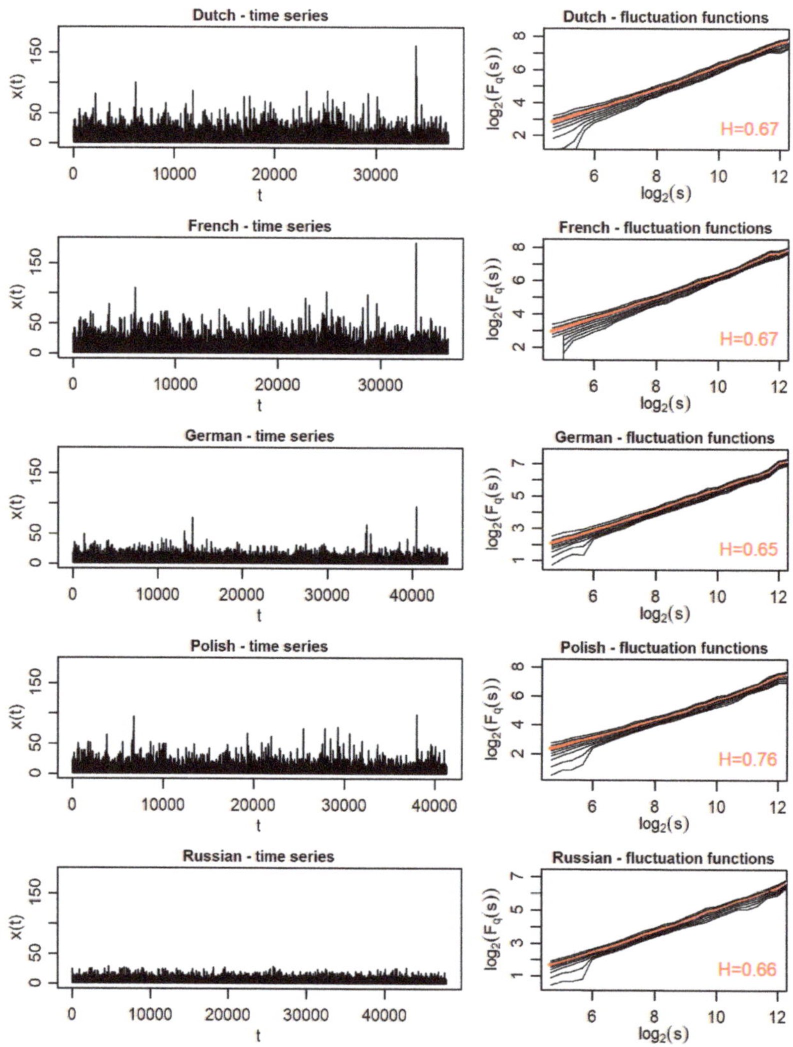

Figure 2. Time series representing the distances between consecutive punctuation marks (**left column**) and the corresponding fluctuation functions (**right column**) in *Finnegans Wake* and its translations. In each fluctuation function plot, the function for $q = 2$ is marked in red, and the Hurst exponent H is provided in the bottom-right corner.

The illustration that complements this section, shown in Figure 3, explicitly demonstrates one of the extreme examples of differences in the punctuation distribution in some translations of *Finnegans Wake*. The plot presents the relative arrangement of punctuation marks in a sample two-paragraph excerpt from the book. The discussed excerpt starts at the beginning of the last paragraph on page 579 and ends at the end of page 580 in the original English version. Clearly, consistent with the previous observations, the first three cases (English, Dutch, and French) differ the least, while the Russian translation differs the most by generating many much shorter IPIs .

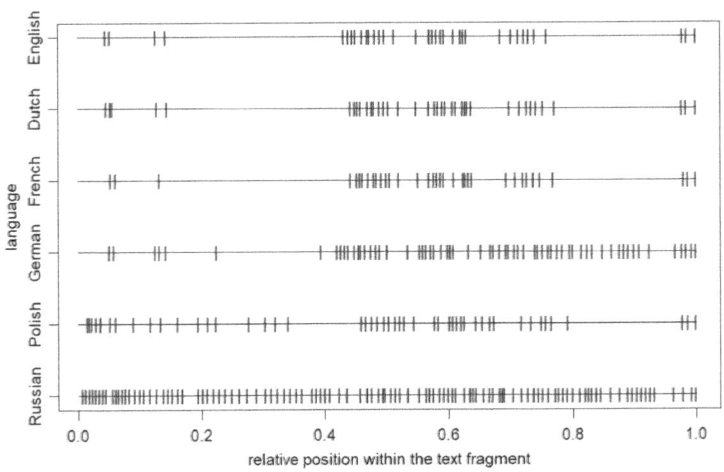

Figure 3. The relative arrangement of punctuation marks corresponding to the six considered cases of the original *Finnegans Wake* and its Dutch, French, German, Polish, and Russian translations in a two-paragraph excerpt from the book (starting at the beginning of the last paragraph on page 579 and ending at the end of page 580 in the original English version).

3.3. Multifractal Sentence Length Variability (SLV)

The distribution of distances between all the punctuation marks clearly follows certain fairly universal rigors, as demonstrated by previous recent studies and confirmed to a large extent even for such an exceptional work as the one examined here, *Finnegans Wake*. At the same time, however, the distributions of the punctuation marks ending sentences (such as periods, question marks, and exclamation marks) have much more freedom in this respect. The distances between them are precisely the lengths of sentences. It has already been shown [53] that *Finnegans Wake* develops a self similar cascade of SLV with exceptionally richly developed multifractal properties. Such patterns appear only in literary works written using the narrative technique of stream of consciousness, and *Finnegans Wake* is the most spectacular in this respect, both in terms of the almost ideal symmetry of the resulting singularity spectrum (Equation (9)) as well as in the sense of the width of this spectrum. The main goal of the present study is to examine to what extent these properties are preserved in the available translations.

The results collected in Figure 4 clearly indicate that, in this respect, the translations largely faithfully preserve the multifractal characteristics of the original, and, in the two cases of French and Polish, this mapping is almost perfect. The patterns of variation in the successive sentence lengths are strikingly similar here, and this applies even to their absolute values, as can be seen on the vertical scales of the upper panels of this figure. Consequently, the total number of sentences is also similar, as is evident from the horizontal scales of these panels. This applies even to the Russian translation, which showed greater discrepancies in the case of full punctuation. The functions $F_q(s)$ show very good scaling that is strongly dependent on q, comparable in all cases. As a consequence, the multifractal spectra $f(\alpha)$ are alike, broad ($\Delta\alpha \approx 1$), and vary only slightly. Minor asymmetries in the spectra can also be observed. For Dutch and Russian, they are right-sided, and for German they are left-sided. This indicates [49] that, in the first two cases, the hierarchy of multifractal correlations is somewhat more developed towards variability in shorter sentence lengths than longer ones, while, in the third case (German), it is the other way around. The multifractal spectra of the French and Polish translations are almost exactly symmetrical, similarly to the English original. All these spectra are also significantly shifted to the right with respect to $\alpha = 0.5$, which signals a clear persistent trend in sentence length variability. This property is also confirmed in terms of the corresponding Hurst exponents H, which are calculated as scaling exponents for $F_{q=2}(s)$, marked in red in these figures. Their corresponding values turn out to be even slightly higher than those when considering the distance variations between all the successive punctuation marks, as indicated in Figure 2.

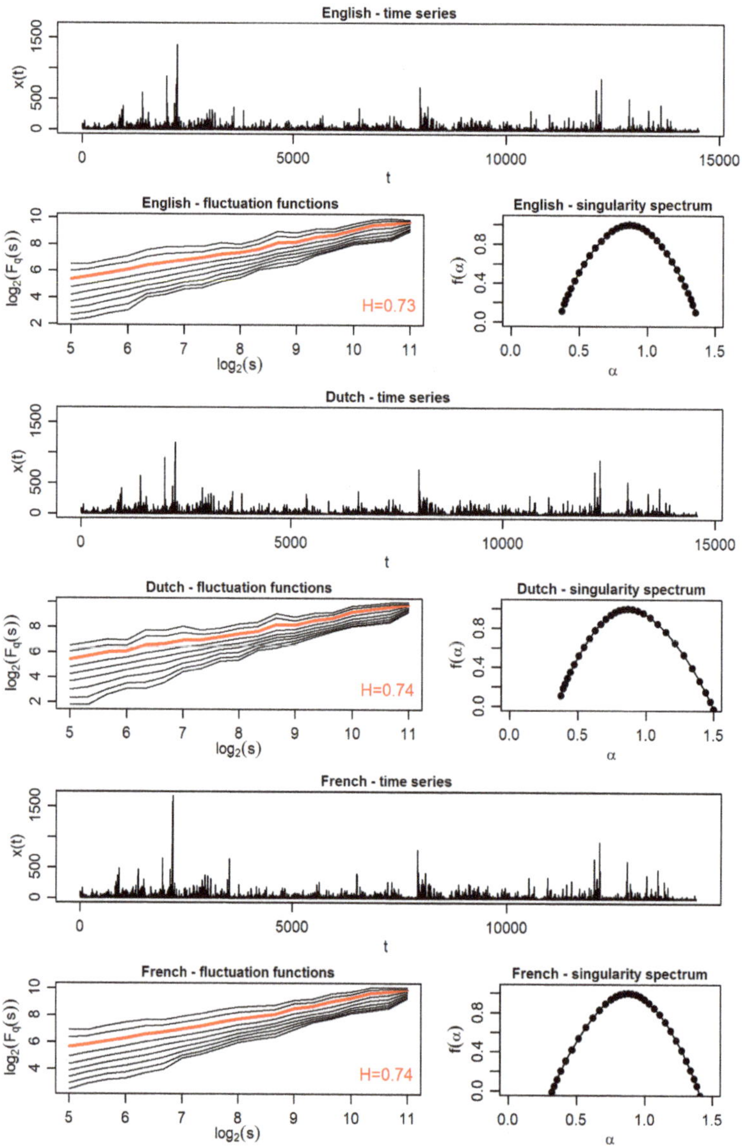

Figure 4. *Cont.*

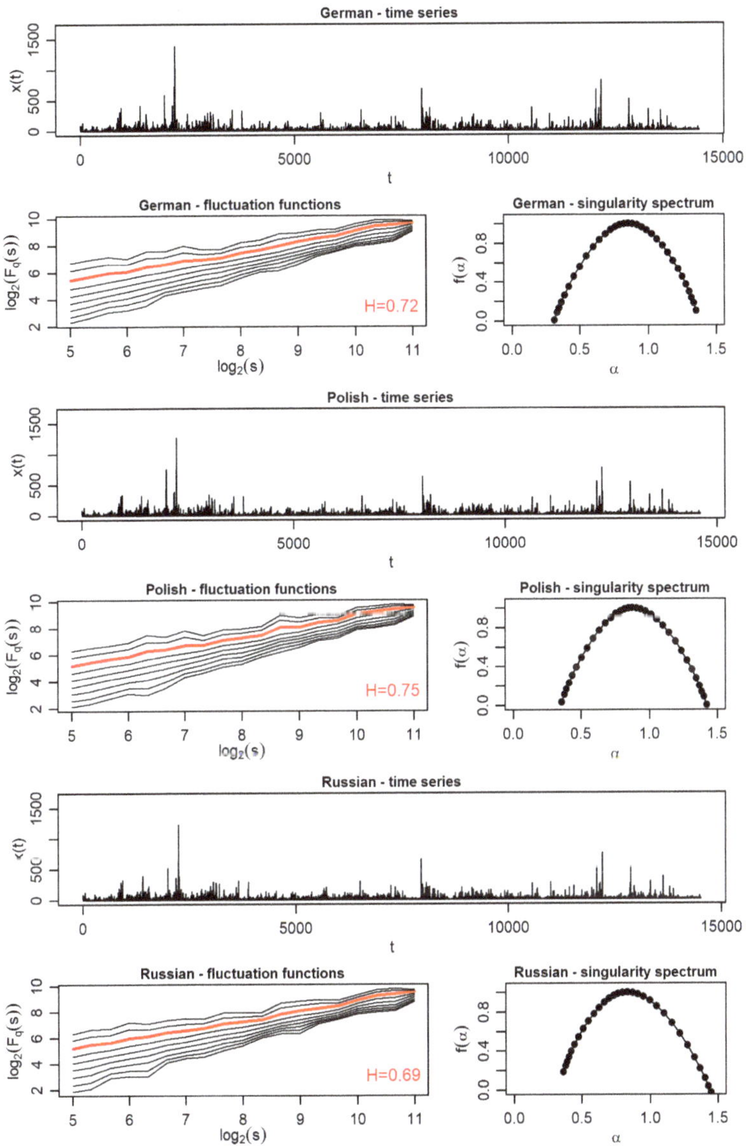

Figure 4. Time series representing sentence lengths, the corresponding fluctuation functions $F_q(s)$, and singularity spectra $f(\alpha)$ in *Finnegans Wake* and its translations. In each of the fluctuation function plots, $F_q(s)$ for $q = 2$ is marked in red, and the Hurst exponent H value is provided in the bottom-right corner.

4. Summary and Conclusions

This study explored the intricate punctuation patterns and sentence length variability within James Joyce's *Finnegans Wake* and its translations into five languages: Dutch, French, German, Polish, and Russian. Through a combination of statistical methods, including the discrete Weibull distribution and multifractal detrended fluctuation analysis (MFDFA), it is identified that the punctuation use in *Finnegans Wake* defies conventional linguistic norms.

Unlike typical texts, its punctuation patterns remain largely translation-invariant, reflecting Joyce's intentional crafting of a translinguistic narrative as one likely factor.

The findings highlight two distinct characteristics. The text exhibits a rare decreasing hazard function in its punctuation intervals, a trait only observed in *Finnegans Wake* among the analyzed works. This uniqueness persists across most translations, with minor variations reflecting linguistic differences, or, which cannot be excluded, some bias related to the translation fidelity. As previously demonstrated [52,53] by the comparison of large corpora of literary texts, the multifractal properties of the sentence lengths in *Finnegans Wake* demonstrate an extraordinary degree of self-similarity and complexity. This trait is preserved remarkably well in translations, particularly in the French and Polish versions, underscoring the robustness of the text's structural organization. These results contribute to a deeper understanding of the interplay between linguistic structure and translation fidelity, affirming the universal complexity embedded in Joyce's work.

The study confirms that *Finnegans Wake* exemplifies a rare literary phenomenon where structural properties transcend linguistic boundaries, maintaining coherence in complexity across translations. This underscores the text's suitability for cross-disciplinary analysis, superimposing linguistics, literature, and complexity science. By demonstrating that punctuation and sentence organization are integral to the text's identity, this research highlights the intricate balance between translation and preservation of artistic intent.

Future research could extend these methods to other experimental literary works to further explore the universality of such translinguistic traits. Additionally, integrating advanced computational techniques and larger datasets might provide new insights into the quantitative characteristics of experimental literature, which explores the deeper layers and possibilities of natural language.

Author Contributions: Conceptualization, K.B., S.D., J.K. and T.S.; Methodology, S.D., J.K. and T.S.; Software, J.K. and T.S.; Validation, K.B., S.D., J.K. and T.S.; Formal analysis, S.D., J.K. and T.S.; Investigation, S.D., J.K. and T.S.; Resources, K.B. and T.S.; Data curation, K.B. and T.S.; Writing—original draft, S.D.; Writing—review and editing, K.B., S.D., J.K. and T.S.; Visualization, J.K. and T.S.; Supervision, S.D. All authors have read and agreed to the published version of the manuscript.

Funding: This research received no external funding.

Institutional Review Board Statement: Not applicable.

Data Availability Statement: The data presented in this study are available on request from the corresponding author. The data are not publicly available due to copyright.

Conflicts of Interest: The authors declare no conflicts of interest.

References

1. Hébert-Dufresne, L.; Allard, A.; Garland, J.; Hobson, E.A.; Zaman, L. The path of complexity. *Npj Complex.* **2024**, *1*, 4. [CrossRef]
2. Kwapień, J.; Drożdż, S. Physical approach to complex systems. *Phys. Rep.* **2012**, *515*, 115–226. [CrossRef]
3. Dębowski, L. *Information Theory Meets Power Laws: Stochastic Processes and Language Models*; Wiley: Hoboken, NJ, USA, 2020. [CrossRef]
4. Takahira, R.; Tanaka-Ishii, K.; Dębowski, L. Entropy Rate Estimates for Natural Language—A New Extrapolation of Compressed Large-Scale Corpora. *Entropy* **2016**, *18*, 364. [CrossRef]
5. Montemurro, M.A.; Zanette, D.H. Universal Entropy of Word Ordering Across Linguistic Families. *PLoS ONE* **2011**, *6*, e19875. [CrossRef]
6. Alvarez-Lacalle, E.; Dorow, B.; Eckmann, J.P.; Moses, E. Hierarchical structures induce long-range dynamical correlations in written texts. *PNAS* **2006**, *103*, 7956–7961. [CrossRef] [PubMed]
7. Liu, J.; Gunn, E.; Youssef, F.; Tharayil, J.; Lansford, W.; Zeng, Y. Fractality in Chinese prose. *Digit. Scholarsh. Humanit.* **2023**, *38*, 604–620. [CrossRef]

8. Sánchez, D.; Zunino, L.; Gregorio, J.D.; Toral, R.; Mirasso, C. Ordinal analysis of lexical patterns. *Chaos* **2023**, *33*, 033121. [CrossRef]
9. Pawłowski, A. Time-Series analysis in linguistics: Application of the ARIMA method to cases of spoken Polish. *J. Quant. Linguist.* **1997**, *4*, 203–221. [CrossRef]
10. Kosmidis, K.; Kalampokis, A.; Argyrakis, P. Language time series analysis. *Phys. A* **2006**, *370*, 808–816. [CrossRef]
11. Cancho, R.F.i.; Solé, R.V. The small world of human language. *Proc. R. Soc. Lond. Ser. B Biol. Sci.* **2001**, *268*, 2261–2265. [CrossRef]
12. Amancio, D.R.; Antiqueira, L.; Pardo, T.A.S.; Da, L.; Costa, F.; Oliveira, O.N.; Nunes, M.G.V. Complex networks analysis of manual and machine translations. *Int. J. Mod. Phys. C* **2008**, *19*, 583–598. [CrossRef]
13. Liu, H.; Xu, C. Can syntactic networks indicate morphological complexity of a language? *Europhys. Lett.* **2011**, *93*, 28005. [CrossRef]
14. Cong, J.; Liu, H. Approaching human language with complex networks. *Phys. Life Rev.* **2014**, *11*, 598–618. [CrossRef]
15. Wachs-Lopes, G.A.; Rodrigues, P.S. Analyzing natural human language from the point of view of dynamic of a complex network. *Expert Syst. Appl.* **2016**, *45*, 8–22. [CrossRef]
16. Kulig, A.; Kwapień, J.; Stanisz, T.; Drożdż, S. In narrative texts punctuation marks obey the same statistics as words. *Inf. Sci.* **2017**, *375*, 98–113. [CrossRef]
17. Akimushkin, C.; Amancio, D.R.; Oliveira, O.N. Text authorship identified using the dynamics of word co-occurrence networks. *PLoS ONE* **2017**, *12*, 0170527. [CrossRef]
18. Stanisz, T.; Kwapień, J.; Drożdż, S. Linguistic data mining with complex networks: A stylometric-oriented approach. *Inf. Sci.* **2019**, *482*, 301–320. [CrossRef]
19. Raducha, T.; Gubiec, T. Predicting language diversity with complex networks. *PLoS ONE* **2018**, *13*, e0196593. [CrossRef] [PubMed]
20. Naranan, S.; Balasubrahmanyan, V. Models for power law relations in linguistics and information science. *J. Quant. Linguist.* **1998**, *5*, 35–61. [CrossRef]
21. Newman, M. Power laws, Pareto distributions and Zipf's law. *Contemp. Phys.* **2005**, *46*, 323–351. [CrossRef]
22. Ausloos, M. Punctuation effects in english and esperanto texts. *Phys. A* **2010**, *389*, 2835–2840. [CrossRef]
23. Piantadosi, S.T. Zipf's word frequency law in natural language: A critical review and future directions. *Psychon. Bull. Rev.* **2014**, *21*, 1112–1130. [CrossRef]
24. Shanahan, M.; McDonell, K.; Reynolds, L. Role play with large language models. *Nature* **2023**, *623*, 493–498. [CrossRef]
25. Zhao, W.X.; Zhou, K.; Li, J.; Tang, T.; Wang, X.; Hou, Y.; Min, Y.; Zhang, B.; Zhang, J.; Dong, Z.; et al. A Survey of Large Language Models. *arXiv* **2023**, arXiv:2303.18223v13.
26. Stanisz, T.; Drożdż, S.; Kwapień, J. Universal versus system-specific features of punctuation usage patterns in major Western languages. *Chaos Solitons Fractals* **2023**, *168*, 113183. [CrossRef]
27. Stanisz, T.; Drożdż, S.; Kwapień, J. Complex systems approach to natural language. *Phys. Rep.* **2024**, *1053*, 1–84. [CrossRef]
28. Dec, J.; Dolina, M.; Drożdż, S.; Kwapień, J.; Stanisz, T. Multifractal Hopscotch in *Hopscotch* Julio Cortázar. *Entropy* **2024**, *26*, 716. [CrossRef] [PubMed]
29. Bartnicki, K. Finnegans Wake as a System of Knowledge Without Primitive Terms: A Proposal Against the Paradigm of Competence in the So-Called Joyce Industry. Ph.D. Thesis, Friedrich Schiller University Jena, Jena, Germay, 2021. [CrossRef]
30. Joyce, J. *Finnegans Wake [Polish: Finneganów Tren]*; Ha!art: Cracow, Poland, 2012.
31. Joyce, J. *Finnegans Wake [Dutch: Finnegans Wake]*; Athenaeum-Polak & Van Gennep: Amsterdam, The Netherlands, 2002.
32. Joyce, J. *Finnegans Wake [French: Veillée Pinouilles]*. Available online: https://archive.org/details/veillee-pinouilles-18-juin-2020-pdf/mode/2up (accessed on 4 February 2025).
33. Joyce, J. *Finnegans Wake [German: Finnegans Wehg]*; Zweitausendeins: Frankfurt am Main, Germany, 1993.
34. Joyce, J. *Finnegans Wake [Russian: Na Pomine Finneganov]*; Rideró: Yekaterinburg, Russia. Available online: http://samlib.ru/r/rene_a/ (accessed on 4 February 2025).
35. Miller, R. *Survival Analysis*; John Wiley & Sons: Hoboken, NJ, USA, 1997.
36. Nakagawa, T.; Osaki, S. The discrete Weibull distribution. *IEEE Trans. Reliab.* **1975**, *R-24*, 300–301. [CrossRef]
37. Johnson, N.L.; Kotz, S.; Balakrishnan, N. *Continuous Univariate Distributions*; Wiley-Interscience: Hoboken, NJ, USA, 1994.
38. Altmann, E.G.; Pierrehumbert, J.B.; Motter, A.E. Beyond Word Frequency: Bursts, Lulls, and Scaling in the Temporal Distributions of Words. *PLoS ONE* **2009**, *4*, e7678. [CrossRef]
39. Padgett, W.J.; Spurrier, J.D. On discrete failure models. *IEEE Trans. Reliab.* **1985**, *34*, 253–256. [CrossRef]
40. Jimenez, J. Intermittency and cascades. *J. Fluid Mech.* **2000**, *409*, 99–120. [CrossRef]
41. Kantelhardt, J.W.; Zschiegner, S.A.; Koscielny-Bunde, E.; Havlin, S.; Bunde, A.; Stanley, H. Multifractal detrended fluctuation analysis of nonstationary time series. *Phys. A* **2002**, *316*. [CrossRef]

42. Oświęcimka, P.; Kwapień, J.; Drożdż, S. Wavelet versus detrended fluctuation analysis of multifractal structures. *Phys. Rev. E* **2006**, *74*, 016103. [CrossRef] [PubMed]
43. Peng, C.K.; Buldyrev, S.V.; Havlin, S.; Simons, M.; Stanley, H.E.; Goldberger, A.L. Mosaic organization of DNA nucleotides. *Phys. Rev. E* **1994**, *49*, 1685. [CrossRef]
44. Hurst, H.E. The long-term storage capacity of reservoir. *Trans. Am. Soc. Civ. Eng.* **1951**, *116*, 2447. [CrossRef]
45. Heneghan, C.; McDarby, G. Establishing the relation between detrended fluctuation analysis and power spectral density analysis for stochastic processes. *Phys. Rev. E* **2000**, *62*, 6103–6110. [CrossRef]
46. Halsey, T.C.; Jensen, M.H.; Kadanoff, L.P.; Procaccia, I.; Shraimant, B.I. Fractal measures and their singularities: The characterization of strange sets. *Phys. Rev. A* **1986**, *33*, 1141–1151. [CrossRef]
47. Ohashi, K.; Amaral, L.A.; Natelson, B.H.; Yamamoto, Y. Asymmetrical singularities in real-world signals. *Phys. Rev. E* **2003**, *68*, 065204. [CrossRef]
48. Cao, G.; Cao, J.; Xu, L. Asymmetric multifractal scaling behavior in the Chinese stock market: Based on asymmetric MF-DFA. *Phys. A* **2013**, *392*, 797–807. [CrossRef]
49. Drożdż, S.; Oświęcimka, P. Detecting and interpreting distortions in hierarchical organization of complex time series. *Phys. Rev. E* **2015**, *91*, 030902(R). [CrossRef]
50. Gómez-Gómez, J.; Carmona-Cabezas, R.; Ariza-Villaverde, A.B.; Gutiérrez de Ravé, E.; Jiménez-Hornero, F.J. Multifractal detrended fluctuation analysis of temperature in Spain (1960–2019). *Phys. A* **2021**, *578*, 126118. [CrossRef]
51. Kwapień, J.; Watorek, M.; Bezbradica, M.; Crane, M.; Mai, T.T.; Drozdz, S. Analysis of inter-transaction time fluctuations in the cryptocurrency market. *Chaos* **2022**, *32*, 083142. [CrossRef]
52. Stanisz, T.; Drożdż, S.; Kwapień, J. Statistics of punctuation in experimental literature—The remarkable case of Finnegans Wake by James Joyce. *Chaos: Interdiscip. J. Nonlinear Sci.* **2024**, *34*, 083124. [CrossRef] [PubMed]
53. Drożdż, S.; Oświęcimka, P.; Kulig, A.; Kwapień, J.; Bazarnik, K.; Grabska-Gradzińska, I.; Rybicki, J.; Stanuszek, M. Quantifying origin and character of long-range correlations in narrative texts. *Inf. Sci.* **2016**, *331*, 32–44. [CrossRef]

entropy

MDPI

Article

Towards Identifying Objectivity in Short Informal Text

Chaowei Zhang [1,*], Cheng Zhao [1], Zewei Zhang [2] and Yuchao Huang [3]

[1] The Department of Information Engineering, Yangzhou University, Yangzhou 225012, China; mz120230977@stu.yzu.edu.cn
[2] The Department of Computer Science and Software Engineering, Auburn University, Auburn, AL 36849, USA; zzw475622@gmail.com
[3] The Future Design Laboratory of Zhejiang University, Hangzhou 310027, China; huangyuchao@zju.edu.cn
* Correspondence: cwzhang@yzu.edu.cn

Abstract: Short informal texts are increasingly prevalent in modern communication, often containing fragmented grammar, personal opinions, and limited context. Traditional NLP tasks for the texts ordinarily focus on the subjective aspect learning, such as sentiment analysis and polarity classification. The study of learning objectivity from the texts is similarly significant, which can benefit many real-world applications including information filtering, content verification, etc. Unfortunately, this study is not being explored. This paper proposes a novel task that aims at identifying objectivity in short informal texts. Inspired by the characteristics of objective statements that normally need complete syntax structures for knowledge expression and delivery, we try to leverage the viewpoint of subjects (U), the tense of predicates (V), and the viewpoint of objects (O) as critical factors for objectivity learning. Upon that, we further propose a two-stage objectivity identification approach: (1) a UVO quantification module is implemented via a proposed OpenIE and large language model (LLM)-based triple feature quantification procedure; (2) an objectivity identification module employs pre-trained base models like BERT or RoBERTa that are constrained with the quantified UVO. The experimental result demonstrates our approach can outperform the base models up to 5.91% in objective-F1 and up to 6.97% in accuracy.

Keywords: short informal texts; objectivity learning; relational triples; large language models

1. Introduction

The proliferation of online instant communication platforms, such as TikTok, has led to the emergence of large volumes of informal short texts [1]. These texts are often found in social media posts, instant messaging, and online comments, which are characterized by their informal language and brevity [2]. Unlike well-organized formal documents, informal short texts often contain slang [3], abbreviations [4], emoticons [5], non-standard grammar usages [6], and ambiguous language [7], which can hinder accurate information extraction and comprehension [8]. For instance, during the COVID-19 pandemic, tweets like "COVID vaccines alter DNA" spread rapidly as pseudo-factual claims, exploiting informal syntax to evade detection by traditional fact-checking tools [9]. Consequently, it has become a considerable challenge for crowds to identify the objective contents that are the carriers of reliable messages and factual statements from such texts [10,11]. Normally, subjective informal short texts are often laden with personal emotions, opinions, and abbreviations and may exhibit incomplete sentence structures, leading to crowds suffering from identifying the relationships between the relational triplet (a.k.a., subject, predicate, and object) under an uncertain context [12]. However, objective informal short texts, despite being able to be used in informal scenarios, are rooted in factual information and provide accurate, clear communication for knowledge expression and convey. In this case, the goal of differentiating objective informal short texts from subjective ones is quite arduous, especially in the domains where misinformation or inappropriate content can spread rapidly [13].

Objectivity detection in informal texts intersects with but critically diverges from other established NLP tasks such as sentiment analysis [14], stance detection [15], fact verification [16], and opinion mining [17]. For example, sentiment analysis aims to identify the polarity of texts as positive, negative, or neutral; stance detection tries to classify support/opposition toward a target; fact verification tries to validate claims against external evidence; and opinion mining aims to extract subjective expressions from statements. Objectivity detection distinguishes factual claims from non-factual content by evaluating whether statements are grounded in verifiable facts, leveraging syntactic markers (e.g., third-person subjects, definitive tenses) as internal indicators of factuality while addressing the unique challenge of informal texts where opinions and factual phrasing are intricately blended, necessitating deeper structural analysis to disentangle objective assertions from subjective framing. One of the primary challenges in identifying objectivity from informal short texts is their lexical sparsity, where the limited number of words in each text hinders the application of traditional frequency-based or context-dependent models [18–20]. These models struggle to capture the subtle linguistic cues that indicate objectivity, particularly in informal short texts [21–24]. Thus, there is an urgent need to explore more effective tactics for objectivity identification in these texts, thereby fully unlocking their potentiality at automating information processing tasks like fact-checking [25–27], information filtering [28,29], and content verification [30,31].

Traditional relational triple-relevant tasks primarily aim to structure information from texts for further analysis and knowledge representation to assist the tasks of information extraction [32], knowledge graph construction [33,34], semantic understanding [35], etc. That means the concept of relational triples maintains a foundational possibility that could support a nuanced approach to understanding content in brief and fragmented statements, especially for informal short texts. Specifically, relational triples help deconstruct these fragmented statements into their core components, which allow for a more nuanced understanding of the underlying objectivity in the statement [36]. Moreover, the latent features of relational triples, such as viewpoints and tenses, are especially valuable as they reveal the essential elements needed to determine whether a statement is objective or subjective [37]. For example, first-person pronouns (e.g., "I believe", "We think") typically signal subjective opinions, while impersonal subjects (e.g., "Research shows", "Statistics indicate") suggest an objective stance. Furthermore, tense is not merely a grammatical marker; in pragmatics, it functions as a tool to frame information in terms of relevance and reliability. Therefore, effective feature representation for relational triples holds significant potential for identifying objectivity from informal short texts.

To explore the effectiveness of the triple features in identifying objectivity, this paper proposes a novel approach that integrates open information extraction (OpenIE) techniques coupled with LLMs aiming to extract these features from informal short texts. Specifically, we extract and quantify the viewpoint of subjects, the tense of predicates, and the viewpoint of objects (UVO) triple features from these texts to provide a structural foundation that can enhance the analysis and comprehension of the texts. This approach prevents the LLM from unintentionally altering the original text's meaning, thereby preserving the integrity of the input while improving the model's performance in identifying objective information. To leverage these features, we embed them into various fine-tunable pre-trained language models, such as RoBERTa and BERT, to verify the effectiveness of the features in detecting objective content within informal short texts.

In summary, this study proposes a novel method aiming at detecting objectivity in informal short texts, which relieves the inherent challenges posed by the informal and sparse nature of these texts. The contribution of this study is summarized as follows.

- This work originally formalizes objectivity detection as a distinct task tailored to the challenges of informal language. Meanwhile, we systematically analyze textual objectivity in platforms like social media and instant messaging.
- We propose a rule-based UVO (subject–predicate–object) quantification framework, which integrates syntactic analysis, tense detection via POS tagging with large-language-model-augmented triple extraction.
- Our two-stage framework synergizes OpenIE, LLMs, and fine-tuned pre-trained encoders. We mitigate hallucination risks inherent in LLMs by decoupling interpretable triple extraction from neural classification via rule-based and LLM voting while preserving input integrity.
- We rigorously evaluate our method across three benchmark datasets and achieve up to 6.97% accuracy improvements over baseline models. The insights revealed in our work provide actionable guidelines for domain-specific adaptation of objectivity detection systems.

2. Related Works

2.1. Early Objectivity Identification

Objective identification could be treated as a derivative of subjectivity learning known as affective computing or sentiment analysis. Differing from subjective learning, the key to objective identification lies in accurately locating fact-based statements, which is crucial for low false tolerance applications. Early research on subjectivity learning primarily focused on lexical or syntax-based methods. In one of the foundational works by [38] on subjectivity classification, an annotation framework was proposed to classify sentences as subjective or objective with the assistance of multiple annotators. These efforts ensured consistency in text annotation and set a solid groundwork for future research. Building on this, Riloff and Wiebe [39] proposed a more systematic learning framework that automatically extracted subjective expressions from texts, which highlighted the complementary nature of objective identification.

As research advanced, objective identification moved beyond simple lexical features to incorporate broader contextual and grammatical information. Afterward, Wiebe et al. [40] introduced a hybrid approach that combined rule-based and statistical methods. This approach analyzed sentence structure, verb types, and noun specificity to more accurately distinguish objective statements from subjective ones, which provided valuable insights for identifying objectivity.

2.2. Objectivity Identification via Artificial Intelligence

With the development of deep learning and machine learning, subjectivity learning evolved into more complex supervised learning models, significantly improving objective identification. For example, Hu and Liu [41] employed a support vector machine (SVM) to classify product reviews as subjective or objective, which effectively vectorizes text features to distinguish opinions from factual descriptions via SVM. Liu et al. [42] used decision trees to build a hierarchical structure for subjective and objective classification, but this method struggled with capturing objective features in complex linguistic patterns. The neural networks, such as CNNs, RNNs, and LSTM also demonstrate excellent performance in objective classification by automatically learning and extracting features from texts. For example, Zhang et al. [43] highlighted the effectiveness of LSTM in capturing logical relationships within long texts, which is beneficial for objective identification in sentences. Additionally, Hu et al. [44] leveraged RNNs to capture contextual information and sequence modeling in text classification, which particularly emphasizes the advantages of RNN in handling sequential data.

Since the rise of Transformer architecture [45], the pre-trained language models, such as BERT [46], RoBERTa [47], T5 [48], BART [49], and especially the series of GPT [50–54], have opened new possibilities for objectivity identification. Qasim et al. [55] demonstrated that fine-tuned BERT models perform exceptionally well in sentiment analysis and other text classification tasks, showing potential for their application in subjectivity classification. These models excel at understanding textual context, which significantly enhances classification accuracy. Larger language models, including GPT-4 [54] and LLaMA [56], have also been applied to text classification tasks. Sun et al. [57] explored the effectiveness of large language models in text classification, showcasing their superiority in capturing context and understanding semantics.

2.3. *Objectivity in Informal Short Texts*

Despite the success of these advanced models, objective identification in short texts presents unique challenges due to informal short texts increasingly incorporating informal and non-standard language [58]. Specifically, informal texts often lack sufficient contextual information and may contain ambiguous information leading to obscuring the traditional subject–object–predicate relationships that underpin many text classification models [59]. This case has prompted researchers to explore new ways to address the challenges posed by the phenomenon of "colloquialism". Recent advances have leveraged pre-trained language models to handle informal texts, which can contextualize sparse and unstructured text to infer the intent and factuality behind statements [60–62]. However, these generative models struggle with short informal texts as they tend to "fill in the gaps" by completing incomplete statements, which may lead to altered meanings and reduced precision in identifying objectivity. Thus, it remains a critical challenge when adopting LLMs in objective identification in short informal texts.

In summary, detecting objectivity in informal short texts remains a challenging task due to their informal nature [63,64] and lexical sparsity. Traditional feature engineering methods [65] have proven inadequate, and while deep learning and pre-trained models have improved performance, they still encounter significant limitations. The ongoing research continues to explore hybrid methods that combine advanced neural networks with syntactic and semantic extraction techniques to better capture the nuances of informal language and improve objectivity detection in short texts [66]. One of the possible solutions aiming at objectivity detection in short informal texts is to investigate the latent information implied in relational triples by combining the nature of objectivity in text. This study is supposed to investigate the effectiveness of the latent features of the relational triples—viewpoints of subjects, tense of predicates, and viewpoints of objects in the task. The details of our methodology design are demonstrated in the following two sections.

3. Feature Extraction and Quantification

In this section, we describe the systematic procedure for describing objective statements at the sentence level. According to the event expression concept proposed by Zhang et al. [67], a sentence (s_i) can be represented by an event set that maintains one or more events (i.e., $E_i = \{e_{i1}, e_{i2}, \ldots, e_{im}\}$), where each *event* is a triple set consisting of a subject, a predicate, and an object ($e_{ij} = \{u_{ij}, v_{ij}, o_{ij}\}, 1 \leq j \leq m$). Thus, we select subjects (u), predicates (v), and objects (o) of sentences in the form of raw elements as a way to represent sentence-level documents. We subsequently apply the viewpoint of a subject, the tense of a predicate, and the viewpoint of an object as fundamental features to investigate the efficacy of these three parameters in identifying objective cues in short texts. Notably, while our framework processes emojis as text tokens, we acknowledge that visual elements like emojis or images in informal texts may carry additional contextual or tonal cues critical for objectivity determination. Future extensions of this work will explore multimodal integration (e.g., emoji semantic embeddings or image captioning) to address this limitation.

3.1. Implementation Overview

A complex sentence may include multiple triples, which poses a greater challenge in terms of analysis. Moreover, it is crucial to select the most suitable triple from the triple candidates to represent the sentence for accurate semantic analysis. To this end, we apply a sequence of advanced NLP techniques for the best triple selection and the triple-feature quantification, as shown in Figure 1.

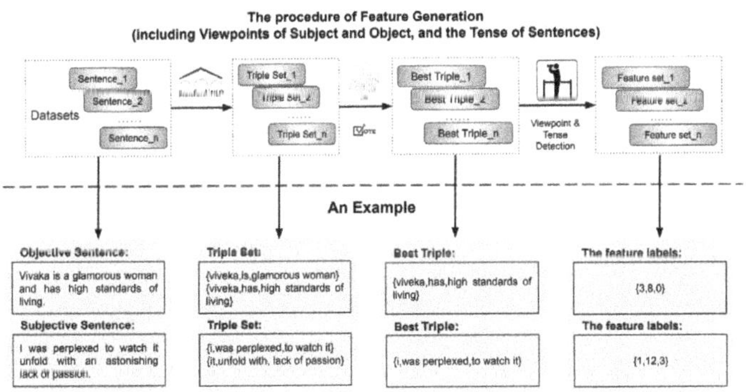

Figure 1. The procedure of UVO feature extraction (including viewpoints of subject and object and the tense of predicate in a short text) is demonstrated using an example.

Specifically, we initially extract triples from a sentence using the OpenIE of Stanford NLP. As the example shown in Figure 1, there are two triples—{vivaka, is, glamorous woman} and *{vivaka, has, high standards of living}*, that can be obtained from the complex objective sentence—*{'Vivaka is a glamorous woman and has high standards of living'}* using Stanford OpenIE. Next, we leverage six LLMs including source-opened *OLlama 3 70B* (https://ollama.com/library/llama3:70b, accessed on 26 April 2025), *Mistral 7B* (https://huggingface.co/mistralai/Mistral-7B-v0.1, accessed on 26 April 2025), *WizardLM 2 7B* (https://huggingface.co/dreamgen/WizardLM-2-7B, accessed on 26 April 2025), *Phi-3 14B* (https://huggingface.co/microsoft/Phi-3-medium-128k-instruct, accessed on 26 April 2025), *Gemma 2 9B* (https://huggingface.co/google/gemma-2-9b, accessed on 26 April 2025), and *Zhipu AI* (https://www.zhipuai.cn/, accessed on 26 April 2025) to select the most representative triple from the set of OpenIE-extracted candidates through a simple voting mechanism. During this selection, LLMs prioritize triples where the predicate's tense aligns with temporal cues identified in the sentence (e.g., flagging mismatches like "Back in the day, I go there daily" for reanalysis). The selected triple best captures the main structure and semantics of the original sentence. These models collectively span a diverse range of capabilities, from instruction tuning to general-purpose dialogue, thereby offering multi-perspective judgment for triple selection. The prompt used to select the best triple from the triple sets is shown in Table 1, where the prompt takes the original sentence s_i and the generated triple set E_i as inputs coupled with a goal and three constraints. Then, we identify the viewpoints of the subject and object and the tense of the predicate embraced in the best triple using our presented rule-based methods. Finally, the viewpoints of the subject and object and the tense of the predicate will be used to facilitate the identification of objective statements.

Table 1. The prompt used in LLMs for best triple selection.

text:{text} triples:{triples} Goal: Choose the best subject-predicate-object triples from the text. Requirements 1: The output format should be: (subject, predicate, object). Requirements 2: Do not rephrase the triples. Requirements 3: Avoid excessive introduction or background description.

3.2. The Detection of Viewpoints and Tense

To reveal the usability of the viewpoints of the subject and object and the tense of the predicate in discovering objective short informal texts, we design a sequence of rule-based methods to quantify these latent features.

To cover all possible subject- and object-related cases of sentences, we introduce a set, $P = \{0, 1, 2, 3, 4\}$, where first viewpoints, second viewpoints, third viewpoints, and the other cases using non-pronoun named entities (including slang terms like "doggo") are represented as 1, 2, 3, and 4, respectively. Value 0 in set P indicates that no subject or object is identified from input sentences caused by the phenomenon of *"Ellipsis"*. Sentence s_i is expressed by the best triple set $e_i = \{u_i, v_i, o_i\}$, where u_i denotes a subject, v_i represents a predicate, and o_i is an object. Function F_{view} returns the viewpoint of subject u_i - p_i^u and the viewpoint of object o_i - p_i^o. Both p_i^u and p_i^o must belong to the value set P. More formally, we have

$$F_{view}(u_i, o_i, L_1, L_2, L_3, NER) = (p_i^u, p_i^o), \{p_i^u, p_i^o\} \in P \quad (1)$$

where L_1, L_2, and L_3 are manually collected word vocabularies that are comprised of all pronouns in first viewpoints, second viewpoints, and third viewpoints, respectively. NER refers to an existing named entity recognition method. If neither a subject nor an object of sentence s_i is initially discovered, then p_i^u or p_i^o is tentatively set to 0. When sentence s_i embraces a subject (an object), the value of p_i^u (p_i^o) should be set to 1, 2, 3, or 4.

3.3. Tense Detection

To quantify the tense of the predicate in a sentence, we extend three basic tenses—'future', 'present', and 'past'—into twelve tenses using three tense-based parameters—'Perfect', 'Continuous', and 'Continuous Perfect'. Such twelve tenses cover all the sentence cases summarized in Table 2 accompanied with the POS tagging schemes.

Let $Q = \{1, 2, 3, 4, 5, 6, 7, 8, 9, 10, 11, 12, 0\}$ denote a set of twelve tenses containing future perfect continuous tense (i.e., 1), future perfect tense (i.e., 2), future continuous tense (i.e., 3), future simple tense (i.e., 4), present perfect continuous tense (i.e., 5), present perfect tense (i.e., 6), present continuous tense (i.e., 7), present simple tense (i.e., 8), past perfect continuous tense (i.e., 9), past perfect tense (i.e., 10), past continuous tense (i.e., 11), past simple tense (i.e., 12), and undetected tense (i.e., 0). We devise function T_{tense}, which returns tense q_i of the best triple (i.e., $e_i = \{u_i, v_i, o_i\}$). To address temporal cues expressed lexically (e.g., "back in the day"), we supplement the POS-based tense detection with a rule-based filter that maps common temporal adverbs/phrases (e.g., "yesterday", "currently") to their implied tenses using a predefined lexicon derived from the TempEval-3 corpus [68]. For instance, the phrase "back in the day" triggers a past-tense override regardless of verb form. Intuitively, the value of q_i is an element in set Q. More specifically, we have

$$q_i = T_{tense}(v_i, schemes), q_i \in Q. \tag{2}$$

where *schemes* is a rule-based POS tagging detector used to discriminate the tense of the predicate v_i, as demonstrated in Table 2. More detailed, we employ the *CoreNLP* POS tagger approach [69], which is nested in the Stanford CoreNLP toolkit to identify the part of speech of words. The tense of q_i will be returned according to the index of the scheme list if predicate v_i has a match within the schemes, To handle non-standard or dialectal verb forms (e.g., "ain't"), we rely on lexical normalization within the POS tagging phase and apply fallback rules leveraging temporal adverbs (e.g., "yesterday", "currently") for tense inference. For slang predicates (e.g., "yeet"), the LLM-aided triple selection (Section 2.1) ensures contextual role inference even if POS tagging mislabels tense; otherwise, q_i is set to 0, meaning that no tense is diagnosed in the case.

Table 2. All twelve tenses accompanied with corresponding POS tagging schemes and labels.

Tenses	POS Schemes	Example	Labels
Future Perfect Continuous	"MD" + "VB" + "VBN" + "VBG"	I will have been doing	1
Future Perfect	"MD" + "VB" + "VBN"	I will have done	2
Future Continuous	"MD" + "VB" + "VBG"	I will be doing	3
Future Simple	"MD"	I will do	4
Present Perfect Continuous	"VBD" + "VBN" + "VBG"	I have been doing	5
Present Perfect	"VBD" + "VBN" + "VBG"	I have done	6
Present Continuous	"VBD" + "VBN"	I am doing	7
Present Simple	"VBD"	I do	8
Past Perfect Continuous	"VBP"/"VBZ" + "VBN" + "VBG"	I had been doing	9
Past Perfect	"VBP"/"VBZ" + "VBN"	I had done	10
Past Continuous	"VBP"/"VBZ" + "VBG"	I was doing	11
Past Simple	"VBP"/"VBZ"	I did	12

4. System Framework

To verify the effectiveness of the above features extracted from the data processing procedure (the viewpoints of subject and object, the tense of predicate) on objectivity identification, we deploy a local short-text learning system using text encoders aiming at objectivity detection. Specifically, the system framework embraces a text representation module, a triple-feature learning module, and a classification module.

4.1. Text Representation

Given a randomly selected sentence s_i, the true label y_i, and the corresponding generated triple-features including the viewpoint of subject p_i^u, the tense of predicate q_i, and the viewpoint of the object p_i^o, we initialize three pre-trained encoders, namely, subject encoder, predicate encoder, and object encoder for the feature representation. More specifically, the subject encoder firstly encodes s_i as a sequential representation $F_i^u = \{h_{i1}^u, h_{i2}^u, \dots, h_{in}^u\}$ (n is the length of sentence s_i) and then acquires the global representation of the subject's viewpoint—f_i^u—via an attention block, same as the predicate encoder and object encoder. Thus, we have

$$\begin{cases} f_i^u = Attention(E_u(s_i)), \\ f_i^v = Attention(E_v(s_i)), \\ f_i^o = Attention(E_o(s_i)), \end{cases} \tag{3}$$

where $E()$ is a specific text encoder (subject encoder E_u, predicate encoder E_v, or object encoder E_o) used for obtaining the sequential representations of text—$\{F_i^u, F_i^v, F_i^o\}$. It is worth noting that the three Encoders are trainable with the fine-tuning technique by complying with the two restrictions—the triple-features (p_i^u, q_i, or p_i^o) extracted from the procedure of UVO feature generation and the original true label of text y_i.

4.2. Triple-Feature Learning

The triple-feature learning module is designed to assist in identifying objective statements. We leverage the three features extracted in the feature generation procedure, including (1) the viewpoint of subject p^u, (2) the tense of predicate q, and (3) the viewpoint of the object p^o to train the three local deployed MLPs. As demonstrated in Figure 2, the MLPs are deployed to learn each of the three features. Moreover, the MLPs take the global representations of text as input and then utilize the value of each feature as the true label for parameter training. The loss function for learning the feature (i.e., the viewpoint of the subject u_i) is presented as follows:

$$\mathcal{L}_u = \text{CrossEntropy}(p^u, \text{MLP}(f^u)) \tag{4}$$

where f^u is the global representation calculated by feeding F^u into the attention block. It is noteworthy that the value of p^u is transformed with one-hot encoding. Moreover, the MLPs' loss for learning the predicate tense and the object's viewpoint are denoted as $\mathcal{L}_v$ and $\mathcal{L}_o$.

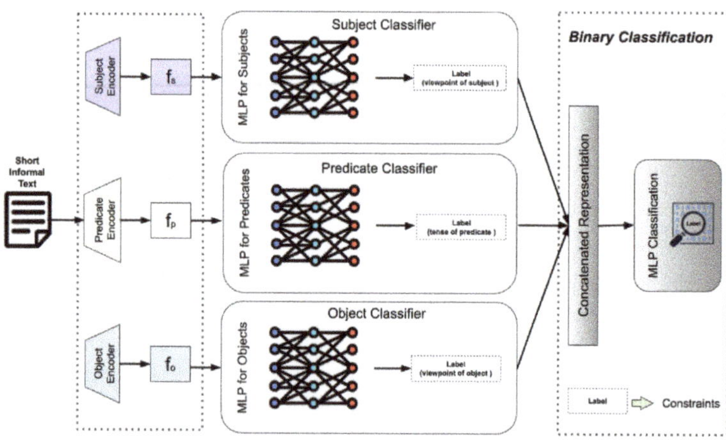

Figure 2. The system framework of objectivity identification. The labels used in three classifiers are generated from the procedure of UVO feature generation (including viewpoints of subject and object and the tense of predicate).

4.3. Binary Classification

Once the forward inference processes of the three MLPs are accomplished, the representations of the last hidden layers from the MLPs are concatenated as the final representation of the input sentence. Formally, the final representation of sentence s is represented as

$f = (f^u \oplus f^v \oplus f^o)$. Finally, we fed f into another MLP to approach the true label of the sentence—y. Thus, the learning objective of the last MLP is denoted as the following loss function.

$$\mathcal{L}_s = \text{MSE}(y, \text{MLP}(f)) \tag{5}$$

where f is the final representation of sentence s, and it goes through an MLP to match the dimensionality of the true label for loss calculation. Thus, the total loss of the entire model is as follows.

$$\mathcal{L}_{\text{final}} = \mathcal{L}_u + \mathcal{L}_v + \mathcal{L}_o + \mathcal{L}_s \tag{6}$$

5. Experimental Results

In this section, we conduct two kinds of experiments by varying the three extracted features, namely, (1) viewpoints of subjects; (2) tenses of predicates; and (3) viewpoints of objects. The first group experiment aims to evaluate the effectiveness of the three-element variable set (p^o, q, p^o) in improving the performance of base models. In the second experiment group, we evaluate the impact of each of these three parameters in distinguishing objective sentences by constructing three one-elemental variable sets (i.e., (p^u),(q), and (p^o)).

5.1. Experimental Configurations

Dataset: In this study, we adopt three existing datasets to conduct our experiment: (1) the movie review dataset that encompasses 5000 movie review snippets from Rottentomatoes (www.rottentomatoes.com, accessed on 26 April 2025) as a subjective data collection and 5000 objective data items from the plot summaries available from the Internet Movie Database (www.imdb.com); (2) the SST dataset (https://nlp.stanford.edu/sentiment/, accessed on 26 April 2025), also known as Stanford Sentiment Treebank dataset, which are labeled in five fold. Specifically, SST consists of 11,855 single sentences extracted from movie reviews including 1510 very negative, 3140 negative, 2242 neutral, 3111 positive, and 1852 very positive sentences; (3) the Twitter US Airline Review dataset from Kaggle that consists of 2349 sentences labeled in positive, 3073 neutral data points, and 9131 negative data points (https://www.kaggle.com/datasets/crowdflower/twitter-airline-sentiment/data, accessed on 26 April 2025). To adapt the setting of our tasks, we treat data points labeled as 'Neutral' as proxies for objective data, while acknowledging that this approach may conflate non-polar opinions with factual objectivity [13]. For instance, subjective but neutral statements (e.g., "I think the event was average") could be mislabeled. To assess label fidelity, we conducted a post hoc analysis on 500 randomly selected samples, finding that 78% of 'neutral' labels aligned with true objectivity (e.g., factual statements like "The meeting starts at 3 PM"). All of the others were subjective data for the SST and the Airline Review datasets.

Baselines: We adopt four pre-trained language models to tier up our proposed method for its effectiveness verification.

- BERT, a bidirectional transformer-based model developed by Google, which is pre-trained using masked language modeling (MLM) and next sentence prediction (NSP) tasks. BERT can read an input sequence from two sides to capture deep contextual information, which allows it to be adaptable for our tasks.
- RoBERTa is developed by Facebook AI, which is an optimized variant of BERT. It removes the next sentence prediction (NSP) task, uses dynamic masking, and is trained on much larger datasets for more epochs than BERT.

- BART is a transformer model also developed by Facebook AI that combines BERT's bidirectional encoding coupled with GPT's autoregressive decoding. Its encoder–decoder architecture makes it versatile for both natural language comprehension and generation.
- T5, developed by Google, is a unified framework that can convert NLP tasks into a text-to-text problem. It uses a transformer-based encoder–decoder architecture to handle diverse tasks. Its text-to-text approach simplifies task design and has achieved strong performance across a variety of benchmarks.
- GPT-2, developed by OpenAI, is a large-scale transformer-based language model trained on a massive corpus of internet text in an unsupervised manner. Unlike encoder–decoder frameworks, GPT-2 uses a unidirectional decoder-only architecture, generating text by predicting the next word in a sequence.

Implementation Details: In our experiments, we set up fixed hyper-parameters, including the learning rate in Adam optimizer (i.e., $1^{-5\times10}$), dropout rate before the classification layer (i.e., 0.3), and the number of epochs (i.e., 5 times) across all the datasets. In more detail, each dataset is divided as 80% for training and 20% for testing. We adopt various metrics among all the methods to evaluate the performance including training loss (Tr-Loss), training accuracy (Tr-Acc), testing loss (Te-Loss), testing accuracy (Te-Acc), and the F1 score for objective data (Obj-F1), as shown in Tables 3 and 4. All the results from the tables are the average score of five epochs. Finally, the objective of this study is to evaluate the effectiveness of the extracted UVO features in identifying objectivity from informal texts by taking advantage of fine-tuning three kinds of pretrained language models.

Table 3. The performance comparison between encoders-only baselines and the triple-feature-based encoders across the three datasets. The triple features, UVO, represent viewpoints of subjects, tenses of predicates, and viewpoints of objects, respectively. The best values in various metrics have been bolded.

Datasets	Methods	Tr-Loss	Tr-Acc	Te-Loss	Te-Acc	Obj-F1
Movie	RoBERTa	0.136	0.953	0.139	0.955	0.954
	RoBERTa + UVO	**0.101**	**0.966**	**0.129**	**0.963**	**0.962**
	Bert	0.101	0.972	0.130	0.958	0.954
	Bert + UVO	**0.086**	**0.975**	**0.117**	**0.965**	**0.963**
	Bart	0.181	0.933	0.133	0.956	0.954
	Bart + UVO	**0.167**	**0.934**	**0.121**	**0.961**	**0.961**
	T5	0.181	0.940	0.122	0.960	0.956
	T5 + UVO	**0.152**	**0.946**	**0.119**	**0.960**	**0.956**
	GPT2	0.565	0.684	0.566	0.702	0.727
	GPT2 + UVO	**0.503**	**0.731**	**0.565**	**0.705**	**0.735**
SST	RoBERTa	0.536	0.765	0.518	0.803	0.889
	RoBERTa + UVO	**0.498**	**0.809**	**0.486**	**0.812**	**0.898**
	Bert	0.520	0.769	0.554	0.786	0.879
	Bert + UVO	**0.450**	**0.816**	**0.464**	**0.817**	**0.898**
	Bart	0.528	0.792	0.554	0.801	0.890
	Bart + UVO	**0.511**	**0.792**	**0.491**	**0.815**	**0.897**
	T5	0.575	0.707	0.551	0.739	0.834
	T5 + UVO	**0.461**	**0.809**	**0.499**	**0.809**	**0.893**
	GPT2	0.497	0.808	0.538	0.784	0.879
	GPT2 + UVO	**0.478**	**0.820**	**0.493**	**0.811**	**0.896**
Twitter	RoBERTa	0.344	0.874	0.353	0.871	0.921
	RoBERTa + UVO	**0.282**	**0.887**	**0.318**	**0.872**	**0.922**
	Bert	0.328	0.875	0.395	0.867	0.919
	Bert + UVO	**0.252**	**0.902**	**0.343**	0.866	**0.919**
	Bart	0.436	0.801	0.401	0.848	0.905
	Bart + UVO	**0.359**	**0.840**	**0.334**	**0.863**	**0.916**
	T5	0.372	0.850	0.335	0.864	0.920
	T5 + UVO	**0.329**	**0.860**	**0.302**	**0.874**	**0.926**
	GPT2	0.471	0.806	0.492	0.808	0.880
	GPT2 + UVO	**0.440**	**0.817**	**0.460**	**0.811**	**0.894**

Table 4. The ablation study we conducted for investigating the role of each of the UVO features in affecting the performance of five pre-trained base models on objectivity detection across the three datasets. The best values in various metrics have been bolded.

Datasets	Models	Constraints			Tr-Loss	Tr-Acc	Te-Loss	Te-Acc	Obj-F1
		U	V	O					
Movie	RoBERTa	✓			**0.131**	**0.959**	0.127	**0.963**	**0.961**
			✓		0.136	0.957	**0.118**	0.961	0.961
				✓	0.142	0.951	0.130	0.958	0.957
	Bert	✓			**0.095**	**0.970**	**0.102**	**0.968**	**0.966**
			✓		0.108	0.969	0.139	0.953	0.952
				✓	0.104	0.970	0.112	0.963	0.962
	BART	✓			**0.190**	0.932	0.156	**0.949**	**0.948**
			✓		0.220	0.920	0.181	0.939	0.939
				✓	0.194	**0.934**	**0.153**	0.947	0.946
	T5	✓			**0.174**	**0.946**	0.135	**0.958**	**0.954**
			✓		0.213	0.926	0.148	0.949	0.944
				✓	0.184	0.939	**0.130**	0.955	0.951
	GPT2	✓			**0.502**	0.731	**0.585**	**0.687**	**0.735**
			✓		0.505	**0.733**	0.619	0.676	0.719
				✓	0.504	0.729	0.601	0.684	0.727
SST	RoBERTa	✓			0.548	0.765	0.524	0.808	0.893
			✓		**0.523**	**0.796**	**0.490**	**0.822**	**0.902**
				✓	0.539	0.773	0.519	0.805	0.890
	Bert	✓			0.557	0.764	0.534	0.800	0.890
			✓		**0.528**	**0.781**	**0.514**	**0.806**	**0.891**
				✓	0.543	0.767	0.528	0.789	0.881
	BART	✓			0.537	0.774	0.545	0.814	0.897
			✓		**0.515**	**0.797**	**0.494**	**0.827**	**0.905**
				✓	0.518	0.796	0.529	0.813	0.897
	T5	✓			0.665	0.603	0.584	0.693	0.780
			✓		**0.521**	**0.766**	**0.506**	**0.800**	**0.887**
				✓	0.526	0.760	0.516	0.798	0.886
	GPT2	✓			0.498	0.806	0.507	0.802	0.891
			✓		**0.486**	**0.814**	**0.480**	**0.816**	**0.899**
				✓	0.492	0.811	0.487	0.803	0.891
Twitter	RoBERTa	✓			**0.317**	**0.877**	**0.321**	**0.873**	**0.923**
			✓		0.349	0.859	0.323	0.868	0.921
				✓	0.360	0.858	0.350	0.863	0.917
	Bert	✓			**0.304**	**0.889**	**0.346**	**0.868**	**0.919**
			✓		0.331	0.877	0.347	0.860	0.915
				✓	0.360	0.860	0.364	0.858	0.914
	BART	✓			0.409	0.806	**0.356**	**0.856**	0.910
			✓		**0.399**	**0.820**	0.362	0.855	**0.913**
				✓	0.450	0.779	0.395	0.834	0.896
	T5	✓			**0.393**	**0.836**	0.350	0.853	0.913
			✓		0.414	0.815	0.366	0.852	0.909
				✓	0.404	0.826	**0.339**	**0.862**	**0.917**
	GPT2	✓			**0.440**	**0.819**	0.455	**0.823**	**0.902**
			✓		0.446	0.810	**0.453**	0.816	0.897
				✓	0.462	0.814	0.466	0.808	0.891

5.2. The First Group Experiment

Table 3 presents the performance comparison between the UVO-added base models and the UVO-free baselines on the three datasets. As shown in the table, UVO-added base models significantly outperform the base models without UVO on every dataset in terms of all the metrics, which demonstrates the effectiveness of triple features in identifying objective sentences. Specifically, we have the following observations from Table 3.

1. Training Efficiency Improvement: In all datasets, the addition of UVO features consistently reduces the training loss as well as yields better training accuracy across all models, which suggests that these features help the models converge better and capture the structure of the data more effectively. In particular, the Bert + UVO model shows the largest reduction in training loss on the Movie dataset, which decreased by 36.69% compared with the base Bert model. Moreover, the Bert + UVO model demonstrated the highest increase in training accuracy percentage increment (6.65%) compared with the corresponding base model on the SST dataset. Notably, even for the GPT2 model—which generally shows higher training loss and lower accuracy compared to other models—the addition of UVO features reduces the training loss significantly across all datasets. For example, in the Movie dataset, the training loss decreased from 0.565 to 0.503, while training accuracy improved from 0.684 to 0.731, indicating better convergence due to the structural cues provided by UVO.
2. Better Generalization on Testing Phase: For all the base pre-trained models, UVO features can help them to achieve both better testing loss and accuracy, which reflects the better generalization capabilities of UVO triple features. This improvement signifies that the UVO-added base models are not overfitting to the training data but learning meaningful patterns that transfer well to the test data. For example, the above-mentioned Bert + UVO model on the Movie dataset also can reduce the testing loss by 1.1% and enhance the testing accuracy by 0.8% under the promise of effective training progress. It is noteworthy that the most significant reduction in testing loss (i.e., 16.18%) is from the UVO-added Bert model on the SST dataset compared with the Bert-only, and the UVO-added Bert model also produces the best testing accuracy improvement (i.e., 4.03%) while tested on the SST dataset. In addition, GPT2 models also exhibit consistent improvements in generalization: on the SST dataset, testing loss decreased from 0.538 to 0.460 and test accuracy increased from 0.784 to 0.811 with UVO. A similar trend is observed on the Twitter dataset, where testing accuracy rose from 0.808 to 0.811 and testing loss slightly reduced, demonstrating that GPT2 also benefits from UVO in learning transferable patterns.
3. Efficient Objectivity Identification: UVO features consistently improve the Obj-F1 score across all datasets and models, demonstrating that the inclusion of the triple features helps all the base models handle objective information better by focusing on syntactic roles (viewpoints of subjects, tenses of predicates, and viewpoints of objects). For instance, the T5 + UVO model achieves an Obj-F1 of 0.956 on the Movie dataset and 0.926 on the Twitter dataset. Similarly, GPT2 + UVO shows marked improvement over the baseline GPT2 in Obj-F1 scores: from 0.727 to 0.735 (+0.8%) on Movie, from 0.873 to 0.896 (+2.3%) on SST, and from 0.880 to 0.894 (+1.4%) on Twitter, reinforcing the conclusion that UVO helps even relatively weaker baselines like GPT2 to better capture objectivity.

5.3. The Ablation Study

To further investigate the impact of each feature in the UVO triple, we conduct two ablation studies by either utilizing or disabling each of the UVO features. We first activate the effect of a single feature on the performance of base models. As shown in Table 4, we verify the efficiency of each of the UVO features on all base pre-trained models across the three datasets. Furthermore, we have the following observations from the table.

1. In the Movie dataset, the viewpoint of subjects (i.e., U) can supply better improvement to the performance in terms of five metrics across all of the base pre-trained models compared with the other two features—the tense of predicates and the viewpoint of objects (i.e., Bert + U achieves the best performance—0.095 training loss, 0.97 training accuracy, 0.102 testing loss, 0.968 testing accuracy, and 0.966 F1 score in objective). Such a phenomenon reflects that the viewpoint of object (O) and the tense of predicate (V) might be less influential than the subject's perspective in movie reviews. While the viewpoint of objects and the tense of movie reviews are important, they are secondary to how the reviewer personally feels and expresses those feelings. Since the subject's expression of opinion is critical in movie reviews, its viewpoint allows the model to focus more on this subjective lens, which is key for improving generalization in objectivity detection.
2. It is observable from the results yielded from the SST dataset that the tense of predicates (i.e., V) promotes the base models with the relatively best performance in perspectives of all five metrics compared with the viewpoints of subjects and objects. That means, in the SST dataset, the tense of predicates is a major clue in determining whether the content is a personal opinion or a more neutral statement of fact. More insightful, since SST data consists of shorter texts, and other contextual information is sparse, the verbs carry a significant proportion of the determination in helping the model to identify objectivity.
3. For the Twitter dataset, various pre-trained base models yield relatively unstable performance compared with the experiment results on the other two datasets. However, the viewpoint of subjects (U) still plays the most important role in distinguishing objective texts. Specifically, the different performance of fine-tuning pre-trained models on the Twitter dataset when applying the U, V, or O constraints reflects their architectural differences and varying pretraining objectives. Models like RoBERTa excel in short text, informal settings like Twitter due to their fine-tuned contextual embeddings, yet BART performs better at tasks involving sequence structure (e.g., tense or object viewpoint detection). It is crucial to understand the differences between models or datasets to detect objectivity in short informal text.

Similar to the first group of ablation study, Table 5 demonstrates the efficiency of each tuple of the UVO features (implemented by disabling each of the UVO features), which reveals that the performance of base models may be varied while each of the UVO features is disabled in different datasets. The analysis for the second ablation study is enumerated as follows.

- In the Movie dataset, combinations of two features—especially U+V and U+O—tend to slightly improve performance compared to using a single constraint. For example, BERT with U+O achieves 0.098 Tr-Loss, 0.969 Tr-Acc, 0.118 Te-Loss, 0.964 Te-Acc, and 0.966 Obj-F1, nearly matching the best single-feature result (BERT+U). This suggests that while subject viewpoint (U) is dominant, integrating it with either tense (V) or object viewpoint (O) can reinforce semantic grounding. However, the marginal improvement indicates that most of the predictive strength still lies in the subject's viewpoint, highlighting the expressive nature of movie reviews.

Table 5. The ablation study we conducted for investigating the role of two of the UVO features in affecting the performance of five pre-trained base models on objectivity detection across the three datasets. The best values in various metrics have been bolded.

Datasets	Models	Constraints			Tr-Loss	Tr-Acc	Te-Loss	Te-Acc	Obj-F1
		U	V	O					
Movie	RoBERTa	✓	✓		**0.125**	**0.957**	**0.122**	**0.963**	**0.964**
		✓		✓	0.138	0.949	0.133	0.954	0.957
			✓	✓	0.152	0.943	0.135	0.953	0.956
	Bert	✓	✓		0.099	0.970	**0.097**	**0.969**	**0.970**
		✓		✓	**0.096**	**0.972**	0.116	0.961	0.964
			✓	✓	0.098	0.969	0.118	0.964	0.966
	BART	✓	✓		**0.169**	**0.935**	0.135	0.955	0.955
		✓		✓	0.179	0.930	**0.123**	**0.956**	**0.959**
			✓	✓	0.173	0.934	0.138	0.951	0.952
	T5	✓	✓		0.159	0.945	0.119	**0.964**	**0.962**
		✓		✓	**0.157**	0.945	**0.114**	0.962	0.959
			✓	✓	0.160	**0.946**	0.121	0.960	0.957
	GPT2	✓	✓		**0.505**	**0.729**	**0.583**	**0.688**	**0.670**
		✓		✓	0.507	0.727	0.615	0.674	0.627
			✓	✓	0.505	0.722	0.598	0.673	0.652
SST	RoBERTa	✓	✓		0.503	**0.806**	0.478	0.820	0.900
		✓		✓	0.518	0.793	0.497	0.810	0.896
			✓	✓	**0.496**	0.803	**0.475**	**0.824**	**0.903**
	Bert	✓	✓		0.497	0.795	0.504	0.802	0.889
		✓		✓	0.488	0.789	0.525	0.798	0.886
			✓	✓	**0.455**	**0.807**	**0.470**	**0.811**	**0.894**
	BART	✓	✓		0.522	0.780	**0.511**	**0.818**	**0.900**
		✓		✓	0.526	0.774	0.539	0.803	0.889
			✓	✓	**0.508**	**0.798**	0.515	0.802	0.889
	T5	✓	✓		0.471	0.807	**0.482**	0.812	0.895
		✓		✓	0.466	0.810	0.491	0.812	0.895
			✓	✓	**0.461**	**0.811**	0.483	**0.813**	**0.896**
	GPT2	✓	✓		0.493	**0.811**	0.496	0.805	0.892
		✓		✓	**0.493**	0.810	0.496	0.804	0.892
			✓	✓	0.494	0.809	0.482	0.815	0.897
Twitter	RoBERTa	✓	✓		**0.296**	**0.883**	**0.302**	**0.878**	**0.927**
		✓		✓	0.304	0.878	0.315	0.877	0.925
			✓	✓	0.335	0.857	0.344	0.861	0.915
	Bert	✓	✓		**0.267**	**0.894**	0.336	0.869	0.920
		✓		✓	0.297	0.880	0.355	0.858	0.913
			✓	✓	0.294	0.879	**0.323**	**0.876**	**0.925**
	BART	✓	✓		0.364	0.844	**0.309**	**0.878**	**0.927**
		✓		✓	0.397	0.813	0.329	0.863	0.916
			✓	✓	**0.360**	**0.847**	0.309	0.874	0.925
	T5	✓	✓		0.359	0.846	0.318	0.869	0.923
		✓		✓	0.385	0.821	0.335	0.862	0.918
			✓	✓	**0.334**	**0.862**	**0.314**	**0.871**	**0.925**
	GPT2	✓	✓		0.465	0.808	0.453	**0.819**	**0.899**
		✓		✓	**0.463**	**0.809**	0.461	0.813	0.895
			✓	✓	0.468	0.807	**0.451**	0.814	0.895

- For SST, the dual-feature combinations V+O and U+V perform slightly better than single features for some models. For instance, RoBERTa with V+O yields a Te-Acc of 0.824 and Obj-F1 of 0.903, higher than U or O alone. This reinforces the earlier observation that verb tense (V) carries the most semantic weight in the SST dataset, and when enhanced with object viewpoint (O), the contextual understanding is deepened. Since SST texts are short and less informative, having both predicate tense and object perspectives can aid in identifying subtle subjectivity markers.
- In Twitter, U+V and U+O combinations generally outperform other pairs or single features. For example, T5 with U+O achieves 0.821 Tr-Acc and 0.915 Obj-F1, higher than V+O. These results imply that in short and informal texts like tweets, integrating subject viewpoint (U) with an additional semantic constraint (V or O) enhances model robustness. However, due to noise and structural variation in tweets, improvements remain dataset- and model-dependent.

6. Conclusions

As key elements in sentence structure, viewpoints of subjects, tenses of predicates, and viewpoints of objects can directly influence the tone and objectivity of the text. Inspired by this, we propose to investigate the impact of the three features on objective detection in short informal text. To achieve this goal, we propose an LLM- and OpenIE-based triple feature extraction procedure to quantify each of the features. Upon that, we construct a pre-trained encoder-based model aiming at objectivity detection by leveraging the extracted triple features as the constraints of model training. Our experimental results demonstrate the superiority of our method in identifying objectivity in datasets that consist of short informal text. Moreover, we conduct the corresponding ablation studies to further explore the impact of each of the three triple features on the task, which reflects that each UVO feature has different influences while varying base models or datasets. Although the proposed UVO quantification framework demonstrates strong performance in detecting objectivity in informal texts, the dependence on OpenIE and rule-based filtering may lead to unstable outputs when processing noisy or structurally complex sentences. This can affect the quality of downstream triple selection. Furthermore, our experiments are conducted primarily on social media and sentiment datasets. The framework's generalizability to highly specialized domains has not yet been evaluated. In the future, our approach will be applied to further investigate the difference in objectivity between human-written text and LLM-generated text.

Author Contributions: Conceptualization, C.Z. (Chaowei Zhang); Methodology, C.Z. (Chaowei Zhang); Software, C.Z. (Cheng Zhao); Validation, C.Z. (Cheng Zhao); Formal analysis, Z.Z.; Resources, Y.H.; Data curation, Z.Z.; Writing—original draft, Y.H. All authors have read and agreed to the published version of the manuscript.

Funding: This study is partially supported by the National Natural Science Foundation of China (62403412), and the Natural Science Foundation of the Higher Education Institutions of Jiangsu Province of China under grant 23KJB520040.

Data Availability Statement: The data presented in this study are available on request from the corresponding author.

Conflicts of Interest: The authors declare no conflicts of interest.

References

1. Laureate, C.D.P.; Buntine, W.; Linger, H. A systematic review of the use of topic models for short text social media analysis. *Artif. Intell. Rev.* **2023**, *56*, 14223–14255. [CrossRef]
2. Tommasel, A.; Godoy, D. Short-text feature construction and selection in social media data: A survey. *Artif. Intell. Rev.* **2018**, *49*, 301–338. [CrossRef]
3. Wuraola, I.; Dethlefs, N.; Marciniak, D. Understanding Slang with LLMs: Modelling Cross-Cultural Nuances through Paraphrasing. In Proceedings of the 2024 Conference on Empirical Methods in Natural Language Processing, Miami, FL, USA, 12–16 November 2024; pp. 15525–15531.
4. Liu, J.; Tian, X.; Tong, H.; Xie, C.; Ruan, T.; Cong, L.; Wu, B.; Wang, H. Enhancing Chinese abbreviation prediction with LLM generation and contrastive evaluation. *Inf. Process. Manag.* **2024**, *61*, 103768. [CrossRef]
5. Qiu, Z.; Qiu, K.; Lyu, H.; Xiong, W.; Luo, J. Semantics Preserving Emoji Recommendation with Large Language Models. In Proceedings of the 2024 IEEE International Conference on Big Data (BigData), Washington, DC, USA, 15–18 December 2024; IEEE: Piscataway, NJ, USA, 2024; pp. 7131–7140.
6. Yadav, A.; Garg, T.; Klemen, M.; Ulcar, M.; Agarwal, B.; Robnik-Sikonja, M. From Translation to Generative LLMs: Classification of Code-Mixed Affective Tasks. *IEEE Trans. Affect. Comput.* **2025**. [CrossRef]
7. Kamath, G.; Schuster, S.; Vajjala, S.; Reddy, S. Scope ambiguities in large language models. *Trans. Assoc. Comput. Linguist.* **2024**, *12*, 738–754. [CrossRef]
8. Pang, B.; Lee, L. Opinion mining and sentiment analysis. *Found. Trends Inf. Retr.* **2008**, *2*, 1–135. [CrossRef]
9. Sachdeva, B.; Rathee, H.; Tiwari, N. Role of social media in causing and mitigating disasters. *Int. J. Resil. Fire Saf. Disasters* **2021**, *1*, 52–58.
10. Fersini, E.; Messina, E.; Pozzi, F.A. Expressive signals in social media languages to improve polarity detection. *Inf. Process. Manag.* **2016**, *52*, 20–35. [CrossRef]
11. Lawrie, D.; Mayfield, J.; Oard, D.W.; Yang, E.; Nair, S.; Galuščáková, P. HC3: A suite of test collections for CLIR evaluation over informal text. In Proceedings of the 46th International ACM SIGIR Conference on Research and Development in Information Retrieval, Taipei, Taiwan, 23–27 July 2023; pp. 2880–2889.
12. Alhussain, A.I.; Azmi, A.M. Automatic story generation: A survey of approaches. *ACM Comput. Surv.* **2021**, *54*, 1–38. [CrossRef]
13. Chenlo, J.M.; Losada, D.E. An empirical study of sentence features for subjectivity and polarity classification. *Inf. Sci.* **2014**, *280*, 275–288. [CrossRef]
14. Wankhade, M.; Rao, A.C.S.; Kulkarni, C. A survey on sentiment analysis methods, applications, and challenges. *Artif. Intell. Rev.* **2022**, *55*, 5731–5780. [CrossRef]
15. Küçük, D.; Can, F. Stance detection: A survey. *ACM Comput. Surv.* **2020**, *53*, 1–37. [CrossRef]
16. Bekoulis, G.; Papagiannopoulou, C.; Deligiannis, N. A review on fact extraction and verification. *ACM Comput. Surv.* **2021**, *55*, 1–35. [CrossRef]
17. Miao, Z.; Li, Y.; Wang, X.; Tan, W.C. Snippext: Semi-supervised opinion mining with augmented data. In Proceedings of the Web Conference 2020, Taipei, Taiwan, 20–24 April 2020; pp. 617–628.
18. Biber, D. *Variation Across Speech and Writing*; Cambridge University Press: Cambridge, UK, 1991.
19. Udochukwu, O.; He, Y. A rule-based approach to implicit emotion detection in text. In Proceedings of the Natural Language Processing and Information Systems: 20th International Conference on Applications of Natural Language to Information Systems, NLDB 2015, Passau, Germany, 17–19 June 2015; Proceedings 20; Springer: Berlin/Heidelberg, Germany, 2015; pp. 197–203.
20. Abirami, M.; Uma, M.; Prakash, M. Sentiment analysis of informal text using a Rule based Model. *J. Chem. Pharm. Sci.* **2016**, *9*, 2854–2858.
21. Androutsopoulos, J. From variation to heteroglossia in the study of computer-mediated discourse. In *Digital Discourse: Language in the New Media*; Oxford University Press: Oxford, UK, 2011; pp. 277–298.
22. Hutto, C.; Gilbert, E. Vader: A parsimonious rule-based model for sentiment analysis of social media text. In Proceedings of the International AAAI Conference on Web and Social Media, Ann Arbor, MI, USA, 1–4 June 2014; Volume 8, pp. 216–225.
23. Zhuang, W.; Zeng, Q.; Zhang, Y.; Liu, C.; Fan, W. What makes user-generated content more helpful on social media platforms? Insights from creator interactivity perspective. *Inf. Process. Manag.* **2023**, *60*, 103201. [CrossRef]
24. Lin, Y.; Wang, W.; Wang, Y. Hot Technology Analysis Method Based on Text Mining and Social Networks. In Proceedings of the 2023 4th International Conference on Machine Learning and Computer Application, Hangzhou, China, 27–29 October 2023; pp. 161–167.
25. Guo, Z.; Schlichtkrull, M.; Vlachos, A. A survey on automated fact-checking. *Trans. Assoc. Comput. Linguist.* **2022**, *10*, 178–206. [CrossRef]

26. Soprano, M.; Roitero, K.; La Barbera, D.; Ceolin, D.; Spina, D.; Demartini, G.; Mizzaro, S. Cognitive Biases in Fact-Checking and Their Countermeasures: A Review. *Inf. Process. Manag.* **2024**, *61*, 103672. [CrossRef]
27. Augenstein, I.; Baldwin, T.; Cha, M.; Chakraborty, T.; Ciampaglia, G.L.; Corney, D.; DiResta, R.; Ferrara, E.; Hale, S.; Halevy, A.; et al. Factuality challenges in the era of large language models and opportunities for fact-checking. *Nat. Mach. Intell.* **2024**, *6*, 852–863. [CrossRef]
28. Liu, B.; Li, C.; Zhou, W.; Ji, F.; Duan, Y.; Chen, H. An attention-based deep relevance model for few-shot document filtering. *ACM Trans. Inf. Syst.* **2020**, *39*, 1–35. [CrossRef]
29. Liu, K.; Wang, Z.; Yang, Y.; Huang, C.; Niu, M.; Lu, Q. Hierarchical Diachronic Embedding of Knowledge Graph Combined with Fragmentary Information Filtering. In Proceedings of the International Conference on Artificial Neural Networks, Heraklion, Greece, 26–29 September 2023; Springer: Berlin/Heidelberg, Germany, 2023; pp. 435–446.
30. Fields, J.; Chovanec, K.; Madiraju, P. A survey of text classification with transformers: How wide? How large? How long? How accurate? How expensive? How safe? *IEEE Access* **2024**, *12*, 6518–6531. [CrossRef]
31. Taha, K.; Yoo, P.D.; Yeun, C.; Homouz, D.; Taha, A. A comprehensive survey of text classification techniques and their research applications: Observational and experimental insights. *Comput. Sci. Rev.* **2024**, *54*, 100664. [CrossRef]
32. Ren, F.; Zhang, L.; Zhao, X.; Yin, S.; Liu, S.; Li, B. A simple but effective bidirectional framework for relational triple extraction. In Proceedings of the Fifteenth ACM International Conference on Web Search and Data Mining, Tempe, AZ, USA, 21–25 February 2022; pp. 824–832.
33. Rosso, P.; Yang, D.; Cudré-Mauroux, P. Beyond triplets: Hyper-relational knowledge graph embedding for link prediction. In Proceedings of the Web Conference 2020, Taipei, Taiwan, 20–24 April 2020; pp. 1885–1896.
34. Zhao, Y.; Zhou, H.; Zhang, A.; Xie, R.; Li, Q.; Zhuang, F. Connecting embeddings based on multiplex relational graph attention networks for knowledge graph entity typing. *IEEE Trans. Knowl. Data Eng.* **2022**, *35*, 4608–4620. [CrossRef]
35. Li, P.; Wang, X.; Liang, H.; Zhang, S.; Zhang, Y.; Jiang, Y.; Tang, Y. A fuzzy semantic representation and reasoning model for multiple associative predicates in knowledge graph. *Inf. Sci.* **2022**, *599*, 208–230. [CrossRef]
36. González, W.J. Semantics of science and theory of reference: An analysis of the role of language in basic science and applied science. In *Language and Scientific Research*; Springer: Berlin/Heidelberg, Germany, 2021; pp. 41–91.
37. Rodrigo-Ginés, F.J.; Carrillo-de Albornoz, J.; Plaza, L. A systematic review on media bias detection: What is media bias, how it is expressed, and how to detect it. *Expert Syst. Appl.* **2024**, *237*, 121641. [CrossRef]
38. Wiebe, J.; Bruce, R.; O'Hara, T.P. Development and use of a gold-standard data set for subjectivity classifications. In Proceedings of the 37th Annual Meeting of the Association for Computational Linguistics, College Park, MD, USA, 20–26 June 1999; pp. 246–253.
39. Riloff, E.; Wiebe, J. Learning extraction patterns for subjective expressions. In Proceedings of the 2003 Conference on Empirical Methods in Natural Language Processing, Sapporo, Japan, 11–12 July 2003; pp. 105–112.
40. Wiebe, J.; Wilson, T.; Bruce, R.; Bell, M.; Martin, M. Learning subjective language. *Comput. Linguist.* **2004**, *30*, 277–308. [CrossRef]
41. Hu, M.; Liu, B. Mining and summarizing customer reviews. In Proceedings of the Tenth ACM SIGKDD International Conference on Knowledge Discovery and Data Mining, Seattle, WA, USA, 22–25 August 2004; pp. 168–177.
42. Liu, B.; Hu, M.; Cheng, J. Opinion observer: Analyzing and comparing opinions on the web. In Proceedings of the 14th International Conference on World Wide Web, Chiba, Japan, 10–14 May 2005; pp. 342–351.
43. Zhang, X.; Zhao, J.; LeCun, Y. Character-level convolutional networks for text classification. *Adv. Neural Inf. Process. Syst.* **2015**, *28*, 1–9.
44. Hu, H.; Liao, M.; Zhang, C.; Jing, Y. Text classification based recurrent neural network. In Proceedings of the 2020 IEEE 5th Information Technology and Mechatronics Engineering Conference (ITOEC), Chongqing, China, 12–14 June 2020; IEEE: Piscataway, NJ, USA, 2020; pp. 652–655.
45. Vaswani, A. Attention is all you need. *Adv. Neural Inf. Process. Syst.* **2017**, *30*, 1–11.
46. Devlin, J. Bert: Pre-training of deep bidirectional transformers for language understanding. *arXiv* **2018**, arXiv:1810.04805.
47. Liu, Y. Roberta: A robustly optimized bert pretraining approach. *arXiv* **2019**, arXiv:1907.11692.
48. Raffel, C.; Shazeer, N.; Roberts, A.; Lee, K.; Narang, S.; Matena, M.; Zhou, Y.; Li, W.; Liu, P.J. Exploring the limits of transfer learning with a unified text-to-text transformer. *J. Mach. Learn. Res.* **2020**, *21*, 1–67.
49. Lewis, M. Bart: Denoising sequence-to-sequence pre-training for natural language generation, translation, and comprehension. *arXiv* **2019**, arXiv:1910.13461.
50. Radford, A. Improving Language Understanding by Generative Pre-Training. 2018. Available online: https://cdn.openai.com/research-covers/language-unsupervised/language_understanding_paper.pdf (accessed on 27 May 2025).
51. Radford, A.; Wu, J.; Child, R.; Luan, D.; Amodei, D.; Sutskever, I. Language models are unsupervised multitask learners. *OpenAI Blog* **2019**, *1*, 9.
52. Brown, T.B. Language models are few-shot learners. *arXiv* **2020**, arXiv:2005.14165.

53. Ouyang, L.; Wu, J.; Jiang, X.; Almeida, D.; Wainwright, C.; Mishkin, P.; Zhang, C.; Agarwal, S.; Slama, K.; Ray, A.; et al. Training language models to follow instructions with human feedback. *Adv. Neural Inf. Process. Syst.* **2022**, *35*, 27730–27744.
54. Achiam, J.; Adler, S.; Agarwal, S.; Ahmad, L.; Akkaya, I.; Aleman, F.L.; Almeida, D.; Altenschmidt, J.; Altman, S.; Anadkat, S.; et al. Gpt-4 technical report. *arXiv* **2023**, arXiv:2303.08774.
55. Qasim, R.; Bangyal, W.H.; Alqarni, M.A.; Ali Almazroi, A. A Fine-Tuned BERT-Based Transfer Learning Approach for Text Classification. *J. Healthc. Eng.* **2022**, *2022*, 3498123. [CrossRef]
56. Touvron, H.; Lavril, T.; Izacard, G.; Martinet, X.; Lachaux, M.A.; Lacroix, T.; Rozière, B.; Goyal, N.; Hambro, E.; Azhar, F.; et al. Llama: Open and efficient foundation language models. *arXiv* **2023**, arXiv:2302.13971.
57. Sun, X.; Li, X.; Li, J.; Wu, F.; Guo, S.; Zhang, T.; Wang, G. Text classification via large language models. *arXiv* **2023**, arXiv:2305.08377.
58. Tang, L.; Shalyminov, I.; Wong, A.W.m.; Burnsky, J.; Vincent, J.W.; Yang, Y.; Singh, S.; Feng, S.; Song, H.; Su, H.; et al. Tofueval: Evaluating hallucinations of llms on topic-focused dialogue summarization. *arXiv* **2024**, arXiv:2402.13249.
59. Laurenzi, E.; Mathys, A.; Martin, A. An LLM-Aided Enterprise Knowledge Graph (EKG) Engineering Process. *Proc. AAAI Symp. Ser.* **2024**, *3*, 148–156. [CrossRef]
60. Zhang, K.; Choi, Y.; Song, Z.; He, T.; Wang, W.Y.; Li, L. Hire a linguist!: Learning endangered languages in LLMs with in-context linguistic descriptions. In Proceedings of the Findings of the Association for Computational Linguistics ACL 2024, Bangkok, Thailand, 11–16 August 2024; pp. 15654–15669.
61. Guo, R.; Xu, W.; Ritter, A. Meta-Tuning LLMs to Leverage Lexical Knowledge for Generalizable Language Style Understanding. In Proceedings of the 62nd Annual Meeting of the Association for Computational Linguistics (Volume 1: Long Papers), Bangkok, Thailand, 11–16 August 2024; pp. 13708–13731.
62. Smith, G.; Fleisig, E.; Bossi, M.; Rustagi, I.; Yin, X. Standard Language Ideology in AI-Generated Language. *arXiv* **2024**, arXiv:2406.08726.
63. Song, G.; Ye, Y.; Du, X.; Huang, X.; Bie, S. Short text classification: A survey. *J. Multimed.* **2014**, *9*, 635–643. [CrossRef]
64. Chen, M.; Jin, X.; Shen, D. Short text classification improved by learning multi-granularity topics. In Proceedings of the Twenty-Second International Joint Conference on Artificial Intelligence, Barcelona, Spain, 16–22 July 2011.
65. Verdonck, T.; Baesens, B.; Óskarsdóttir, M.; vanden Broucke, S. Special issue on feature engineering editorial. *Mach. Learn.* **2024**, *113*, 3917–3928. [CrossRef]
66. Nayak, T.; Majumder, N.; Goyal, P.; Poria, S. Deep neural approaches to relation triplets extraction: A comprehensive survey. *Cogn. Comput.* **2021**, *13*, 1215–1232. [CrossRef]
67. Zhang, C.; Gupta, A.; Kauten, C.; Deokar, A.V.; Qin, X. Detecting Fake News for Reducing Misinformation Risks Using Analytics Approaches. *Eur. J. Oper. Res.* **2019**, *279*, 1036–1052. [CrossRef]
68. UzZaman, N.; Llorens, H.; Derczynski, L.; Allen, J.; Verhagen, M.; Pustejovsky, J. SemEval-2013 Task 1: TempEval-3: Evaluating Time Expressions, Events, and Temporal Relations. In Proceedings of the Second Joint Conference on Lexical and Computational Semantics (*SEM), Volume 2: Proceedings of the Seventh International Workshop on Semantic Evaluation (SemEval 2013), Atlanta, GA, USA, 13–14 June 2013; Manandhar, S., Yuret, D., Eds.; Association for Computational Linguistics: Stroudsburg, PA, USA, 2013; pp. 1–9.
69. Manning, C.D.; Surdeanu, M.; Bauer, J.; Finkel, J.R.; Bethard, S.; McClosky, D. The Stanford CoreNLP natural language processing toolkit. In Proceedings of the 52nd Annual Meeting of the Association for Computational Linguistics: System Demonstrations, Baltimore, MD, USA, 23–24 June 2014; pp. 55–60.

entropy

MDPI

Article

Scaling Laws in Language Families

Maelyson Rolim Fonseca dos Santos [1,2,*,†] and Marcelo Andrade de Filgueiras Gomes [2,*,†]

1 Instituto Federal de Educação, Ciência e Tecnologia de Pernambuco, Barreiros 55560-000, PE, Brazil
2 Departamento de Física, Universidade Federal de Pernambuco, Recife 50670-901, PE, Brazil
* Correspondence: maelyson.rolim@barreiros.ifpe.edu.br (M.R.F.d.S.); mafg@ufpe.br (M.A.d.F.G.)
† These authors contributed equally to this work.

Abstract: This article investigates scaling laws within language families using data from over six thousand languages and analyzes emergent patterns observed in Zipf-like classification graphs. Both macroscopic (based on the number of languages by family) and microscopic (based on the number of speakers by language within a family) aspects of these classifications are examined. Particularly noteworthy is the discovery of a distinct division among the fourteen largest contemporary language families, excluding Afro-Asiatic and Nilo-Saharan languages. These families are found to be distributed across three language family quadruplets, each characterized by significantly different exponents in the Zipf graphs. This finding sheds light on the underlying structure and organization of major language families, revealing intriguing insights into the nature of linguistic diversity and distribution.

Keywords: language families; scaling laws; Zipf distributions

1. Introduction

Complex systems are extensively characterized by scaling symmetry and power law distributions. These phenomena are evident in a wide range of studies, such as those on cities [1], growth models [2], cellular automata [3], and cognitive sciences [4], among many others. Language also exhibits characteristics of complex systems [5–7] and currently several statistical laws related to linguistic studies are well established [8].

An intriguing aspect in the analysis of complex systems is the role of entropy as a measure of disorder and diversification. In the linguistic context, historical processes of migration, fragmentation, and interaction between populations could be analogously interpreted as an "entropic flow", where ancestral languages diversify and spread across the Earth's surface, akin to the entropy of mixing in fluids [9]. In thermodynamics, mixing entropy quantifies the increase in disorder when two gases or liquids irreversibly interpenetrate. Similarly, linguistic diversification might be viewed as a process that maximizes a "cultural entropy", where the fragmentation of proto-languages into diverse families and subsequent contact and hybridization processes generate emergent statistical patterns [10,11]. The Zipf-like scaling laws observed in this study, with characteristic exponents for different family groupings, may reflect distinct stages of this entropic process. For instance, higher exponents could correspond to linguistically "more mixed" systems, where prolonged interactions and homogenizing demographic pressures reduce relative diversity, while lower exponents might indicate families in early stages of fragmentation, preserving more pronounced hierarchies. This entropic perspective provides a conceptual bridge between the historical dynamics of languages and physical formalisms of self-organization in non-equilibrium systems [12,13].

Notedly interesting on the interface between statistical studies and linguistics, Zipf's law for written texts relates the word frequency to their corresponding rank and points to a robust power-law dependence between these variables with a scaling exponent close to unity [14]. In recent decades, the expansion of available corpora and computational power has facilitated the study of Zipf's law across various language systems [15–19].

Additionally, other types of non-Zipfian scaling laws along several decades of variability were also found in the study of the distribution of living languages by Gomes et al. [20], as well as by Santos and Gomes [21]. Although these studies have explored the relations between linguistic diversity and certain geographical, demographic, and economic factors, they have not delved into the investigation of scaling within language families.

More than two decades ago, Zanette analyzed the distribution of language family sizes across seventeen families [22]. Wichmann later studied the distribution of family sizes using data from the fourteenth edition of Ethnologue [23], while Hammarstrom conducted a similar study based on his own language family classification from the sixteenth edition of Ethnologue [24]. In this paper, we first examine the distribution of language family sizes using a more recent dataset. Differently from all previous studies, we then present an original analysis of the distribution of language sizes within each family, focusing on the fourteen largest language families.

The structure of this paper is as follows: in Section 2 we discuss the materials and methods. Section 3 introduces a scaling law emerging from the classification of language families according to the number of languages. Moreover, in Section 3 we study and we present the distribution of language sizes as measured by the number of speakers of 14 largest contemporary language families and discuss the appearance of quadruplets of language families. Section 4 ends with a brief summary of our conclusions.

2. Materials and Methods

We analyzed data from the digital twentieth edition of Ethnologue [25], one of the most comprehensive and longstanding catalogs of the world's languages, first published in 1951. Over the past decades, Ethnologue has been widely used in linguistic and interdisciplinary research. Although our analysis is based on a more recent edition, a critical review of earlier versions of this database was conducted by Hammarström [26]. This edition classifies 6711 living languages into 141 families. The data for this study were manually extracted from the online version of Ethnologue; each language entry was individually accessed and recorded to ensure fidelity to the original classification and demographic information. In our analysis presented in this section, similar to that carried out by Wichmann [23], we excluded 388 languages categorized as constructed languages, creoles, sign languages, language isolates, mixed languages, pidgins, and unclassified languages. These languages were excluded because they do not consistently follow historical or genealogical patterns that define language families, making their inclusion problematic for comparative statistical analysis. For instance, constructed and pidgin languages are often the result of artificial or contact-driven processes rather than common ancestry, while isolates and unclassified languages lack sufficient data to be confidently grouped into families. Their inclusion could distort the scaling relations we aim to investigate.

An important feature to highlight is the wide range in the number of languages per family, varying from a single language, as in the Carajá family, to over a thousand, as in the Niger–Congo family. Regarding this variation in family sizes, Greenhill proposed five possible explanations: family age, population size, technology (agriculture/language dispersal hypothesis), geography and ecology, and social factors [27].

We began by ranking the language families in decreasing order of size, following an approach similar to Zipf's ranking of words in *Human Behavior and the Principle of Least Effort* [14]. Thus, we assign rank $r = 1$ to the Niger–Congo family, composed of 1526 languages; rank $r = 2$ to the Austronesian family, with 1224 languages; rank $r = 3$ to the Trans-New Guinea family, with 478 languages; and so on. In this way, we can express the cardinal size—that is, the number of languages—N_F in a family of rank r as

$$N_F \sim r^{-\theta}. \quad (1)$$

Observing that only ten families have more than one hundred languages, it is worth asking whether a similar pattern is also found within each family. If so, this would imply that, when classifying individual languages by their number of speakers N, we should observe

$$N \sim r^{-\kappa}, \quad (2)$$

where r is the language rank and κ is the corresponding exponent. In the next section, we present the results obtained from this classification.

3. Results and Discussion

3.1. Macroscopic Aspects

The resulting graph of ranking the language families in decreasing order of size is shown in Figure 1. The choice of double logarithmic axes in this figure is justified so that the plotted points can be visualized linearly [28].

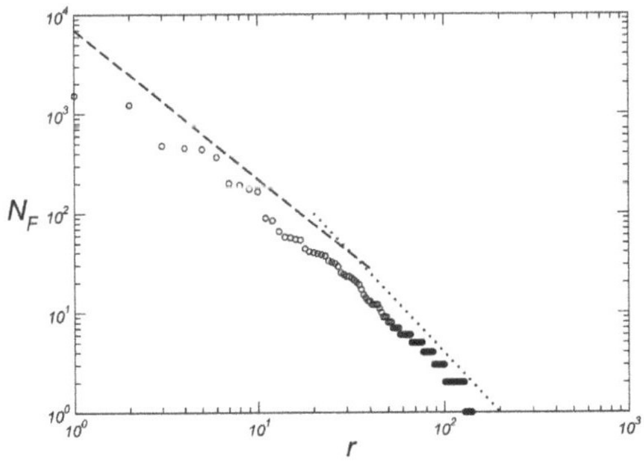

Figure 1. Number of languages N_F of each language family as a function of their rank r. The dotted (dashed) line with slope -2.0 (-1.5) corresponds to scaling behavior associated with stable distributions [29]. The scaling exponents describe the data along approximately one decade of variability in the values of r.

In Figure 1 can be viewed two different scaling behaviors associated with two distinct values for the exponent theta (Equation (1)): a first region with $\theta = 1.5$ for intermediary values of rank and a second region with $\theta = 2.0$ for large values of rank. The difference in exponents shown in Figure 1 may reflect the fact that larger families tend to be older and/or have more complex demographic histories. Importantly, the dotted and dashed lines in Figure 1 (and subsequent figures) do not represent statistical fits to the data. Instead, these lines act as qualitative guides to highlight proposed scaling regimes that capture the main observed trends. Such visual representations are common in exploratory analyses of complex systems, where strict parametric fitting may not adequately capture crossover behaviors or data heterogeneity. The distribution of the size of language families that we report here resembles the so-called Hollow Curve [27], a characteristic right-skewed distribution commonly observed in biological taxa, where a few groups are very large while most are small.

For large values of r, the exponent $\theta = 2.0$ referred in Figure 1 is close to the value ($\theta = 1.905$) previously obtained from the fourteenth edition of *Ethnologue* [23] but greater than the value ($\theta = 1.38$) obtained by Hammarström from their own language family classification from the sixteenth edition of *Ethnologue* [24]. This last value, however, seems more akin to the slope 1.5 observed for the intermediary rank values in Figure 1. Our result contradicts Zanette [22] who, analyzing a set of seventeen families from *A Guide to the World's Languages* by Ruhlen [30], proposed that the number of languages would decrease exponentially with the family rank.

Hammarström points out that since linguistic differentiation occurs mainly through human migration, the cardinal size of a family can be considered a measure of the diffusive spread of a family [24] . From this perspective, we can understand that only a small number of language families are spread over large areas of the Earth's surface. At the same time, it is possible to understand that most families have a small geographical reach.

3.2. Microscopic Aspects

Table 1 shows the kappa value for the fourteen largest language families according to the number of speakers (Equation (2)). In the following paragraphs we will discuss each of these families specifically and will show that, with the exception of the Afro-Asian and Nilo-Saharan families, the remaining twelve families are distributed in three quadruplets of language families grouped according to the exponent of Zipf's distributions. At large rank values, the curves shown in Figures 2–4 deviate from the classical Zipfian form, primarily due to the dominance of languages with very small speaker populations. This deviation likely arises from reduced viability or incomplete recording of languages with very few speakers.

Table 1. Fourteen largest language families according to the number of languages N_F. The value r indicates the classification of the language according to this ordering and r_s indicates the classification of the family according to the linguistic population N. The value κ is the exponent of the scaling law $N \sim r^{-\kappa}$.

r	Family	N_F	N (in Millions)	r_s	κ
01	Niger–Congo	1526	458.90	03	1.2
02	Austronesian	1224	324.88	05	1.6
03	Trans-New Guinea	478	3.55	21	1.1
04	Sino-Tibetan	452	1355.71	02	1.7
05	Indo-European	440	3077.11	01	1.7
06	Afro-Asiatic	366	444.85	04	2.6
07	Nilo-Saharan	201	50.33	12	1.4
08	Australian	193	0.04	51	1.6
09	Otomanguean	176	1.68	24	1.1
10	Austro-Asiatic	166	104.99	09	2.0
11	Tai-Kadai	90	80.1	10	1.2
12	Dravidian	85	228.1	06	2.0
13	Tupian	66	6.2	19	2.1
14	Uto-Astecan	58	1.9	22	2.1

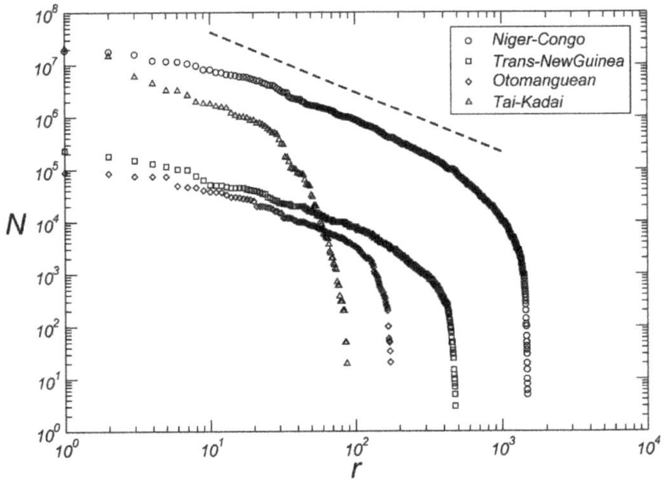

Figure 2. Number of speakers N by language as a function of rank r for the Niger–Congo, Trans-New Guinea, Otomanguean and Tai-Kadai families. The dashed line provides guidance for eye adjustments $N \sim r^{-\kappa}$ with $\kappa = 1.15$.

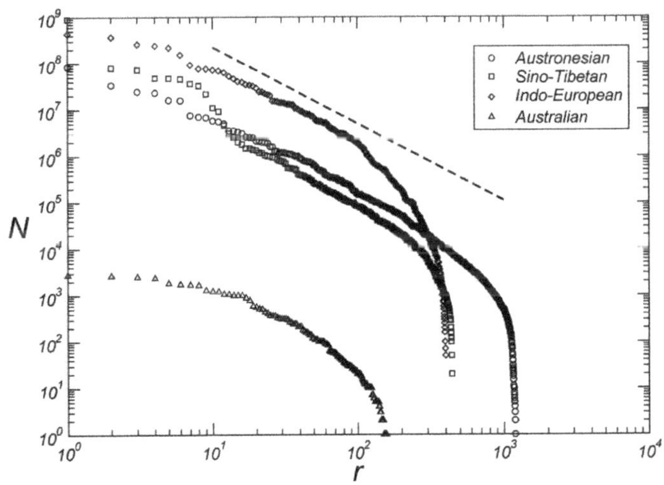

Figure 3. Number of speakers N by language as a function of rank r for the Austronesian, Sino-Tibetan, Indo-European, and Australian families. The dashed line provides guidance for eye adjustments $N \sim r^{-\kappa}$ with $\kappa = 1.65$.

The largest language family, the Niger–Congo, is composed of 1526 languages and encompasses nearly all native languages in Africa below the Sahara and is characterized by $\kappa = 1.2$. Another family, consisting of over a thousand languages, is the Austronesian, with its 1224 languages scattered from Indonesia through the island of New Guinea to Easter Island. This family, originating from the region of Taiwan, has $\kappa = 1.6$. New Guinea is also home to the Trans-New Guinea family, composed of 478 languages and characterized by $\kappa = 1.1$. The Sino-Tibetan family has 452 languages but boasts over 380 times more speakers than the Trans-New Guinea family and it has $\kappa = 1.7$. The Indo-European family includes almost all languages in Europe and many languages in the Asian continent.

Among the top twenty languages globally, eleven belong to this family, totaling over three billion speakers distributed across 440 languages with $\kappa = 1.7$.

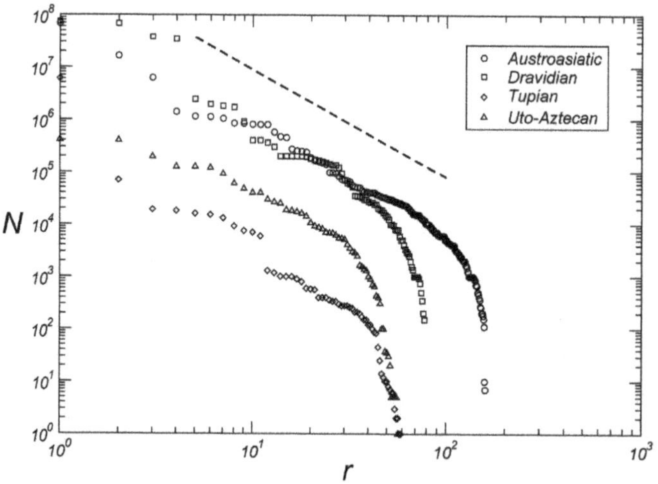

Figure 4. Number of speakers N by language as a function of rank r for the Austroasiatic, Dravidian, Tupian and Uto-Aztecan families. The dashed line provides guidance for eye adjustments $N \sim r^{-\kappa}$ with $\kappa = 2.05$.

The 366 languages that make up the Afro-Asiatic family likely descend from the language spoken by human groups that migrated from the African continent to the Middle East over 50,000 years ago. It was in Phoenician, a language of this family, that the first phonetic alphabet was constructed. Unlike the five largest families characterized by two exponent values ($\kappa = 1.15 \pm 0.05$ and $\kappa = 1.65 \pm 0.05$), the Afro-Asiatic family has $\kappa = 2.6$, making this value the highest among the fourteen major language families. South of the Afro-Asiatic languages region lies another family whose κ value is also distinct from those two characteristic of the five largest families: comprising over 200 languages, the Nilo-Saharan family has $\kappa = 1.4$.

The Australian family, although very diverse in the number of languages (193 in total), has experienced a significant decline in the number of speakers over the past centuries, making it the smallest family by this criterion among those discussed here. Australian languages with fewer than ten thousand speakers have $\kappa = 1.6$, which is the same value reported for the Austronesian family. The Otomanguean family, characteristic of the current Mexican territory, also experienced a population reduction after colonization processes and presents $\kappa = 1.1$. This value is identical to that reported for the Trans-New Guinea family.

The Austro-Asiatic language family, spoken from Southeast Asia to East India, ranks as the ninth-largest language family globally, with $\kappa = 2.0$. Southeast Asia is also home to the Tai-Kadai family, boasting about half the number of languages as the Austro-Asiatic family, and it is characterized by $\kappa = 1.2$. Moving to the Dravidian family, typical in South Asia, it boasts a linguistic population exceeding two hundred million speakers. Many of these languages persisted despite the expansion and dominance of Indo-European languages in the Indian region. The Dravidian family shares the same exponent as the Austro-Asiatic family, with $\kappa = 2.0$.

The processes resulting from colonization irreversibly affected the distribution of languages on the American continent. Two of the language families most affected were the

Tupian, in South America, and the Uto-Aztecan, in North America. These two families are characterized by $\kappa = 2.1$. Together with the Austro-Asiatic and Dravidian families, these two families constitute a quadruplet characterized by $\kappa = 2.05 \pm 0.05$.

Therefore, as discussed in the preceding paragraphs, with the exception of the Afro-Asiatic and Nilo-Saharan families, we have three quadruplets of language families grouped according to the exponent of the Zipf distributions. The first is composed of the Trans-New Guinea, Otomanguean, Niger–Congo, and Tai-Kadai families with $\kappa = 1.15 \pm 0.05$, as seen in Figure 2. With $\kappa = 1.65 \pm 0.05$, the second quadruplet comprises the families Austronesian, Australian, Sino-Tibetan, and Indo-European, as seen in Figure 3. And finally the third quadruplet with the families Austro-Asiatic, Dravidian, Tupian, and Uto-Aztecan has $\kappa = 2.05 \pm 0.05$, as seen in Figure 4.

4. Conclusions

Here we present power laws emerging from the distributions of both the size of linguistic families according to the number of languages and the size of languages according to their respective number of speakers within each of the fourteen largest families, based on methods for classifying and ordering data from more than six thousand languages. In contrast to previous studies, we show in Figure 1 that the cardinal size of a language family is related to the rank of the family by two exponents (1.5 and 2.0). The first for intermediate values of rank and the second for large values of rank. We then show that for twelve language families (Niger–Congo, Trans-New Guinea, Otomanguean, Tai-Kadai; Austronesian, Sino-Tibetan, Indo-European, Australian; Austroasiatic, Dravidian, Tupian, Uto-Aztecan) out of the fourteen largest families, we have three quadruplets of language families grouped according to the exponent of the Zipf distributions, namely, $\kappa = 1.15 \pm 0.05$ (Figure 2), $\kappa = 1.65 \pm 0.05$ (Figure 3), and $\kappa = 2.05 \pm 0.05$ (Figure 4). We believe that future research, with particular emphasis on detailed human migratory processes, should seek to understand (*i*) why these twelve language families form statistically well-characterized quadruplets according to the values of the exponents, and (*ii*) why the Afro-Asiatic and Nilo-Saharan families, both from the African continent, have different exponent values from those of the aforementioned quadruplets. Regarding this last aspect, it seems clear to us to expect that the continent where the greatest diversity of languages was originally generated should also present the greatest number of different emerging classes of statistical distributions that help to characterize this same diversity. Assessing whether these results hold across alternative datasets represents an important direction for future investigation.

Author Contributions: All authors contributed equally to this work. All authors have read and agreed to the published version of the manuscript.

Funding: This research was funded by Coordenação de Aperfeiçoamento de Pessoal de Nível Superior (CAPES) PROEX 23038.003069/2022-87, Grant #0041/2022.

Data Availability Statement: No new data were created or analyzed in this study. The data used are publicly available through the online platform of the 20th edition of *Ethnologue* (https://www.ethnologue.com/20 (accessed on 14 July 2021)).

Acknowledgments: M. R. F. Santos acknowledges CAPES for doctoral scholarship at the beginning of this research, Instituto Federal de Educação, Ciência e Tecnologia de Pernambuco for license for the completion of the doctorate and Stanisław Drożdż for stimulating discussions at the CCS 2023.

Conflicts of Interest: The authors declare no conflicts of interest.

References

1. Batty, M. The size, scale, and shape of cities. *Science* **2008**, *319*,769–771. [CrossRef] [PubMed]
2. Barabási, A.L.; Stanley, H.E. *Fractal Concepts in Surface Growth;* Cambridge University Press: Cambridge, UK, 1995.
3. Christensen, K.; Olami, Z. Scaling, phase transitions, and nonuniversality in a self-organized critical cellular-automaton model. *Phys. Rev. A* **1992**, *46*, 1829–1838. [CrossRef] [PubMed]
4. Kello, C.T.; Brown, G.D.A.; Ferrer-i Cancho, R.; Holden, J.G.; Linkenkaer-Hansen, K.; Rhodes, T.; Van Orden, G.C. Scaling laws in cognitive sciences. *Trends in Cognitive Sciences* **2010**, *14*, 223–232. [CrossRef] [PubMed]
5. Balasubrahmanyan, V.K.; Naranan, S. Quantitative linguistics and complex system studies. *J. Quant. Linguist.* **1996**, *3*, 177–228. [CrossRef]
6. Kretzschmar, W.A. *Language and Complex Systems;* Cambridge University Press: Cambridge, UK, 2015.
7. Stanisz, T.; Drożdż, S.; Kwapień, J. Complex systems approach to natural language. *Phys. Rep.* **2024**, *1053*, 1–84. [CrossRef]
8. Altmann, E.G.; Gerlach, M. Statistical laws in linguistics. In *Creativity and Universality in Language. Lecture Notes in Morphogenesis;* Degli Esposti, M., Altmann, E., Pachet, F., Eds.; Springer: Cham, Switzerland, 2016.
9. Galam, S. *Sociophysics;* Springer: New York, NY, USA, 2012.
10. Bentley, R.A.; Hahn, M.W.; Shennan, S.J. Random drift and culture change. *Proc. R. Soc. Lond. Ser. B Biol. Sci.* **2024**, *271*, 1443–1450. [CrossRef] [PubMed]
11. Castellano, C.; Fortunato, S.; Loreto, V. Statistical physics of social dynamics. *Rev. Mod. Phys.* **2009**, *81*, 591–646. [CrossRef]
12. Gell-Mann, M. *The Quark and the Jaguar: Adventures in the Simple and the Complex;* Macmillan: Shipley, UK, 1995.
13. Shannon, C.E. A mathematical theory of communication. *Bell Syst. Tech. J.* **1948**, *27*, 379–423. [CrossRef]
14. Zipf, G.K. *Human Behavior and the Principle of Least Effort;* Addison-Wesley: Cambridge, UK, 1949.
15. Ferrer i Cancho, R.; Solé, R.V. Zipf's law and random texts. *Adv. Complex Syst.* **2002**, *5*, 1–6. [CrossRef]
16. Ferrer i Cancho, R. The variation of Zipf's law in human language. *Eur. Phys. J. B* **2005**, *44*, 249–257. [CrossRef]
17. Jayaram, B.D.; Vidya, M.N. Zipf's law for indian languages. *J. Quant. Linguist.* **2008**, *15*, 293–317. [CrossRef]
18. Baixeries, J.; Elvevåg, B.; Ferrer-i Cancho, R. The evolution of the exponent of Zipf's law in language ontogeny. *PLoS ONE* **2013**, *8*, e53227. [CrossRef] [PubMed]
19. Moreno-Sánchez, I.; Font-Clos, F.; Corral, Á. Large-scale analysis of Zipf's law in english texts. *PLoS ONE* **2016**, *11*, e0147073. [CrossRef] [PubMed]
20. Gomes, M.A.F.; Vasconcelos, G.L.; Tsang, I.J.; Tsang, I.R. Scaling relations for diversity of languages. *Phys. A Stat. Mech. Its Appl.* **1999**, *271*, 489–495. [CrossRef]
21. Santos, M.R.F.; Gomes, M.A.F. Revisiting scaling relations for linguistic diversity. *Phys. A Stat. Mech. Its Appl.* **2019**, *532*, 121821. [CrossRef]
22. Zanette, D.H. Self-similarity in the taxonomic classification of human languages. *Adv. Complex Syst.* **2001**, *4*, 281–286. [CrossRef]
23. Wichmann, S. On the power-law distribution of language family sizes. *J. Linguist.* **2005**, *41*, 117–131. [CrossRef]
24. Hammarström, H. A full-scale test of the language farming dispersal hypothesis. *Diachronica* **2010**, *27*, 197–213. [CrossRef]
25. Simons, G.F.; Fennig, C.D. *Ethnologue: Languages of the World,* 20th ed.; SIL International: Dallas, TX, USA, 2017.
26. Hammarströmm, H. Ethnologue 16/17/18th editions: A comprehensive review. *Language* **2015**, *91*, 723–737. [CrossRef]
27. Greenhill, S.J. Demographic correlates of language diversity. In *The Routledge Handbook of Historical Linguistics;* Bowern, C., Evans, B., Eds.; Routledge: Amsterdam, The Netherlands, 2014; pp. 557–578.
28. Newman, M.E.J. Power laws, Pareto distributions and Zipf's law. *Contemp. Phys.* **2005**, *46*, 323–351. [CrossRef]
29. Takayasu, H. *Fractals in the Physical Sciences;* Manchester University Press: Manchester, UK, 1990.
30. Ruhlen, M. *A Guide to the World's Languages;* Classification; Stanford University Press: Redwood City CA, USA, 1987; Volume I.

entropy

MDPI

Article

Long-Range Dependence in Word Time Series: The Cosine Correlation of Embeddings

Paweł Wieczyński [1,†] and Łukasz Dębowski [2,*,†]

1 Independent Researcher, 80-180 Gdańsk, Poland; wieczynskipawel@gmail.com
2 Institute of Computer Science, Polish Academy of Sciences, 01-248 Warsaw, Poland
* Correspondence: ldebowsk@ipipan.waw.pl
† These authors contributed equally to this work.

Abstract: We analyze long-range dependence (LRD) for word time series, understood as a slower than exponential decay of the two-point Shannon mutual information. We achieve this by examining the decay of the cosine correlation, a proxy object defined in terms of the cosine similarity between word2vec embeddings of two words, computed by an analogy to the Pearson correlation. By the Pinsker inequality, the squared cosine correlation between two random vectors lower bounds the mutual information between them. Using the Standardized Project Gutenberg Corpus, we find that the cosine correlation between word2vec embeddings exhibits a readily visible stretched exponential decay for lags roughly up to 1000 words, thus corroborating the presence of LRD. By contrast, for the Human vs. LLM Text Corpus entailing texts generated by large language models, there is no systematic signal of LRD. Our findings may support the need for novel memory-rich architectures in large language models that exceed not only hidden Markov models but also Transformers.

Keywords: word embeddings; mutual information; cosine similarity; power laws; stretched exponential; long-range dependence

1. Introduction

Consider a time series $(W_i)_{i\in\mathbb{Z}}$ such as a text in natural language, a sequence of real numbers, or a sequence of vectors. Let $I(W_i; W_{i+n})$ be the Shannon mutual information between two random variables separated by n positions. By short-range dependence (SRD), we understand an asymptotic exponential bound for the decay of this dependence measure,

$$I(W_i; W_{i+n}) = O(\exp(-\delta n)), \quad \delta > 0. \tag{1}$$

By long-range dependence (LRD), we understand any sort of decay of the dependence measure that does not fall under (1). In particular, under LRD, we may have a power-law decay of the dependence measure,

$$I(W_i; W_{i+n}) \sim n^{-\gamma}, \quad \gamma > 0, \tag{2}$$

which resembles a more standard definition of LRD for the autocorrelation function by Beran [1], or we may have a stretched exponential decay thereof,

$$I(W_i; W_{i+n}) \sim \exp(-\delta n^{\beta}), \quad \delta > 0, \quad 0 < \beta < 1. \tag{3}$$

The SRD is characteristic of mixing Markov and hidden Markov processes ([2], Theorem 1), which assume that the probability of the next token depends only on a finite number of preceding tokens or on a bounded memory. Hence, the observation of LRD for sufficiently large lags implies that the time series generation cannot be modeled by a mixing Markov process of a relatively small order or—via the data-processing inequality ([3], Chapter 2.8)—by a mixing hidden Markov process with a small number of hidden states.

By contrast, it has often been expressed that texts in natural language exhibit LRD [2,4–11]. Several empirical studies analyzing textual data at different linguistic levels, such as characters [2,4], words [9], or punctuation [11], have indicated that correlations in natural language persist over long distances. This persistent correlation suggests that dependencies in human language extend far beyond adjacent words or short phrases, spanning across entire paragraphs or even longer discourse structures.

The LRD should be put on par with other statistical effects signaling that natural language is not a finite-state hidden Markov process, a theoretical linguistic claim that dates back to [12–14]. Let us write blocks of words $W_j^k := (W_j, W_{j+1}, \dots, W_k)$. A power-law growth of the block mutual information

$$I(W_1^n; W_{n+1}^{2n}) \sim n^{\beta}, \quad 0 < \beta < 1, \tag{4}$$

is known as Hilberg's law or as the neural scaling law [15–17]. Another observation [18] is a power-law logarithmic law of the maximal repetition length

$$L(W_1^n) \sim (\log n)^{\alpha}, \quad \alpha > 1, \tag{5}$$

where we denote the maximal repetition length

$$L(W_1^n) := \max\left\{k \geq 1 : W_{i+1}^{i+k} = W_{j+1}^{j+k} \text{ for some } 0 \leq j < i \leq n-k\right\}. \tag{6}$$

The long-range dependence (2) or (3), Hilberg's law (4), and the maximal repetition law (5) have been all reported for natural language, whereas it can be mathematically proved that none of them is satisfied by finite-state hidden Markov processes [19,20].

The LRD, Hilberg's law, and the maximal repetition law independently—and for different reasons—support the necessity of using complex memory architectures in contemporary large language models (LLMs). Neural networks designed for natural language processing must incorporate mechanisms capable of mimicking these laws. The older generation n-gram models struggle with this requirement for reasons that can be analyzed mathematically. By contrast, it is has been unclear whether Transformers [21], with their attention-based mechanisms, can leverage these extensive relationships. Understanding the nature of the LRD, Hilberg's law, and the maximal repetition law in textual data may shed some light onto neural architectures that can progress on language modeling tasks.

Various smoothing techniques were proposed to discern LRD at the character or phoneme level [2,4,6,7]. Under no advanced estimation, the power-law decay of the Shannon mutual information between two characters dissolves into noise for lags up to 10 characters [4]. By contrast, Lin and Tegmark [2] considered sophisticated estimation techniques and reported the power-law decay of the Shannon mutual information between two characters for much larger lags.

Because of the arbitrariness of word forms relative to the semantic content of the text, we are not convinced that the results by Lin and Tegmark [2] are not an artifact of their estimation method. For this reason, following the idea of Mikhaylovskiy and Churilov [9], we have decided to seek the LRD on the level of words. We have supposed that pairs of words rather than pairs of characters better capture the long-range semantic coherence of the text. For this reason, we have expected that the LRD effect extends for a larger distance on the level of words than on the level of characters. Indeed, in the present study, we report a lower bound on the Shannon mutual information between two words that is salient for lags up to 1000 words, which is four decades of magnitude larger than the unsmoothed effect for characters.

A modest goal of this paper is to systematically explore a simple measure of dependence to check whether texts in natural language and those generated by large language models exhibit the LRD. Rather than directly investigating the Shannon mutual information, which is difficult to estimate for large alphabets and strongly dependent sources, we elect a measure of dependence called the cosine correlation. This object is related to the cosine similarity of two vectors and somewhat resembles the Pearson correlation. Formally, the cosine correlation between two random vectors U and V equals

$$\mathrm{CC}(U;V) := \mathbf{E}\,\frac{U}{\|U\|}\cdot\frac{V}{\|V\|} - \mathbf{E}\,\frac{U}{\|U\|}\cdot\mathbf{E}\,\frac{V}{\|V\|}, \tag{7}$$

where $\mathbf{E}\,X$ is the expectation of random variable X, $U \cdot V$ is the dot product, and $\|U\| := \sqrt{U \cdot U}$ is the norm. By contrast, the cosine similarity of two non-random vectors u and v is

$$\cos(u;v) := \frac{u}{\|u\|}\cdot\frac{v}{\|v\|}. \tag{8}$$

In order to compute the cosine correlation or the cosine similarity for actual word time series, we need a certain vector representation of words. As a practical vector representation of words, one may consider word2vec embeddings used in large language models [22,23]. Word embeddings capture semantic relationships between words by mapping them into continuous spaces, allowing for a more meaningful measure of similarity between distant words in a text. In particular, Mikhaylovskiy and Churilov [9] observed an approximate power-law decay for the expected cosine similarity $\mathbf{E}\cos(U;V)$ of word embeddings.

The paper by Mikhaylovskiy and Churilov [9] lacked, however, the following important theoretical insight. As a novel result of this paper, we demonstrate that the cosine correlation $\mathrm{CC}(U;V)$, rather than the expected cosine similarity $\mathbf{E}\cos(U;V)$, provides a lower bound for the Shannon mutual information $I(U;V)$. Applying the Pinsker inequality [24,25], we obtain the bound

$$I(U;V) \ge \frac{\mathrm{CC}(U;V)^2}{2}. \tag{9}$$

This approach provides an efficient alternative to direct statistical estimation of mutual information, which is often impractical due to the sparse nature of natural language data. In particular, a slower than exponential decay of the cosine correlation implies LRD. Thus, a time series with a power-law or stretched exponential decay of the cosine correlation is not a Markov process or a hidden Markov process.

Indeed, on the experimental side, we observe a stretched exponential decay of the cosine correlation, which is clearly visible roughly for lags up to 1000 words—but only for natural texts. By contrast, artificial texts do not exhibit this trend in a systematic way. Our source of natural texts is the Standardized Project Gutenberg Corpus [26], a diverse collection of literary texts that offers a representative sample of human language usage. Our source of artificial texts is the Human vs. LLM Text Corpus [27]. To investigate the effect of semantic correlations, we also consider the cosine correlation between moving sums of neighboring embeddings, a technique that we call pooling. Curiously, pooling does not make the stretched exponential decay substantially slower. The lack of a prominent LRD signal was already noticed for the previous generation of language models by Takahashi and Tanaka-Ishii [6,7].

Our observation of the slow decay of the cosine correlation in general confirms the prior results of Mikhaylovskiy and Churilov [9] and supports the hypothesis of LRD. We notice that Mikhaylovskiy and Churilov [9] did not try to fit the stretched exponential decay to their data and their power-law model was not visually very good. Both theoretical and experimental findings of this paper contribute to the growing body of statistical evidence proving that natural language is not a finite-state hidden Markov process.

What is more novel is that our findings may support the view that natural language cannot be either generated by Transformer-based large language models—in view of no systematic decay trend of the cosine correlation for the Human vs. LLM Text Corpus. As mentioned, the LRD, Hilberg's law, and the maximal repetition law independently substantiate the necessity of sophisticated memory architectures in modern computational linguistic applications. These results open avenues for further research into the theoretical underpinnings of language structure, potentially informing the development of more effective models for language understanding and generation.

The organization of the article is as follows. Section 2 presents the theoretical results. Section 3 discusses the experiment. In particular, Section 3.1 presents our data. Section 3.2 describes the experimental methods. Section 3.3 presents the results. Section 3.4 offers the discussion. Section 4 contains the conclusion.

2. Theory

Similarly as Mikhaylovskiy and Churilov [9] but differently than Li [4], Lin and Tegmark [2], and Takahashi and Tanaka-Ishii [6,7], we will seek for LRD on the level of words rather than on the level of characters or phonemes. The Shannon mutual information between words is difficult to estimate for large alphabets and strongly dependent sources. Thus, we consider its lower bound defined via the cosine correlation of word2vec embeddings [22,23].

Let $\mathbf{E}\, X := \int X dP$ denote the expectation of a real random variable X. Let $\ln x$ be the natural logarithm of x and let $H(X) := \mathbf{E}[-\ln p(X)]$ be the Shannon entropy of a discrete random variable X, where $p(X)$ is the probability density of X with respect to a reference measure ([3], Chapters 2.1 and 8.1). The Shannon mutual information between variables X and Y ([3], Chapters 2.4 and 8.5) equals

$$I(X;Y) := H(X) + H(Y) - H(X,Y) \tag{10}$$

By contrast, the Pearson correlation between real random variables X and Y is defined as

$$\mathrm{Corr}(X;Y) := \frac{\mathrm{Cov}(X;Y)}{\sqrt{\mathrm{Var}(X)\,\mathrm{Var}(Y)}}. \tag{11}$$

where we denote the covariance $\mathrm{Cov}(X;Y) := \mathbf{E}\, XY - \mathbf{E}\, X\, \mathbf{E}\, Y$ and the variance $\mathrm{Var}(X) := \mathrm{Cov}(X;X)$. By the Cauchy–Schwarz inequality, we have $|\mathrm{Corr}(X;Y)| \le 1$.

We will introduce an analog of the Pearson correlation coefficient for vectors, which we call the cosine correlation. First, let us recall three standard concepts. For vectors $u = (u_1, u_2, ..., u_d)$ and $v = (v_1, v_2, ..., v_d)$, we consider the dot product

$$u \cdot v := \sum_{k=1}^{d} u_k v_k, \tag{12}$$

the norm $\|u\| := \sqrt{u \cdot u}$, and the cosine similarity

$$\cos(u; v) := \frac{u}{\|u\|} \cdot \frac{v}{\|v\|}. \tag{13}$$

By the Cauchy–Schwarz inequality, we have $|\cos(u; v)| \leq 1$.

Now, we consider something less standard. For vector random variables U and V, we define the cosine correlation

$$\begin{aligned} \mathrm{CC}(U; V) &:= \mathbf{E} \frac{U}{\|U\|} \cdot \frac{V}{\|V\|} - \mathbf{E} \frac{U}{\|U\|} \cdot \mathbf{E} \frac{V}{\|V\|} \\ &= \mathbf{E} \left[\frac{U}{\|U\|} - \mathbf{E} \frac{U}{\|U\|} \right] \cdot \left[\frac{V}{\|V\|} - \mathbf{E} \frac{V}{\|V\|} \right]. \end{aligned} \tag{14}$$

If U and V are discrete and we denote the difference of measures

$$\Delta(u, v) := P(U = u, V = v) - P(U = u) P(V = v) \tag{15}$$

then we may write

$$\mathrm{CC}(U; V) = \sum_{u,v} \Delta(u, v) \cos(u; v). \tag{16}$$

We observe that if random variables U and V are unidimensional, then $\cos(U, V) = 1$ with probability 1 and $\mathrm{CC}(U; V) = 0$. Similarly, $\mathrm{CC}(U; V) = 0$ if $\cos(U, V)$ is constant with probability 1 or if U and V are independent.

To build some more intuitions, let us notice the following three facts. First of all, the cosine correlation between two copies of a random vector lies in the unit interval.

Theorem 1. *We have*

$$0 \leq \mathrm{CC}(U; U) \leq 1. \tag{17}$$

Proof. Let us write $U' := U / \|U\|$. We have

$$\begin{aligned} \mathrm{CC}(U; U) &= \mathbf{E}\, U' \cdot U' - \mathbf{E}\, U' \cdot \mathbf{E}\, U' = 1 - \sum_{k=1}^{d} (\mathbf{E}\, U'_k)^2 \\ &\geq 1 - \sum_{k=1}^{d} \mathbf{E} (U'_k)^2 = 1 - \mathbf{E} \sum_{k=1}^{d} (U'_k)^2 = 1 - \mathbf{E}\, U' \cdot U' = 0. \end{aligned} \tag{18}$$

Hence, the claim follows. □

Second, the cosine correlation satisfies a version of the Cauchy–Schwarz inequality.

Theorem 2. *We have*

$$|\mathrm{CC}(U; V)| \leq \sqrt{\mathrm{CC}(U; U)\, \mathrm{CC}(V; V)} \leq 1. \tag{19}$$

Proof. Let us write $U' := U/\|U\|$ and $V' := V/\|V\|$. By the Cauchy–Schwarz inequalities $|\mathrm{Cov}(X;Y)| \le \sqrt{\mathrm{Var}(X)\,\mathrm{Var}(Y)}$ for random scalars X and Y and $\sum_{k=1}^{d} \sqrt{x_k y_k} \le \sqrt{\sum_{k=1}^{d} x_k}\sqrt{\sum_{k=1}^{d} y_k}$ for real numbers $x_k, y_k \ge 0$, we obtain

$$\begin{aligned}|\mathrm{CC}(U;V)| &\le \sum_{k=1}^{d} |\mathrm{Cov}(U'_k;V'_k)| \le \sum_{k=1}^{d} \sqrt{\mathrm{Var}(U'_k)\,\mathrm{Var}(V'_k)} \\ &\le \sqrt{\sum_{k=1}^{d} \mathrm{Var}(U'_k)}\sqrt{\sum_{k=1}^{d} \mathrm{Var}(V'_k)} = \sqrt{\mathrm{CC}(U;U)\,\mathrm{CC}(V;V)}. \end{aligned} \tag{20}$$

Hence, the claim follows (17). □

Third, we will show that cosine correlation $\mathrm{CC}(U;V)$ provides a lower bound for mutual information $I(U;V)$.

Theorem 3. *We have*

$$I(U;V) \ge \frac{\mathrm{CC}(U;V)^2}{2}. \tag{21}$$

Proof. Let us recall the Pinsker inequality

$$\sum_x p(x) \ln \frac{p(x)}{q(x)} \ge \frac{1}{2}\left(\sum_x |p(x) - q(x)|\right)^2 \tag{22}$$

for two discrete probability distributions p and q [24,25]. By the Pinsker inequality (22), the Cauchy–Schwarz inequality $|\cos(u;v)| \le 1$, and identity (16), we obtain

$$\begin{aligned} I(U;V) &\ge \frac{1}{2}\left(\sum_{u,v} |\Delta(u,v)|\right)^2 \ge \frac{1}{2}\left(\sum_{u,v} |\Delta(u,v)||\cos(u,v)|\right)^2 \\ &\ge \frac{1}{2}\left(\sum_{u,v} \Delta(u,v)\cos(u,v)\right)^2 = \frac{\mathrm{CC}(U;V)^2}{2}. \end{aligned} \tag{23}$$

Hence, the claim follows. □

We note in passing that the Pinsker inequality can be modified as the Bretagnolle–Huber bound

$$\sum_x p(x) \ln \frac{p(x)}{q(x)} \ge -\ln\left(1 - \frac{1}{4}\left(\sum_x |p(x) - q(x)|\right)^2\right) \tag{24}$$

for probability distributions p and q [28,29]. Respectively, we obtain

$$I(U;V) \ge -\ln\left(1 - \frac{\mathrm{CC}(U;V)^2}{4}\right). \tag{25}$$

This bound is weaker than (21) since $|\mathrm{CC}(U;V)| \le 1$.

Let $(W_i)_{i\in\mathbb{Z}}$ be the text in natural language treated as a word time series. Let $\phi(w) = (\phi_1(w), \phi_2(w), \dots, \phi_d(w))$ be an arbitrary vector representation of word w, such as word2vec embeddings [22,23], and let $F_i := (F_{i1}, F_{i2}, \dots, F_{id}) := \phi(W_i)$. In particular, since embeddings $F_i = \phi(W_i)$ are functions of words W_i, by the data-processing inequality ([3], Chapter 2.8) and by the cosine correlation bound (21), we obtain

$$I(W_i; W_j) \geq I(F_i; F_j) \geq \frac{CC(F_i; F_j)^2}{2}. \tag{26}$$

Wrapping up, a slow decay of cosine correlation $CC(F_i; F_{i+n})$ implies a slow decay of mutual information $I(W_i; W_{i+n})$. Since $I(W_i; W_{i+n})$ is damped exponentially for any mixing Markov or hidden Markov process $(W_i)_{i \in \mathbb{Z}}$ by Theorem 1 of Lin and Tegmark [2], observing a power-law or a stretched exponential decay of cosine correlation $CC(F_i; F_{i+n})$ is enough to demonstrate that process $(W_i)_{i \in \mathbb{Z}}$ is not a mixing Markov or hidden Markov process.

The framework that we have constructed in this section has its prior in the literature. We remark that Mikhaylovskiy and Churilov [9] investigated estimates of expectation $\mathbf{E} \cos(F_i; F_{i+n})$ rather than cosine correlation $CC(F_i; F_{i+n})$. That approach required estimation and subtraction of the asymptotic constant term. Mikhaylovskiy and Churilov [9] observed an approximate power-law decay but they did not mention the cosine correlation bound (21) in their discussion explicitly.

3. Experiment

3.1. Data

Our data consisted of three elements: a dictionary of embedding vectors for a subset of human languages, a corpus of texts written by humans in these languages, and a corpus of texts in English created by artificial intelligence. The considered set of human languages included 17 languages. Originally, we planned to use 20 languages with the largest text counts in the considered corpora but three of them, Esperanto, Chinese, and Tagalog, had to be excluded because the embedding dictionary did not cover these languages.

In particular, the source of pretrained word embeddings was chosen as the NLPL repository [23]. To provide a uniform baseline across languages, for all considered languages, we used 100-dimensional embedding vectors trained on the CoNLL17 corpora with the same algorithm, being the word2vec continuous skipgram algorithm. None of these embedding vector spaces includes lemmatization. The vocabulary sizes of the embedding spaces for the considered 17 languages are presented in Table 1.

Table 1. Vocabulary sizes of the chosen embedding spaces.

Language	Vocabulary Size
Catalan (ca)	799,020
Danish (da)	1,655,886
German (de)	4,946,997
Greek (el)	1,183,194
English (en)	4,027,169
Spanish (es)	2,656,057
Finnish (fi)	2,433,286
French (fr)	2,567,698
Hungarian (hu)	2,702,663
Italian (it)	2,469,122
Japanese (ja)	3,989,605
Latin (la)	555,381
Dutch (nl)	2,610,658
Norwegian (no)	223,763
Polish (pl)	4,420,598
Portuguese (pt)	2,536,452
Swedish (sv)	3,010,472

As the source of texts written by humans, we chose the Standardized Project Gutenberg Corpus (SPGC) [26]. The corpus provides texts after some preprocessing and tokenization, as detailed in [26]. We filtered the SPGC to obtain a more manageable yet representative subset of texts. As we have mentioned, we restricted the corpus to 17 languages with the largest text counts simultaneously covered by the applied NLPL embedding dictionary. Moreover, we filtered out files of the size above 1000 KB and we sampled up to 100 texts (or fewer if not available) per language in order to achieve roughly balanced subsets across particular languages.

To provide a comparison with texts generated by artificial intelligence, we also considered the Human vs. LLM Text Corpus (HLLMTC) [27]. All texts in the HLLMTC are in English. To make this corpus more easily computationally tractable, we sampled 1000 human written texts and 6000 LLM generated texts, where we chose 1000 texts per each of the six selected large language models. To convert these texts into word time series, we used off-the-shelf tokenizer [30].

Table 2 provides the summary statistics of the obtained subsets of the Standardized Project Gutenberg Corpus and the Human vs. LLM Text Corpus. In particular, we report the token counts and the coverage of the sampled texts, i.e., the fraction of word tokens of texts that appear in the respective NLPL embedding dictionary.

Table 2. Summary of the used subset of corpora.

Source	# of Texts	# of Tokens		Coverage	
		Mean	Std	Mean	Std
Standardized Project Gutenberg Corpus:					
Catalan (ca)	32	36,827.44	20,376.28	0.97	0.02
Danish (da)	66	51,832.92	30,748.01	0.97	0.02
German (de)	100	41,532.84	32,192.86	0.97	0.02
Greek (el)	100	29,487.20	17,775.53	0.91	0.03
English (en)	100	47,362.22	41,189.09	1.00	0.01
Spanish (es)	100	62,873.16	37,040.17	0.98	0.02
Finnish (fi)	100	35,095.02	28,948.42	0.94	0.03
French (fr)	100	53,948.66	39,585.75	0.96	0.01
Hungarian (hu)	100	50,510.30	31,976.38	0.95	0.02
Italian (it)	100	54,386.01	37,917.55	0.95	0.03
Japanese (ja)	20	268.05	371.40	0.78	0.19
Latin (la)	76	26,769.57	31,199.93	0.93	0.07
Dutch (nl)	100	43,055.05	31,465.48	0.98	0.01
Norwegian (no)	19	39,497.00	24,798.87	0.93	0.03
Polish (pl)	29	14,225.28	16,859.88	0.96	0.06
Portuguese (pt)	100	18,485.80	18,533.72	0.96	0.01
Swedish (sv)	100	37,474.97	29,310.51	0.97	0.02
Human vs. LLM Text Corpus:					
GPT-3.5	1000	444.40	278.66	0.9996	0.0025
GPT-4	1000	628.69	228.94	0.9996	0.0023
Human	1000	666.63	881.15	0.9980	0.0055
LLaMA-13B	1000	437.87	268.76	0.9987	0.0058
LLaMA-30B	1000	404.65	261.39	0.9988	0.0053
LLaMA-65B	1000	369.19	252.73	0.9988	0.0061
LLaMA-7B	1000	489.27	263.58	0.9986	0.0070

3.2. Methods

In this section, we briefly describe what we measured and in what way. We supposed that the LRD on the level of words is due to semantic coherence of the text over longer distances. In particular, mutual information between two words is large as long as the text around these words concerns a similar topic. We supposed that the embedding of this local topic can be roughly estimated as the sum of embeddings of all words in the neighborhood, called a pooled embedding. Let F_i be the embedding of the i-th word in the text. The pooled embeddings are defined as

$$F_i^{(k)} := \sum_{j=0}^{k-1} F_{i+j} \tag{27}$$

for the pooling order $k \geq 1$. In particular, pooled embeddings for $k = 1$ equal word embeddings, $F_i^{(1)} = F_i$.

The object that we wanted to measure was the cosine correlation for pooled embeddings, namely

$$C(n|k) := \mathrm{CC}(F_i^{(k)}; F_{i+n}^{(k)}). \tag{28}$$

Function $C(n|k)$ is substantially larger for $0 \leq n < k$ since the summations for variables $F_i^{(k)}$ and $F_{i+n}^{(k)}$ range partly over overlapping embeddings F_i. Thus, if one wants to estimate the functional form of the decay of $C(n|k)$, it makes sense to fit the respective function exclusively to data points where $n \geq k$.

Let us proceed to the estimation of function $C(n|k)$. Let $\phi(w)$ be the embedding of word w according to the considered word2vec dictionary. From each text, we removed all word tokens that did not have an embedding in the dictionary. In this way, we obtained a collection of word time series $(W_1, W_2, ..., W_N)$, corresponding word embeddings $F_i = \phi(W_i)$, and pooled embeddings $F_i^{(k)}$ given by formula (27). We estimated the expectations as the averages over the times series. That is, we computed the estimator of $C(n|k)$ defined as

$$\hat{C}(n|k) := \frac{1}{N-n} \sum_{i=1}^{N-n} U_i^{(k)} \cdot U_{i+n}^{(k)}, \tag{29}$$

where we used the auxiliary time series

$$U_i^{(k)} := \frac{F_i^{(k)}}{\left\| F_i^{(k)} \right\|} - \frac{1}{N} \sum_{j=1}^{N} \frac{F_j^{(k)}}{\left\| F_j^{(k)} \right\|}. \tag{30}$$

We observe that $F_{i+1}^{(k)} = F_i^{(k)} - F_i + F_{i+k}$. Therefore, the computational complexity of estimator $\hat{C}(n|k)$ for fixed n and k is of order $O(Nd)$, where N is the text length and d is the dimension of embeddings F_i.

For each text $(W_1, W_2, \ldots, W_N)$, we computed estimators $\hat{C}(n|k)$ for lags $n \in A \cap [1, N]$, where

$$A := \left\{1, \lceil 1.1 \rceil, \left\lceil 1.1^2 \right\rceil, \left\lceil 1.1^3 \right\rceil, \ldots \right\}, \tag{31}$$

and pooling orders $k \in \{1, 3, 3^2, 3^3\}$. We observed that the plot of the absolute value $\left|\hat{C}(n|k)\right|$ for considered texts usually dissolved into random noise around $n = 1000$ and there was a hump for $n < k$, as expected. Hence, to estimate the functional form of the decay of $\left|\hat{C}(n|k)\right|$, we restricted the fitting procedure to range $k \leq n \leq 1000$.

The parameter estimation was performed using the `curve_fit` function from the SciPy library [31], which employs the trust region reflective algorithm. We selected this method due to its compatibility with bounded constraints. We estimated parameters of two functions: the power-law decay

$$f(n|c,\gamma) := cn^{-\gamma}, \quad c \in \mathbb{R}, \quad \gamma > 0, \tag{32}$$

and the stretched exponential decay

$$f(n|b,\delta,\beta) := \exp(-\delta n^{\beta} + b), \quad b \in \mathbb{R}, \quad \delta > 0, \quad 0 < \beta < 1, \tag{33}$$

with parameters γ, δ, and β implicitly depending on the pooling order k. As a goodness-of-fit metric, we calculated the sum of squared logarithmic residuals

$$\text{SSLR} := \sum_{n \in A \cap [k,1000]} \left(\log\left|\hat{C}(n|k)\right| - \log f(n|\ldots)\right)^2 \tag{34}$$

divided by the number of the degrees of freedom (ndf) equal to $|A \cap [k, 1000]|$ minus the number of parameters of $f(n|\ldots)$.

We investigated the dependence of the results on the source, understood as the particular language for human-written texts or the particular language model for LLM-generated texts. To check whether there are significant differences of the distribution of a parameter $\alpha \in \{\gamma, \delta, \beta\}$ across particular sources, we used the non-parametric Kruskal–Wallis test with the null hypothesis

$$H_0 : P_1 = P_2 = \ldots = P_J, \tag{35}$$

where P_j is the distribution of parameter α for the j-th source. To further explore differences among different sources, we employed the post-hoc Dunn test with the Bonferroni correction for multiple comparisons.

3.3. Results

Visually, the decay of the absolute cosine correlation estimates $\left|\hat{C}(n|k)\right|$ for $k \leq n \leq 1000$ usually follows a stretched exponential form rather than the exact power-law decay for human-written text. By contrast, no systematic decay for $k \leq n$ can be detected for LLM-generated texts. This tendency can be seen in Figure 1, which is a diagnostic plot of the absolute cosine correlation estimates $\left|\hat{C}(n|k)\right|$ for two texts: *Cecilia: A Story of Modern Rome* in English from the SPGC corpus and *Text no. 702* by GPT 3.5, which is the longest LLM-generated text in the sampled subset of the HLLMTC corpus.

In Tables 3–7, we report the means and the standard deviations of the fitted parameters c and γ of the power-law model (32) and b, δ, and β of the stretched exponential model (33). The values are reported as they depend on a particular language for human-written texts or on a particular language model for LLM-generated texts. When fitting the models, the optimization algorithm did not converge sometimes. The failure rates and the overall goodness of fit are reported in Table 8. Despite the visual appeal of the stretched exponential model, the mean SSLR given by Formula (34) is less for the power-law model. This does not necessarily mean that the power-law model is better, however, since the standard deviation of the SSLR is greater than the mean for the stretched exponential model.

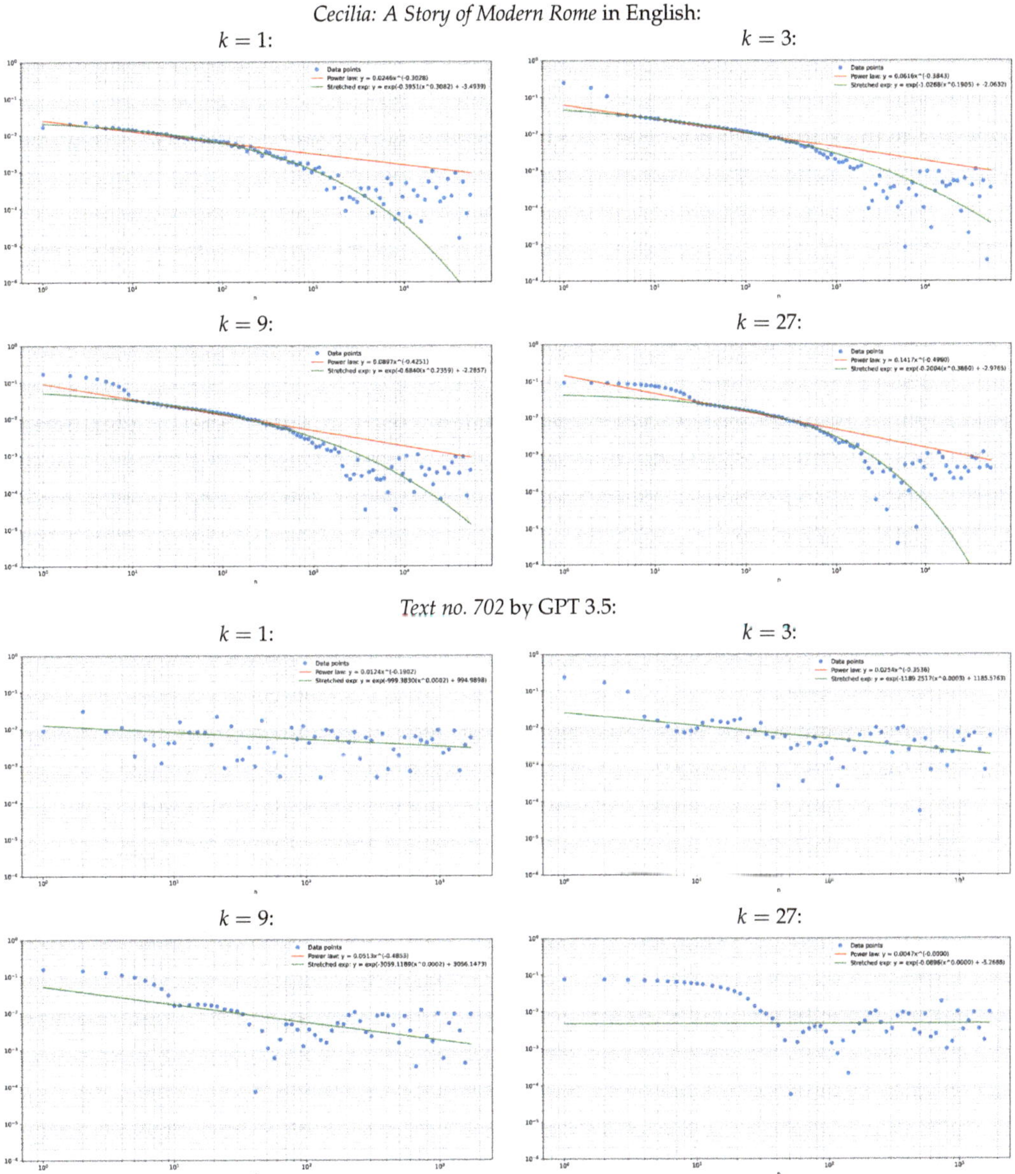

Figure 1. Estimates $\left|\hat{C}(n|k)\right|$ for two diagnostic texts.

Table 3. Means and standard deviations of parameter c.

Source	Pooling Order			
	$k=1$	$k=3$	$k=9$	$k=27$
	Standardized Project Gutenberg Corpus:			
ca	0.0248 ± 0.0035	0.069 ± 0.013	0.099 ± 0.024	0.17 ± 0.12
da	0.0203 ± 0.0045	0.052 ± 0.013	0.080 ± 0.024	0.150 ± 0.091
de	0.0286 ± 0.0079	0.070 ± 0.021	0.108 ± 0.047	0.27 ± 0.67
el	0.031 ± 0.012	0.072 ± 0.024	0.100 ± 0.041	0.19 ± 0.20
en	0.0258 ± 0.0092	0.070 ± 0.026	0.108 ± 0.053	0.20 ± 0.18
es	0.033 ± 0.022	0.088 ± 0.037	0.120 ± 0.049	0.19 ± 0.12
fi	0.0501 ± 0.0082	0.100 ± 0.019	0.144 ± 0.046	0.24 ± 0.13
fr	0.033 ± 0.013	0.087 ± 0.033	0.127 ± 0.076	0.27 ± 0.57
hu	0.0353 ± 0.0058	0.086 ± 0.016	0.119 ± 0.030	0.21 ± 0.18
it	0.0327 ± 0.0093	0.085 ± 0.025	0.117 ± 0.047	0.19 ± 0.16
ja	0.148 ± 0.094	0.45 ± 0.39	3.5 ± 9.6	$21{,}259 \pm 72{,}964$
la	0.067 ± 0.031	0.154 ± 0.076	0.24 ± 0.16	0.63 ± 0.96
nl	0.0257 ± 0.0078	0.071 ± 0.026	0.113 ± 0.057	0.24 ± 0.27
no	0.0136 ± 0.0051	0.040 ± 0.012	0.068 ± 0.019	0.110 ± 0.041
pl	0.046 ± 0.011	0.118 ± 0.025	0.179 ± 0.077	0.42 ± 0.48
pt	0.026 ± 0.011	0.087 ± 0.034	0.16 ± 0.12	0.32 ± 0.50
sv	0.027 ± 0.012	0.065 ± 0.022	0.101 ± 0.039	0.19 ± 0.14
	Human vs. LLM Text Corpus:			
GPT-3.5	0.024 ± 0.022	0.05 ± 0.74	2 ± 57	$5.4 \times 10^{30} \pm 1.7 \times 10^{31}$
GPT-4	0.029 ± 0.019	0.1 ± 3.2	0.02 ± 0.52	-0.1 ± 1.2
Human	0.028 ± 0.021	0.039 ± 0.061	3 ± 59	$1.0 \times 10^{7} \pm 3.1 \times 10^{8}$
LLaMA-13B	0.027 ± 0.027	0.045 ± 0.088	2 ± 31	$1.0 \times 10^{30} \pm 3.1 \times 10^{31}$
LLaMA-30B	0.027 ± 0.024	0.040 ± 0.048	0.3 ± 4.8	$2.6 \times 10^{19} \pm 8.0 \times 10^{20}$
LLaMA-65B	0.026 ± 0.023	0.04 ± 0.18	1 ± 23	$1.7 \times 10^{7} \pm 4.4 \times 10^{8}$
LLaMA-7B	0.029 ± 0.027	0.044 ± 0.051	0.3 ± 4.0	$6.2 \times 10^{11} \pm 1.9 \times 10^{12}$

Table 4. Means and standard deviations of parameter γ.

Source	Pooling Order			
	$k=1$	$k=3$	$k=9$	$k=27$
	Standardized Project Gutenberg Corpus:			
ca	0.449 ± 0.055	0.523 ± 0.063	0.546 ± 0.083	0.62 ± 0.13
da	0.373 ± 0.067	0.442 ± 0.092	0.49 ± 0.11	0.58 ± 0.17
de	0.440 ± 0.079	0.49 ± 0.11	0.53 ± 0.14	0.62 ± 0.23
el	0.405 ± 0.083	0.48 ± 0.13	0.51 ± 0.16	0.57 ± 0.23
en	0.330 ± 0.067	0.418 ± 0.098	0.47 ± 0.12	0.54 ± 0.19
es	0.373 ± 0.090	0.44 ± 0.11	0.45 ± 0.12	0.50 ± 0.16
fi	0.574 ± 0.084	0.552 ± 0.099	0.57 ± 0.13	0.62 ± 0.17
fr	0.415 ± 0.079	0.47 ± 0.11	0.49 ± 0.13	0.55 ± 0.20
hu	0.442 ± 0.075	0.482 ± 0.096	0.49 ± 0.11	0.55 ± 0.17
it	0.421 ± 0.089	0.47 ± 0.12	0.48 ± 0.13	0.52 ± 0.19
ja	0.34 ± 0.19	0.57 ± 0.39	0.79 ± 0.69	2.3 ± 1.6
la	0.40 ± 0.17	0.47 ± 0.23	0.51 ± 0.24	0.60 ± 0.33
nl	0.390 ± 0.074	0.46 ± 0.11	0.51 ± 0.13	0.60 ± 0.20
no	0.347 ± 0.041	0.451 ± 0.060	0.522 ± 0.073	0.59 ± 0.10
pl	0.550 ± 0.075	0.63 ± 0.12	0.65 ± 0.18	0.73 ± 0.26
pt	0.416 ± 0.071	0.57 ± 0.16	0.63 ± 0.22	0.70 ± 0.47
sv	0.407 ± 0.084	0.459 ± 0.095	0.51 ± 0.12	0.60 ± 0.17
	Human vs. LLM Text Corpus:			
GPT-3.5	0.31 ± 0.36	0.19 ± 0.36	0.9 ± 1.9	1.2 ± 2.0
GPT-4	0.47 ± 0.30	0.39 ± 0.42	0.6 ± 1.3	0.8 ± 1.6
Human	0.28 ± 0.26	0.29 ± 0.38	0.7 ± 1.3	0.9 ± 1.3
LLaMA-13B	0.19 ± 0.20	0.25 ± 0.27	0.6 ± 1.3	0.8 ± 1.4
LLaMA-30B	0.20 ± 0.20	0.25 ± 0.35	0.6 ± 1.3	0.8 ± 1.3
LLaMA-65B	0.19 ± 0.20	0.23 ± 0.27	0.6 ± 1.2	0.8 ± 1.2
LLaMA-7B	0.21 ± 0.20	0.25 ± 0.24	0.6 ± 1.0	0.7 ± 1.1

Table 5. Means and standard deviations of parameter b.

Source	Pooling Order			
	$k = 1$	$k = 3$	$k = 9$	$k = 27$
	Standardized Project Gutenberg Corpus:			
ca	0.9 ± 6.1	192 ± 524	26 ± 114	24 ± 127
da	119 ± 552	107 ± 490	69 ± 322	131 ± 571
de	137 ± 586	56 ± 246	134 ± 531	89 ± 480
el	126 ± 501	63 ± 206	128 ± 481	80 ± 335
en	93 ± 472	39 ± 157	108 ± 529	36 ± 183
es	147 ± 615	124 ± 334	34 ± 188	50 ± 255
fi	26 ± 40	151 ± 406	97 ± 419	127 ± 504
fr	81 ± 333	110 ± 478	121 ± 461	106 ± 466
hu	10 ± 28	40 ± 158	21 ± 130	53 ± 197
it	114 ± 483	86 ± 231	83 ± 402	108 ± 527
ja	195 ± 652	449 ± 992	252 ± 902	479 ± 1239
la	406 ± 1044	176 ± 672	231 ± 701	375 ± 1029
nl	4 ± 31	19 ± 101	14 ± 129	47 ± 257
no	271 ± 745	5 ± 27	-0.7 ± 3.9	84 ± 356
pl	157 ± 416	387 ± 1071	364 ± 725	610 ± 1189
pt	353 ± 915	338 ± 884	496 ± 1207	588 ± 1174
sv	114 ± 455	106 ± 534	28 ± 150	69 ± 455
	Human vs. LLM Text Corpus:			
GPT-3.5	1575 ± 1629	709 ± 1156	421 ± 877	198 ± 590
GPT-4	2520 ± 1489	1749 ± 1510	1127 ± 1333	325 ± 793
Human	1211 ± 1462	724 ± 1144	624 ± 1126	581 ± 1149
LLaMA-13B	513 ± 951	502 ± 959	662 ± 1129	522 ± 1029
LLaMA-30B	569 ± 1002	469 ± 893	664 ± 1108	494 ± 993
LLaMA-65B	554 ± 982	479 ± 912	606 ± 1090	458 ± 987
LLaMA-7B	553 ± 1010	506 ± 946	719 ± 1173	526 ± 1062

Table 6. Means and standard deviations of parameter δ.

Source	Pooling Order			
	$k = 1$	$k = 3$	$k = 9$	$k = 27$
	Standardized Project Gutenberg Corpus:			
ca	4.7 ± 6.0	195 ± 524	29 ± 114	27 ± 127
da	123 ± 552	110 ± 490	72 ± 322	134 ± 571
de	141 ± 586	59 ± 246	137 ± 531	92 ± 480
el	130 ± 501	66 ± 206	131 ± 481	83 ± 335
en	97 ± 472	42 ± 157	111 ± 529	39 ± 183
es	151 ± 615	127 ± 334	37 ± 188	52 ± 255
fi	29 ± 40	153 ± 406	100 ± 419	130 ± 504
fr	84 ± 333	112 ± 478	123 ± 461	108 ± 466
hu	13 ± 28	43 ± 158	23 ± 130	56 ± 197
it	118 ± 483	88 ± 231	86 ± 402	111 ± 527
ja	198 ± 652	451 ± 992	254 ± 900	480 ± 1238
la	409 ± 1044	178 ± 672	234 ± 701	377 ± 1028
nl	8 ± 31	21 ± 101	17 ± 129	50 ± 257
no	275 ± 744	8 ± 27	2.6 ± 3.8	87 ± 356
pl	160 ± 416	389 ± 1071	367 ± 725	612 ± 1188
pt	356 ± 915	341 ± 884	498 ± 1207	590 ± 1173
sv	118 ± 455	109 ± 534	31 ± 150	72 ± 455
	Human vs. LLM Text Corpus:			
GPT-3.5	1579 ± 1628	714 ± 1155	426 ± 876	203 ± 589
GPT-4	2523 ± 1488	1753 ± 1509	1132 ± 1331	330 ± 792
Human	1214 ± 1462	728 ± 1144	627 ± 1125	584 ± 1148
LLaMA-13B	517 ± 951	506 ± 958	665 ± 1129	526 ± 1027
LLaMA-30B	573 ± 1001	473 ± 893	667 ± 1107	498 ± 992
LLaMA-65B	558 ± 982	483 ± 911	610 ± 1089	462 ± 986
LLaMA-7B	557 ± 1009	510 ± 946	723 ± 1172	530 ± 1060

Table 7. Means and standard deviations of parameter β.

Source	Pooling Order			
	$k = 1$	$k = 3$	$k = 9$	$k = 27$
Standardized Project Gutenberg Corpus:				
ca	0.16 ± 0.11	0.090 ± 0.093	0.16 ± 0.15	0.24 ± 0.21
da	0.23 ± 0.17	0.17 ± 0.11	0.21 ± 0.16	0.23 ± 0.19
de	0.14 ± 0.15	0.14 ± 0.12	0.18 ± 0.15	0.27 ± 0.19
el	0.18 ± 0.13	0.14 ± 0.15	0.18 ± 0.17	0.30 ± 0.24
en	0.28 ± 0.15	0.17 ± 0.14	0.21 ± 0.15	0.31 ± 0.23
es	0.17 ± 0.12	0.11 ± 0.15	0.15 ± 0.15	0.23 ± 0.20
fi	0.071 ± 0.067	0.067 ± 0.065	0.12 ± 0.11	0.21 ± 0.15
fr	0.14 ± 0.14	0.13 ± 0.15	0.16 ± 0.19	0.24 ± 0.23
hu	0.086 ± 0.069	0.093 ± 0.090	0.18 ± 0.16	0.26 ± 0.25
it	0.119 ± 0.095	0.10 ± 0.13	0.13 ± 0.15	0.20 ± 0.18
ja	0.61 ± 0.45	0.55 ± 0.49	0.59 ± 0.48	0.49 ± 0.48
la	0.20 ± 0.22	0.32 ± 0.25	0.40 ± 0.30	0.52 ± 0.37
nl	0.19 ± 0.15	0.15 ± 0.13	0.19 ± 0.15	0.27 ± 0.21
no	0.30 ± 0.18	0.22 ± 0.11	0.187 ± 0.081	0.23 ± 0.18
pl	0.070 ± 0.068	0.090 ± 0.089	0.15 ± 0.17	0.24 ± 0.24
pt	0.19 ± 0.16	0.13 ± 0.16	0.18 ± 0.20	0.23 ± 0.25
sv	0.16 ± 0.12	0.17 ± 0.13	0.21 ± 0.15	0.29 ± 0.24
Human vs. LLM Text Corpus:				
GPT-3.5	0.07 ± 0.24	0.11 ± 0.30	0.18 ± 0.37	0.28 ± 0.44
GPT-4	0.02 ± 0.12	0.04 ± 0.19	0.06 ± 0.22	0.11 ± 0.30
Human	0.10 ± 0.27	0.19 ± 0.35	0.27 ± 0.41	0.32 ± 0.44
LLaMA-13B	0.24 ± 0.38	0.26 ± 0.39	0.25 ± 0.40	0.32 ± 0.45
LLaMA-30B	0.21 ± 0.36	0.24 ± 0.38	0.24 ± 0.40	0.31 ± 0.44
LLaMA-65B	0.20 ± 0.36	0.23 ± 0.38	0.26 ± 0.42	0.33 ± 0.45
LLaMA-7B	0.25 ± 0.39	0.27 ± 0.39	0.27 ± 0.41	0.30 ± 0.44

Table 8. Failure rates and goodness of fit for power-law (PL) decay and stretched exponential (SE) decay.

Pooling Order	PL Failure Rate (%)	PL Avg. SSLR	SE Failure Rate (%)	SE Avg. SSLR
Standardized Project Gutenberg Corpus:				
1	0.00	0.43 ± 0.47	14.08	2 ± 29
3	0.00	0.28 ± 0.43	29.06	0.4 ± 2.6
9	0.00	0.26 ± 0.44	20.27	0.5 ± 3.2
27	0.37	0.25 ± 0.48	17.21	0.6 ± 4.6
Human vs. LLM Text Corpus:				
1	0.00	1.4 ± 1.1	0.03	11 ± 262
3	0.00	1.3 ± 4.6	0.13	7 ± 220
9	0.00	4 ± 36	0.33	1.6 ± 7.7
27	4.20	1 ± 12	4.99	2 ± 37

3.4. Discussion

Similarly as Mikhaylovskiy and Churilov [9] but differently than Li [4], Lin and Tegmark [2], and Takahashi and Tanaka-Ishii [6,7], we have sought for the LRD on the level of words rather than on the level of characters or phonemes. We have hypothesized that word-level dependencies yield a more prominent effect due to semantic coherence of lexical units over longer distances as compared to phoneme-level correlations, which tend to decay faster, in view of the arbitrariness of word forms.

Indeed, analyzing the cosine similarity of word embeddings, like Mikhaylovskiy and Churilov [9], or their cosine correlation, in the present study, one observes a clearly visible LRD effect for natural, i.e., human-written texts. Mikhaylovskiy and Churilov [9] reported a rough power-law decay without considering an alternative model. By contrast, we have considered both a power-law model and a stretched exponential model and both natural texts and LLM-generated texts.

We report that the slow decay of the cosine correlation extends up to 1000 words for natural texts, whereas it is dominated by noise for LLM-generated texts—as it was already observed for the previous generation of language models [6,7]. These effects can be seen in the diagnostic Figure 1 and independently witnessed by Tables 3–8, where fitting to the random noise results in highly unstable estimates and outliers pumping up the standard deviations beyond the means. Curiously, the decay of the cosine correlation does not change systematically as the pooling order k increases, despite our prior expectation that the cosine correlation would increase monotonically with k.

The distributions of fitted parameters c and γ of the power-law model (32) and b, δ, and β of the stretched exponential model (33) vary significantly across different human languages and different large language models ($p < 0.01$ for the Kruskall–Wallis tests). It means that the cosine correlation decays at different source-specific rates. At the moment, we are unable to state clearly what the cause for this variation may be.

For example, the Japanese language seems an outlier in many categories but this need not be directly caused by language typology. We notice that the available texts in Japanese are very short and their coverage in terms of embeddings is much lower than for other languages. Maybe our experimental methodology fails for very short texts in general. This might be an alternative explanation of the poor fitting results for LLM-generated texts which are very short, as well, as shown in Tables 2 and 8.

4. Conclusions

In this paper, we have provided an empirical support for the claim that texts in natural language exhibit long-range dependence (LRD), understood as a slower than exponential decay of the two-point mutual information. Similar claims have been reiterated in the literature [2,4–11] but we hope that we have provided more direct and convincing evidence.

First, as a theoretical result, we have shown that the squared cosine correlation lower bounds the Shannon mutual information between two vectors. Under this bound, a power-law or a stretched exponential decay of the cosine correlation implies the LRD. In particular, the vector time series which exhibits such a slow decay of the cosine correlation cannot be not a mixing Markov or hidden Markov process by Theorem 1 of Lin and Tegmark [2].

Second, using the Standardized Project Gutenberg Corpus [26] and vector representations of words taken from the NLPL repository [23], we have shown experimentally that the estimates of the cosine correlation of word embeddings follow a stretched exponential decay. This decay extends for lags up to 1000 words without any smoothing, which is four decades of magnitude larger than the unsmoothed, presumably LRD, effect for characters [4].

Third, the stability of this decay suggests that the LRD is a fundamental property of natural language, rather than an artifact of specific preprocessing methods or statistical estimation techniques. The observation of the slow decay of the cosine correlation for natural texts not only supports the hypothesis of LRD but also reaffirms the prior results of Mikhaylovskiy and Churilov [9], who reported a rough power-law decay of the expected cosine similarity of word embeddings.

Fourth, like Takahashi and Tanaka-Ishii [6,7], we have observed the LRD only for natural data. We stress that, as we were able to observe, artificial data do not exhibit the LRD in a systematic fashion. Our source of artificial texts was the Human vs. LLM Text Corpus [27]. We admit that texts in this corpus may be too short to draw firm conclusions and further research on longer LLM-generated texts is necessary to confirm our early claim.

As we have mentioned in the introduction, non-Markovianity effects such as the LRD, Hilberg's law [15–17], and the maximal repetition law [18] may have implications for understanding the limitations and capabilities of contemporary language models. The presence of such effects in natural texts in contrast to texts generated by language models highlights the indispensability of complex memory mechanisms, potentially showing that state-of-the-art architectures such as Transformers [21] are insufficient.

Future research might explore whether novel architectures could capture quantitative linguistic constraints such as the LRD more effectively [32]. Further studies may also explore alternative embeddings or dependence measures and their impact on the stability of the LRD measures such as the stretched exponential decay parameters. Investigating other linguistic corpora, text genres, and languages could also provide valuable insights into the universality of these findings.

Author Contributions: Conceptualization, Ł.D. and P.W.; methodology, Ł.D.; software, P.W.; validation, Ł.D. and P.W.; formal analysis, Ł.D.; investigation, Ł.D. and P.W.; resources, P.W.; data curation, P.W.; writing—original draft preparation, Ł.D.; writing—review and editing, Ł.D. and P.W.; visualization, P.W.; supervision, Ł.D. All authors have read and agreed to the published version of the manuscript.

Funding: This research received no external funding.

Institutional Review Board Statement: Not applicable.

Data Availability Statement: The code and instructions to reproduce the experiments are available at https://github.com/pawel-wieczynski/long_range_dependencies (accessed on 6 June 2025). More figures are available at https://github.com/pawel-wieczynski/long_range_dependencies/tree/main/figures (accessed on 6 June 2025).

Acknowledgments: During the preparation of this manuscript, the authors used ChatGPT 4o for the purpose of drafting some passages of the text in the introduction and conclusion. The authors have reviewed and heavily edited the output and take full responsibility for the content of this publication.

Conflicts of Interest: The authors declare no conflicts of interest. The funders had no role in the design of the study; in the collection, analyses, or interpretation of data; in the writing of the manuscript; or in the decision to publish the results.

Abbreviations

The following abbreviations are used in this manuscript:

CC	cosine correlation
HLLMTC	Human vs. LLM Text Corpus
LLM	large language model
LRD	long-range dependence
SPGC	Standardized Project Gutenberg Corpus
SSLR	sum of squared logarithmic residuals
SRD	short-range dependence

References

1. Beran, J. *Statistics for Long-Memory Processes*; Chapman & Hall: New York, NY, USA, 1994.
2. Lin, H.W.; Tegmark, M. Critical Behavior in Physics and Probabilistic Formal Languages. *Entropy* **2017**, *19*, 299. [CrossRef]
3. Cover, T.M.; Thomas, J.A. *Elements of Information Theory*, 2nd ed.; Wiley & Sons: New York, NY, USA, 2006.
4. Li, W. Mutual Information Functions versus Correlation Functions. *J. Stat. Phys.* **1990**, *60*, 823–837. [CrossRef]
5. Altmann, E.G.; Pierrehumbert, J.B.; Motter, A.E. Beyond Word Frequency: Bursts, Lulls, and Scaling in the Temporal Distributions of Words. *PLoS ONE* **2009**, *4*, e7678. [CrossRef] [PubMed]

6. Takahashi, S.; Tanaka-Ishii, K. Do neural nets learn statistical laws behind natural language? *PLoS ONE* **2017**, *12*, e0189326. [CrossRef] [PubMed]
7. Takahashi, S.; Tanaka-Ishii, K. Evaluating computational language models with scaling properties of natural language. *Comput. Linguist.* **2019**, *45*, 481–513. [CrossRef]
8. Tanaka-Ishii, K. *Statistical Universals of Language: Mathematical Chance vs. Human Choice*; Springer: New York, NY, USA, 2021.
9. Mikhaylovskiy, N.; Churilov, I. Autocorrelations Decay in Texts and Applicability Limits of Language Models. *arXiv* **2023**, arXiv:2305.06615. Available online: https://arxiv.org/abs/2305.06615 (accessed on 6 June 2025).
10. Stanisz, T.; Drożdż, S.; Kwapień, J. Complex systems approach to natural language. *Phys. Rep.* **2024**, *1053*, 1–84. [CrossRef]
11. Bartnicki, K.; Drożdż, S.; Kwapień, J.; Stanisz, T. Punctuation Patterns in "Finnegans Wake" by James Joyce Are Largely Translation-Invariant. *Entropy* **2025**, *27*, 177. [CrossRef] [PubMed]
12. Chomsky, N. Three models for the description of language. *IRE Trans. Inf. Theory* **1956**, *2*, 113–124. [CrossRef]
13. Chomsky, N. *Syntactic Structures*; Mouton & Co.: The Hague, The Netherlands, 1957.
14. Chomsky, N. A Review of B. F. Skinner's Verbal Behavior. *Language* **1959**, *35*, 26–58. [CrossRef]
15. Hilberg, W. Der bekannte Grenzwert der redundanzfreien Information in Texten—Eine Fehlinterpretation der Shannonschen Experimente? *Frequenz* **1990**, *44*, 243–248. [CrossRef]
16. Takahira, R.; Tanaka-Ishii, K.; Dębowski, Ł. Entropy Rate Estimates for Natural Language—A New Extrapolation of Compressed Large-Scale Corpora. *Entropy* **2016**, *18*, 364. [CrossRef]
17. Kaplan, J.; McCandlish, S.; Henighan, T.; Brown, T.B.; Chess, B.; Child, R.; Gray, S.; Radford, A.; Wu, J.; Amodei, D. Scaling Laws for Neural Language Models. *arXiv* **2020**, arXiv:2001.08361. Available online: https://arxiv.org/abs/2001.08361 (accessed on 6 June 2025).
18. Dębowski, Ł. Maximal Repetitions in Written Texts: Finite Energy Hypothesis vs. Strong Hilberg Conjecture. *Entropy* **2015**, *17*, 5903–5919. [CrossRef]
19. Dębowski, Ł. Maximal Repetition and Zero Entropy Rate. *IEEE Trans. Inf. Theory* **2018**, *64*, 2212–2219. [CrossRef]
20. Dębowski, Ł. A Refutation of Finite-State Language Models Through Zipf's Law for Factual Knowledge. *Entropy* **2021**, *23*, 1148 [CrossRef] [PubMed]
21. Vaswani, A.; Shazeer, N.; Parmar, N.; Uszkoreit, J.; Jones, L.; Gomez, A.N.; Kaiser, L.; Polosukhin, I. Attention is All you Need. In Proceedings of the Advances in Neural Information Processing Systems 30: Annual Conference on Neural Information Processing Systems 2017, Long Beach, CA, USA, 4–9 December 2017; pp. 5998–6008.
22. Mikolov, T.; Sutskever, I.; Chen, K.; Corrado, G.S.; Dean, J. Distributed Representations of Words and Phrases and their Compositionality. In Proceedings of the Advances in Neural Information Processing Systems 26: 27th Annual Conference on Neural Information Processing Systems 2013, Lake Tahoe, NV, USA, 5–8 December 2013; pp. 3111–3119.
23. Fares, M.; Kutuzov, A.; Oepen, S.; Velldal, E. Word vectors, reuse, and replicability: Towards a community repository of large-text resources. In Proceedings of the 21st Nordic Conference on Computational Linguistics, Gothenburg, Sweden, 22–24 May 2017; pp. 271–276.
24. Pinsker, M.S. *Information and Information Stability of Random Variables and Processes*; Holden-Day: San Francisco, CA, USA, 1964.
25. Csiszár, I.; Körner, J. *Information Theory: Coding Theorems for Discrete Memoryless Systems*; Cambridge University Press: Cambridge, UK, 2011.
26. Gerlach, M.; Font-Clos, F. A Standardized Project Gutenberg Corpus for Statistical Analysis of Natural Language and Quantitative Linguistics. *Entropy* **2020**, *22*, 126. [CrossRef] [PubMed]
27. Grinberg, Z. Human vs. LLM Text Corpus. 2024. Available online: https://www.kaggle.com/datasets/starblasters8/human-vs-llm-text-corpus (accessed on 13 May 2025).
28. Bretagnolle, J.; Huber, C., Estimation des densités : Risque minimax. In *Séminaire de Probabilités XII*; Lecture Notes in Mathematics; Springer: New York, NY, USA, 1978; Volume 649, pp. 342–363.
29. Canonne, C. A short note on an inequality between KL and TV. *arXiv* **2022**, arXiv:2202.07198. Available online: https://arxiv.org/abs/2202.07198 (accessed on 6 June 2025).
30. Řehůřek, R.; Sojka, P. Software Framework for Topic Modelling with Large Corpora. In Proceedings of the LREC 2010 Workshop on New Challenges for NLP Frameworks, Valletta, Malta, 22 May 2010; pp. 45–50. Available online: http://is.muni.cz/publication/884893/en (accessed on 6 June 2025).
31. Virtanen, P.; Gommers, R.; Oliphant, T.E.; Haberland, M.; Reddy, T.; Cournapeau, D.; Burovski, E.; Peterson, P.; Weckesser, W.; Bright, J.; et al. SciPy 1.0: Fundamental Algorithms for Scientific Computing in Python. *Nat. Methods* **2020**, *17*, 261–272. [CrossRef] [PubMed]
32. Behrouz, A.; Zhong, P.; Mirrokni, V. Titans: Learning to Memorize at Test Time. *arXiv* **2025**, arXiv:2501.00663. Available online: https://arxiv.org/abs/2501.00663 (accessed on 6 June 2025).

entropy

MDPI

Article

Can Informativity Effects Be Predictability Effects in Disguise?

Vsevolod Kapatsinski

Department of Linguistics, University of Oregon, Eugene, OR 97403-1290, USA; vkapatsi@uoregon.edu

Abstract: Recent work in corpus linguistics has observed that informativity predicts articulatory reduction of a linguistic unit above and beyond the unit's predictability in the local context, i.e., the unit's probability given the current context. Informativity of a unit is the inverse of average (log-scaled) predictability and corresponds to its information content. Research in the field has interpreted effects of informativity as speakers being sensitive to the information content of a unit in deciding how much effort to put into pronouncing it or as accumulation of memories of pronunciation details in long-term memory representations. However, average predictability can improve the estimate of local predictability of a unit above and beyond the observed predictability in that context, especially when that context is rare. Therefore, informativity can contribute to explaining variance in a dependent variable like reduction above and beyond local predictability simply because informativity improves the (inherently noisy) estimate of local predictability. This paper shows how to estimate the proportion of an observed informativity effect that is likely to be artifactual, due entirely to informativity improving the estimates of predictability, via simulation. The proposed simulation approach can be used to investigate whether an effect of informativity is likely to be real, under the assumption that corpus probabilities are an unbiased estimate of probabilities driving reduction behavior, and how much of it is likely to be due to noise in predictability estimates, in any real dataset.

Keywords: informativity; predictability; reduction; sound change; phonetics; usage-based linguistics; corpus linguistics

1. Introduction

Research in corpus linguistics has found that the informativity of a word predicts its degree of articulatory reduction above and beyond the unit's predictability given the current context [1–11]. These findings have been interpreted as evidence for speakers being sensitive to a word's information content in deciding how to pronounce it and for long-term memory representations of words accumulating information about its predictability or pronunciation details or both [1–23]. The effects of informativity are remarkably consistent in that all studies that have found an effect of local predictability in a model that included informativity also found an effect of informativity, suggesting that whenever speakers are sensitive to the probability of a word given a particular context, this information also accumulates in memory in association with that word. Consequently, these effects have been argued to provide support for "rich memory" views of the mental lexicon, in which every episode of producing or perceiving a word leaves a trace [13–23].

However, the predictability of a word in a context is not independent of its average predictability across contexts, which is the inverse of informativity and is therefore perfectly correlated with it [24]. Furthermore, when a context is rare, the word's predictability given that context is unreliable and can be improved by taking the word's average predictability into account. To take an extreme example, suppose that the word *inched* occurs only once in the corpus. Then all words following it would have a probability of zero, except for the word that happened to occur after it (let us say *closer*), which would have a probability of 1. These estimates are highly unreliable, since they are based on a sample size of 1. By taking into account probabilities of words in other contexts, we might be able to estimate, for example, that *up* is a lot more probable than *higher* following *inched*. This is especially helpful for a bigram like this, where the words being predicted (*up* and *higher*) are frequent enough to frequently occur in other contexts where their predictabilities can be computed, and the present context (*inched*) is rare so predictability given the current context is unreliable. If instead we have a rare word in a frequent context that co-occurs with it, e.g., *by Kapatsinski*, then informativity will not be of much help as the predictability given the current context is reliable and informativity likely aligns with it. At the limit, if the target word only occurs in the current context, its informativity is its predictability times minus one.

In computational linguistics, this idea is behind the technique of using back-off, where the next word is predicted from a long preceding context but if this context is absent from the training data or less frequent than a predetermined cutoff, the model backs off to using a shorter context, as first proposed in [25]. Jaeger [26] has shown that the estimated effect of predictability on reduction increases when predictability estimates are improved by using back off.

In inferential statistics, this observation is behind the idea of hierarchical/mixed-effects models, which make use of adaptive partial pooling of evidence [27]. Hierarchical regression is preferrable to back-off because it does not use an arbitrary frequency cutoff above which only local context is used and below which it is not used at all. The cutoff falsely assumes that there is zero information (i.e., only noise) in contextual predictability when the context frequency is below the cutoff and that out-of-context probabilities cannot improve in-context probability when context frequency is above the cutoff (for empirical evidence that these assumptions are false, see [28,29]). Ideally, one would take into account both in-context and out-of-context probabilities at all times but optimally weigh them based on how reliable they are in each context. This is what adaptive partial pooling is designed to do. Computational linguistics approaches that have replaced n-gram models with back-off, like recurrent networks and transformers, also implicitly make use of partial pooling. We return to this point in the Discussion.

In estimating some variable in some context, partial pooling relies on observations from that context to the extent that they are available, because they are more relevant than observations from other contexts. But the fewer such observations that are available, the more the model's estimate is based on observations from other contexts. It can be shown that the estimate resulting from partial pooling is generally more accurate than an estimate that does not take other contexts into account (for simulations in the context of estimating reduction probabilities, see [28]). In corpus data, context frequency is 1 for a large proportion of contexts. For example, if context is operationalized as the preceding or following word, as in [1], context frequency would be 1 for ~50% of contexts in any corpus, as shown in [30]. Therefore, partial pooling would yield substantial benefits in estimating predictability from naturalistic language input compared to relying completely on observed predictability in a given context. If speakers estimate predictability in context, it would therefore behoove them to rely on partial pooling to do so [17].

How does partial pooling work? In hierarchical regression [27], the partial pooling estimate for some characteristic ($\hat{\alpha}$) of a word (w) in a context (cx) would be estimated by Equation (1):

$$\hat{\alpha}_{w|cx} = \frac{\frac{n_{cx}}{\sigma^2_{cx}}}{\frac{n_{cx}}{\sigma^2_{cx}} + \frac{1}{\sigma^2_w}}\alpha_{w|cx} + \frac{\frac{1}{\sigma^2_w}}{\frac{n_{cx}}{\sigma^2_{cx}} + \frac{1}{\sigma^2_w}}\alpha_w \quad (1)$$

Here, n_{cx} is the number of observations (frequency) of the context cx, σ^2_{cx} is the observed variance of the characteristic (α) across words in the context cx, and σ^2_w is the variance of average α_w across words (α_w being averaged across all contexts in which each word occurs). Equation (1) says that the estimate based on the current context (α_{cx}, here predictability) is weighted more when the context is frequent (n_{cx} is high), and words tend not to vary in α within that context. Conversely, the estimate based on other contexts (informativity) is weighted more when words (within the relevant class) tend not to differ in (α). If $\alpha_{w|cx}$ is predictability of the word given context, the variance parameters are likely to be relatively unimportant because word predictability distributions within and across contexts are (approximately) scale-free with an exponent of ~1, i.e., Zipfian [30,31], so σ^2_w and σ^2_{cx} tend not to vary dramatically, while n_{cx}, which also follows a scale-free Zipfian distribution, varies enormously across contexts. Therefore, the main influence on the importance of observed local predictability ($\alpha_{w|cx}$) vs. observed informativity (α_w) in estimating local predictability ($\hat{\alpha}_{w|cx}$) given a context should be the frequency of that context.

The present paper demonstrates that, as expected from the reasoning above, the apparent effect of informativity is larger in rare contexts than in frequent ones, while the effect of local, contextual predictability is smaller in rare contexts. This follows directly from the fact that contextual predictability estimates are less reliable in rare contexts and that informativity can help improve these estimates via partial pooling.

Previous simulation work on this question by Cohen Priva and Jaeger [24] has investigated whether the *segmental* informativity effect observed in [32] could be spurious, concluding that it is highly unlikely to be. However, this work did not try to simulate the effects of predictability and informativity on an observable dependent variable like duration, instead asking how correlated informativity is with predictability and vice versa as a function of sample size. This does not allow one to determine how large spurious effects of the two correlated variables on a dependent variable would be. Because the dependent variable and the sizes of the effects of informativity and predictability were not simulated, [24] also does not allow a researcher to use the simulation approach to infer whether an effect of informativity they have observed is attributable to predictability.

The present paper extends this simulation approach by modeling the dependent variable. In addition, it extends it by modeling lexical rather than segmental predictability and informativity. In [24], Cohen Priva and Jaeger mention conducting simulations of informativity and predictability at the lexical level in the discussion section but do not describe the results beyond stating that the "results of those additional studies were similar, except that the problems observed here were observed more strongly and even for larger lexical databases when words, instead of segments, were analyzed. This is presumably due to the overall much larger number of word types, compared to segment types. For example, spurious frequency effects were found when predicting word predictability or word informativity, even for lexical databases with 10,000,000 words" (p.11). As noted in this quote, sampling noise is far less of a concern on the segmental level than on the lexical level, because even the rarest segments like [ð] or [ʒ] in English are very frequent relative to the average word. At the same time, most studies that have argued for an effect of informativity have been conducted at the lexical level [1–11]. Given that spurious effects are more likely at the lexical level, it is important to have a methodology to determine whether any particular lexical informativity effect might be a predictability effect in disguise.

We show, by simulation, that a significant effect of informativity and its greater magnitude in rare contexts could sometimes be due to sampling noise in the predictability estimates. Furthermore, we provide a simple simulation approach to estimate *how much* of an observed effect of informativity is spurious, i.e., due to sampling noise. For our actual example, much of the effect of informativity is not due to sampling noise in predictability estimates. That is, although corpus data could show a significant effect of informativity in the same direction even if speakers showed no real sensitivity to informativity, our simulations show that the magnitude of that spurious effect would rarely be as large as the effect in the real data.

2. Materials and Methods

We begin by demonstrating that predictability and informativity behave as one would expect if (some of) the effect of informativity were due to partial pooling, i.e., informativity helping reduce noise in predictability estimates. Because there is more noise in smaller samples, this involves showing that these predictors interact with context frequency, such that the effect of predictability is stronger in frequent contexts and the effect of informativity is stronger in rare ones. Previous work in [26] has shown that the effect of predictability is stronger in frequent contexts but has not examined whether the effect of informativity interacts with context frequency.

The data we model come from the Switchboard Corpus [33], using the revised word-to-speech alignments from [34]. To ensure that the results generalize across the informativity and predictability continua, I selected two datasets that are opposite of each other in terms of informativity and predictability. One dataset (n = 5146) consists of the words that followed repetition disfluencies in [35], which are mostly nouns. These words are relatively high in informativity and low in predictability, as these factors predict the occurrence of a disfluency before a word [35]. The other dataset (n = 5187), helpfully provided by T. F. Jaeger, consists of determiners (*the, a, an, that, this, these, those, no, every, some, each, another, both, any, all, either, one, many, neither*, or *half*) in prepositional phrases, extracted from the Penn Treebank II Switchboard corpus. None of the determiners are adjacent to disfluencies or pauses. These words are relatively low in informativity and high in predictability. In both datasets, context was operationalized as the preceding word in the utterance, so predictability is $\log(p(w_n|w_{n-1}))$, i.e., forward bigram surprisal with a flipped sign, and informativity is surprisal averaged across contexts.

For the post-disfluency dataset, LSTM [36] probabilities are also available, as described in [35]. However, bigram surprisal is preferrable to surprisal from neural models like [36] for the present purpose of comparing predictability and informativity because part of the power of neural models is that they implicitly implement adaptive partial pooling, relying largely on the specific local context when it is familiar and generalizing over similar contexts when the context is new [37]. Therefore, surprisal/predictability estimates from neural models are influenced by informativity through partial pooling to an unknown extent, since we do not have the equivalent of Equation (1) for neural models. Furthermore, the bigram operationalization makes it easier to determine which two contexts are "the same" for the purposes of calculating context frequency. That is, words that end in the same word are "the same context" for the present study.

Both datasets contain word tokens that are relatively controlled for other factors that influence durations. For the post-disfluency dataset, they are mostly nouns following high-frequency function words (prepositions or articles) [35,38], and surrounding words outside of the disfluency itself have little influence on their durations [38]. This set of words is also of theoretical interest because there is no previous data investigating whether the durations of words following a disfluency are influenced by predictability, and one might expect the interruption in the flow of speech constituting the disfluency to reset the context for speech planning so that predictability given the word preceding the interruption would not influence duration of the word following the interruption. Below, we show that this is not the case. The determiner dataset is controlled for local syntactic and prosodic context: all determiners are in prepositional phrases, and the dataset contains annotation for prosodic word boundaries. All tokens with adjacent boundaries were excluded to make the dataset maximally different from the other.

The two maximally different datasets are intended to illustrate that the choice of the dataset matters little, as nothing in the argument hinges on the choice of the dataset. The predictions follow simply from the mathematical relationship between predictability and informativity, which ensures that the interactions with context will arise as a side effect of the Law of Large Numbers. Thus, these interactions should always be observed given enough power. The simulations below make no assumptions about what the data distribution looks like and bootstrap it from whatever data the researcher is hoping to simulate. For example, this methodology has also been applied to word durations in a Polish corpus, based on guidance from the present author serving as a reviewer (see [11]). Furthermore, the interaction between predictability and context frequency was also seen in [26].

We fit a regression model, predicting word duration (log scaled to reduce skew, as is standard in the literature) from predictability, informativity and their interactions with context. We fit a simple least squares linear regression model (Table 1), using lm() in base R [39], and a mixed-effects lmer() model using the lme4 package [40], version 1.1-35.5, with random intercepts for target words whose durations we model. The random intercepts allow us to control for uncontrolled variables that vary across words along with informativity and thus might spuriously strengthen or weaken the informativity effect independently of its relationship with predictability. I also tried adding a random slope for predictability by word. This model did not converge in the post-disfluency dataset because 64% of the words in the sample occur only once and thus cannot be used to test the random slope. Therefore, the random-effects structure for that dataset consists only of the random intercept for word. The random-effects structure for the determiners model in Table 2 is maximal: it also includes a random slope for predictability and the correlation between the two. p values for linear mixed-effects models were generated using Satterthwaite's method for approximating degrees of freedom for t tests in mixed-effects models [41]. Note that, because the random-effects structure is more complex for determiners, there are more degrees of freedom in the post-disfluency model for the same n. Therefore, the same t value corresponds to a lower p value in the post-disfluency model compared to the determiner model. For the models with interactions in Table 3, the most complex random-effects structure that converged for both datasets included only random intercepts, thus random slopes were not included. Marginal and conditional R^2 values for the mixed-effects models were calculated using the r.squared_GLMM() function in the MuMIn package [42].

To simulate how much of the effect of informativity might be due to predictability, we repeatedly generate data samples in which predictability has the bivariate relationship with duration observed in the real data while informativity has no effect, i.e., the variance not accounted for by predictability is random observation-level noise (if the data are generated from a fixed-effects-only model) or observation-level noise plus the effects of lexical identity (estimated from the observed real data by the mixed-effects model with predictability as a fixed effect and word identity as a random intercept). The true magnitude of the predictability effect is constant across contexts, regardless of context frequency. We then fit the same model we fit to the real data, with predictability and informativity as predictors, with or without context frequency and interactions with it. We ask: (1) how often do we get a "false alarm" on informativity: how often do we observe a significant effect of informativity (at $p < 0.05$) in the same direction as in the real data, even though the data are generated by a process in which there is no true effect of informativity, and (2) how often is that effect at least as large as the effect in the real data.

Specifically, the probability of each word given each context, $p(w_n|w_{n-1})$, feeds into a binomial distribution with the generated context frequencies serving as the number of "trials" in each context. For each "trial", i.e., token of the preceding word, a biased coin is tossed, which comes up heads with the probability given by w_n's probability given the preceding word in the actual sample, $p(w_n|w_{n-1})$. I have also tried varying context frequencies across simulations by sampling context tokens from a multinomial distribution fit to the probabilities in the sample, but this does not appear to change the likelihood of observing a spurious effect of informativity compared to simply using the observed frequencies, thus the simpler approach is taken here.

The binomial sampling process inevitably produces sampling noise. For example, if you toss a coin that has a 50% probability of coming up heads 5 times, the coin will not come up heads on exactly 2.5 occasions and sometimes will come up heads in all 5 trials 3% of the time. The effect of this sampling noise is what adding informativity to the regression model can alleviate. Furthermore, the fewer times you toss the coin, i.e., the lower the context frequency, the more variable probabilities across coin tosses will be given that the underlying probability of the coin coming up heads is kept the same. For example, if you toss the coin once, the coin will generate a probability of 0 or 1 in each sample regardless of underlying probability. If you toss it 5000 times, the average will be close to the underlying probability. This is why we expect predictability effects to be stronger in frequent contexts and why we expect the spurious component of the informativity effect to be largest in rare contexts: in a rare context, estimated predictability is on average farther from the true predictability that influenced duration, and therefore more of the effect of true predictability remains to be captured by informativity.

Predictability of a word given context times the coefficient for predictability in the model without informativity plus the random effect of a word generates the mean of a normal distribution of log duration for that word in that context, and the standard deviation of the residuals of the model serves as the standard deviation of the normal. The simulated duration of each word token is then generated by sampling from this distribution.

The post-disfluency dataset, analysis and simulation code are available on OSF: https://osf.io/c8aws, accessed on 4 July 2025. I do not have permission to redistribute the determiner dataset.

3. Results

3.1. Predictability, Informativity and Context Frequency

Tables 1 and 2 present the linear regression model results. Table 1 presents an analysis in which predictability and informativity jointly predict duration, as in previous empirical work in [1–11]. The two predictors together explain 42.8% of variance in duration of post-disfluency words and 48.2% of variance in duration of determiners in prepositional phrases (based on adjusted R^2). In both datasets, the two predictors are collinear but much more so in the post-disfluency dataset ($r = -0.91$ for the post-disfluency dataset vs. $r = -0.65$ for the determiner dataset). This difference in correlation strength follows from the greater frequencies of determiners, which will often be observed outside of the current context, on which predictability is based. If a word is only observed in the current context, which is true for many post-disfluency words but not true for determiners, predictability is necessarily informativity times minus 1. In both datasets, informativity explains more variance than predictability, but especially so in the determiner dataset: removing informativity from the model in Table 1 reduced adjusted R^2 by 3% in the post-disfluency dataset and by 23% in the determiner dataset, while removing predictability reduces it by <1% in both datasets.

Table 1. Linear regression results predicting log word duration (log seconds) from predictability and informativity. Adjusted $R^2 = 0.439$ for post-disfluency words and = 0.482 for determiners.

Post-Disfluency Words	***b***	***se* (*b*)**	***t***	***p***
Intercept	−2.27	0.017	−132.98	<0.00001
Predictability	−0.05	0.007	−6.89	<0.00001
Informativity	0.13	0.007	18.83	<0.00001
Determiners in PPs	***b***	***se* (*b*)**	***t***	***p***
Intercept	−0.05	0.002	−18.91	<0.00001
Predictability	−0.004	0.001	−6.41	<0.00001
Informativity	0.05	0.001	46.47	<0.00001

Adding the random effects to the models without interactions somewhat reduces the informativity and predictability effect magnitudes, but the qualitative pattern of results remains unchanged (Table 2): both effects are significant and the informativity effect is stronger than the predictability effect.

Table 2. Linear mixed-effects regression results predicting log word duration (log seconds) from predictability and informativity, as in previous work. Marginal $R^2 = 0.27$ for post-disfluency words and = 0.31 for determiners. Conditional $R^2 = 0.47$ for post-disfluency words and 0.61 for determiners.

Post-Disfluency Words	***b***	***se* (*b*)**	***t***	***p***
Intercept	−1.81	0.03	−132.98	<0.00001
Predictability	−0.02	0.005	−4.13	=0.00004
Informativity	0.09	0.006	13.90	<0.00001
Determiners in PPs	***b***	***se* (*b*)**	***t***	***p***
Intercept	−0.02	0.038	−0.59	<0.00001
Predictability	−0.003	0.001	−4.20	=0.0123
Informativity	0.04	0.007	5.31	=0.00003

The model in Table 3 adds interactions with context frequency predicted by partial pooling to the model in Table 2 (a model omitting random effects produced qualitatively identical results). Figures 1 and 2 plot these interactions for the two datasets. Figure 1 shows the effects on durations of post-disfluency words, while Figure 2 shows the effects on durations of determiners in prepositional phrases.

Despite these word types being on the opposite ends of the informativity, predictability, frequency and duration distributions and the difference in context fluency, the interactions of predictability and informativity with context frequency are almost identical. When the context is new/observed only once in the corpus, there is no effect of predictability given the context (context frequency = 1; left panels), as seen from all the lines being on top of each other, and a strong effect of informativity (steep line slope). When the context is very frequent (right panels), there is a strong effect of predictability given that context (the lines are well separated) and a much reduced if not zero effect of informativity (line slopes are shallow and the confidence regions for determiners include a slope of 0 corresponding to no effect of informativity).

Table 3. Linear mixed-effects regression results predicting log word duration (log seconds) from predictability, informativity and their interactions with context (preceding word) frequency. Adjusted $R^2 = 0.448$.

Post-Disfluency Words	*b*	*se* (*b*)	*t*	*df*	*p*
Intercept	−1.91	0.043	−44.53	2577.2	<0.00001
Context Frequency	0.0004	0.00002	2.20	5033.6	0.028
Predictability	0.004	0.011	0.36	4434.3	0.72
Informativity	0.136	0.013	10.59	5076.0	<0.00001
Predictability:Context Frequency	−0.0002	0.0001	−2.63	4698.3	0.00864
Informativity:Context Frequency	−0.0003	0.0001	−4.02	4890.4	0.00006
Determiners in PPs	*b*	*se* (*b*)	*t*	*df*	*p*
Intercept	−6.92	0.039	−177.12	18.9	<0.00001
Context Frequency	0.0001	0.0001	2.08	4802.5	0.038
Predictability	−0.0006	0.001	−0.59	4793.8	0.56
Informativity	0.042	0.008	5.59	21.55	0.00001
Predictability:Context Frequency	−0.0001	0.00001	−5.39	4794.5	<0.00001
Informativity:Context Frequency	−0.0001	0.00002	−5.46	4802.8	<0.00001

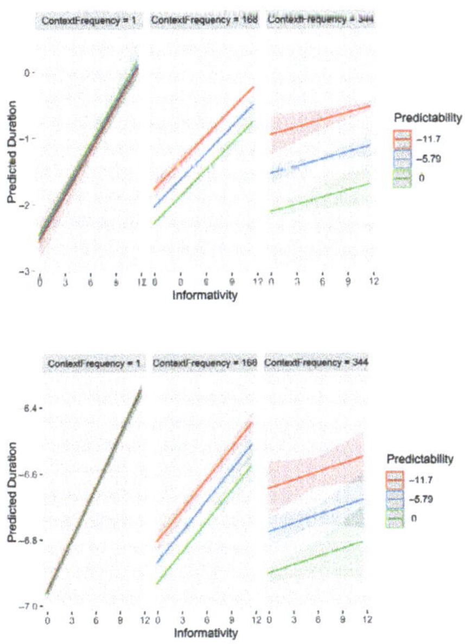

Figure 1. Fixed-effects-only model predictions showing interactions of predictability (x axis) and informativity (lines) with the square root of context frequency (panel). Top row: Post-disfluency words. Bottom row: Determiners in prepositional phrases. The square root is chosen because it scales linearly with standard error and therefore should be a good measure of the amount of information about predictability in each context. The three lines and three panels show the effect for the minimum, mean and maximum (in the dataset) value of each predictor in the Post-disfluency dataset. Since the interactions are linear, the lines for the intervening values would fall between the lines shown. Figure generated using the plot_model() function (type = "emm") from the sjPlot package [43].

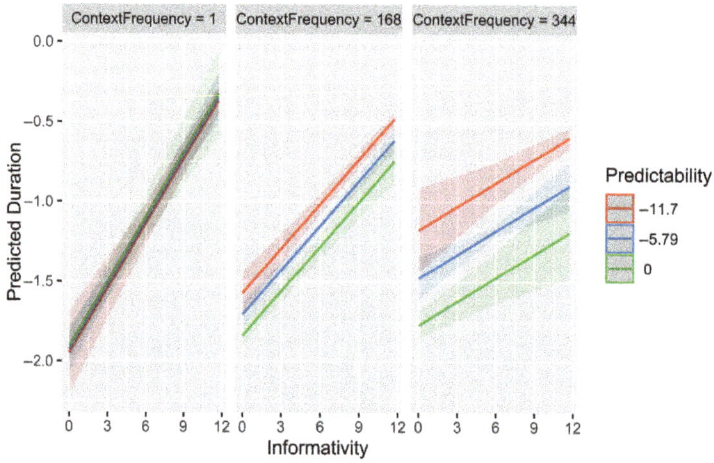

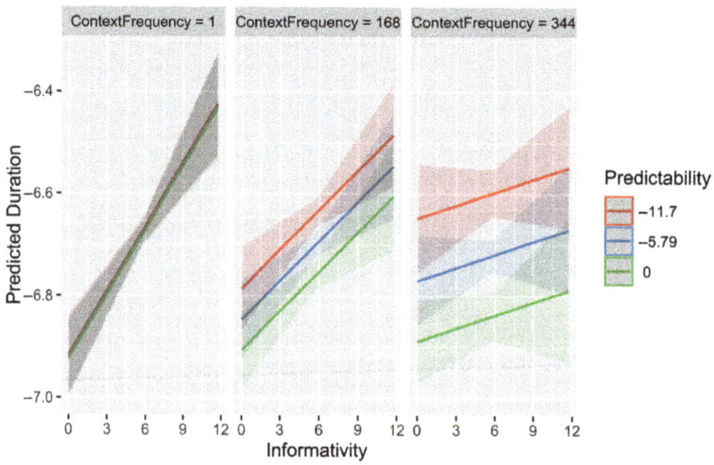

Figure 2. Mixed-effects model predictions showing interactions of predictability (x axis) and informativity (lines) with the square root of context frequency (panel). Top row: Post-disfluency words. Bottom row: Determiners in prepositional phrases. The three lines and three panels show the effect for the minimum, mean and maximum (in the dataset) value of each predictor in the Post-disfluency dataset.

The results in Figures 1 and 2 are exactly as we would expect from adaptive partial pooling: predictability has the largest effect in frequent contexts, where it can be reliably estimated (see also [26]). In turn, informativity has the largest effect in rare contexts, where predictability estimates are least reliable. This is expected either because 1) our/researchers' estimates of predictability are off in rare contexts, and are more accurate in frequent contexts, or 2) because speakers' pronunciations are driven by predictability in the current context (or in-context memories of pronunciations) to the extent that the predictability estimate is reliable (or in-context pronunciation tokens are numerous). To put it another way, how much of the effect of informativity and its interaction with context frequency is an artifact of sampling noise in the data and how much is the result of the speakers' adaptive behavior?

3.2. *Simulating What Would Happen if There Were No Informativity Effect*

As described in the Materials and Methods, we created 1000 simulated datasets in which there was no real effect of informativity, and no real correlation between the effect of predictability and context frequency, but the main effect of predictability, the distribution of context frequencies and the amount of noise were the same as in the real data. Predictability was generated by sampling from a binomial distribution, i.e., for each context, flipping a coin with probability corresponding to true predictability n times, and then noting the resulting sample predictability value.

For these simulations, I report the results for analyzing the data with both a mixed-effects model and a fixed-effects-only model because both analysis methods are common in the literature, and there is some discussion on whether adding a random effect of item makes sense for corpus data in which most words are observed only once (as is the case in the post-disfluency sample). Adding the interactions with context frequency reduces the false alarm rate (i.e., the likelihood of detecting a spurious but significant effect of informativity), as does adding a random intercept by word. However, as discussed below, the likelihood of a false alarm on informativity can nonetheless be high depending on the true size of the predictability effect: the larger the effect of predictability, the greater the likelihood of detecting a spurious effect of informativity.

Figure 3 shows that true predictability and sample predictability are highly correlated, with $r > 0.99$.

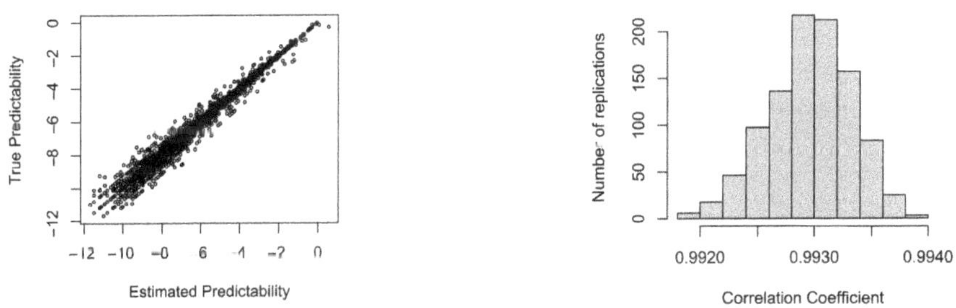

(**a**) (**b**)

Figure 3. (**a**) True predictability vs. estimated predictability based on the sample generated by binomial sampling with probability proportional to true predictability. (**b**) A histogram of the correlation coefficients between true predictability and estimated predictability across the 10,000 replications.

Figure 4 shows data generated from a fixed-effects-only model with only a main effect of predictability. The data are generated from true predictability × b(Pred) + noise in Table 1, with noise matching the standard deviation of the residuals, but estimated predictability and estimated informativity are then used as predictors in the analysis. Despite the strong correlation, the divergence between true predictability and sample predictability is sufficient for informativity to have a significant (but spurious) effect at the 0.05 alpha level more than 9% of the time when probability and informativity are estimated from the sample (Figure 4a). Furthermore, this spurious effect of informativity is usually in the same, theoretically motivated, direction as in the real data (73% of the coefficients in Figure 4a are positive and 91% of significant coefficients are). Of course, if this test were perfectly calibrated, then it should have come out significant by chance 5% of the time and be positive or negative about equally often. This is what we observe in Figure 4b, where the (unobtainable in practice) true predictability and informativity values are used as predictors instead of their values in the simulated data sample. Thus, the small amount of sampling noise seen in Figure 3 is enough to double the false alarm rate, if we simply accept that an observed effect of informativity is a true effect as long as it is significant in a regression model that also includes predictability.

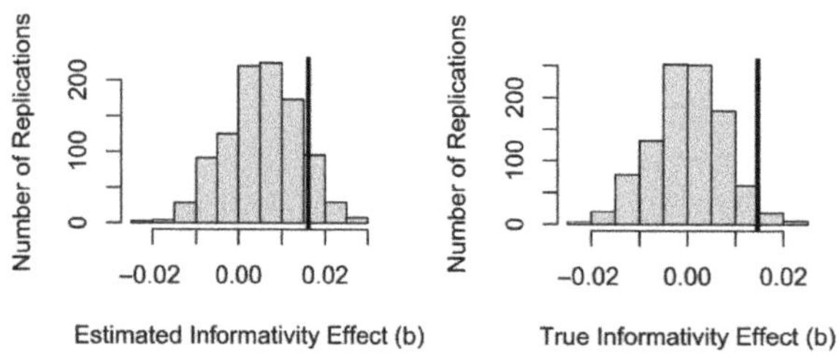

(a) (b)

Figure 4. The distribution of informativity coefficients across 1000 replications of sampling from a population in which there is no real effect of informativity. In (**a**), predictability and informativity used to predict durations are estimated from the observed sample. In (**b**), true predictability and informativity are used. Coefficients to the right of the thick vertical lines are significant at the 0.05 level in the expected direction.

However, Figure 5 shows that the observed size of the informativity coefficient, shown by the vertical dashed line, is much larger than the coefficients obtained in the simulated samples, where there is no true effect of informativity. Thus, although a significant effect of informativity could be spurious, an effect of the size we observed is unlikely to be due to sampling noise alone. Since we made 1000 simulations, $p < 1/1000$ or 0.001.

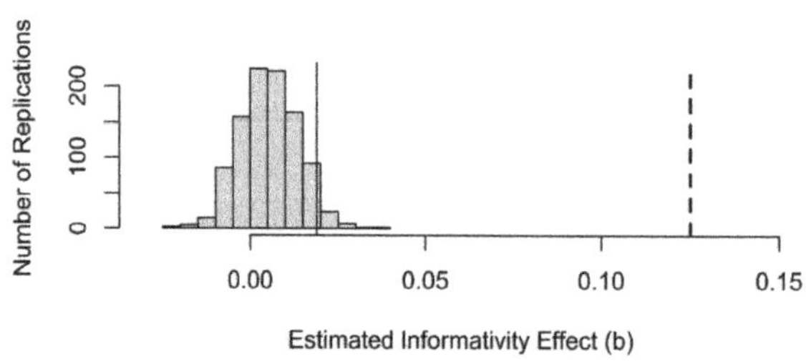

Figure 5. The distribution of informativity coefficients from Figure 4a relative to the informativity coefficient in the real data, shown by the dashed vertical line. The data-generating process behind Figure 3a has a 5% chance of generating a coefficient for informativity to the right of the thin solid vertical line.

A mixed-effects model with a main effect of predictability and a random effect of word will attribute some of the variance attributed to informativity in Table 2 to the random effect of word instead. This is the part of the informativity effect that is attributable to uncontrolled differences between words. Therefore, when the data to be fit are generated from a mixed-effects model instead of the fixed-effects-only model in Figure 5, there will be an elevated chance of finding a spurious effect of informativity in these data even if "true informativity" is used as a predictor, i.e., the equivalent of Figure 4b would still show a >2.5% chance of informativity effect in the expected direction. However, this combined effect should still be smaller than the effect of informativity observed in Table 2.

Figure 6 demonstrates that the effect of informativity in the mixed-effects model is closer to the distribution of simulated informativity effects and that the simulated informativity coefficients are always positive (informative words being longer). Indeed, about a third of the informativity effect in the data is attributable to the random effect of word identity and the correlation of informativity with predictability. However, the coefficient in the real data is still larger than in the simulated data, suggesting that some of the effect of informativity in Table 3 is real. The generating model can of course be made more complex, and realistic, by including additional predictors beyond the random intercept of word, as in [1–11], which may change this conclusion.

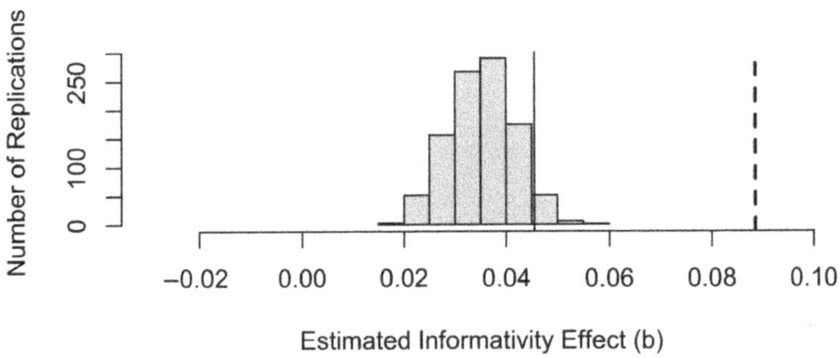

Figure 6. The distribution of informativity coefficients generated from a mixed-effects model in which there is no effect of informativity but there is the same coefficient for predictability and the same random effect of word as in the real data. The data-generating process has a 5% chance of generating a coefficient for informativity to the right of the thin solid vertical line. The dashed vertical line shows the informativity coefficient from the model of real data in Table 2.

The size and likelihood of a false detection of the informativity effect depend on the size of the predictability effect, and the model structure, as shown in Table 4. The likelihood of a false positive on informativity increases as the effect of predictability grows. Informativity captures some of the effect of true predictability, which is not captured by estimated predictability because of the sampling noise causing true and estimated predictability to diverge. The larger the effect of true predictability, the more variance in duration is left over for informativity to capture, and therefore the greater the likelihood that the spurious effect of informativity will be large enough to be significant. Unsurprisingly, the likelihood of a false alarm decreases when interactions with context frequency or the random effect of word are included.

Table 4. Effect sizes and false alarm rates for a spurious effect of informativity (probability of detecting a significant informativity effect at the 5% level in the expected direction when its true effect is zero). The proportions parentheses would be 2.5% (5%/2) if the probability of a false alarm were truly 5%. True model consists of a main effect of predictability (Pred) and a random intercept for word.

Size of the True Predictability Effect	Model	Mean Spurious Effect of Informativity (Probability of a Spurious but Significant Result)
−0.04	Pred + Info	0.05 (100%)
	(Pred + Info)*CxFreq	0.084 (85%)
	Pred + Info + (1 | Word)	0.024 (96%)
	(Pred + Info)*CxFreq + (1 | Word)	0.036 (82%)
−0.08 (as in Table 2)	Pred + Info	0.10 (100%)
	(Pred + Info)*CxFreq	0.17 (100%)
	Pred + Info + (1 | Word)	0.035 (99.8%)
	(Pred + Info)*CxFreq + (1 | Word)	0.05 (96%)

3.3. Interactions with Context Frequency

So far, we have examined whether the main effect of informativity in Tables 1 and 2 could be spurious. What of the predictability:context.frequency and informativity:context.frequency interactions in Table 3? How likely are they to be spurious, and to what extent are these interactions due to partial pooling helping address sampling noise in predictability? For simplicity, all simulations in this section assume that there is a real random effect of word in the data, because this is the standard assumption in the literature. However, we vary whether the model applied to the dataset includes the random effect of word (a random intercept).

Table 5 shows that, when the random effect of word is in the model, the likelihood of a false alarm on the interactions with context frequency decreases when there is a true main effect of informativity. This is not the case if the random intercept for words is omitted. When the random effect of word is (falsely) omitted, spurious interactions between informativity and context frequency are almost always detected when there is a main effect of informativity.

Table 5. The probability of false alarms in fixed-effects-only and mixed-effects models that include interactions with context frequency, on the effect of informativity and the interactions with context. The dataset has a true effect of predictability (varying in size as shown) and a true random intercept of word (of a constant size across the rows) but no true effect of informativity and no true interactions with context frequency. Nominal false alarm level = 0.025.

Model	*b* (Pred)	*b* (Info)	False Alarm or Power on Info	False Alarm on Pred:CxFreq	False Alarm on Info:CxFreqs
Fixed	−0.02	0	0.94	0.06	0.32
	−0.08	0	1	0.50	1
	−0.02	0.044	1	0.14	0.99
	−0.08	0.044	1	0.36	1
Mixed	−0.02	0	0.50	0.05	0.11
	−0.08	0	0.96	0.07	0.21
	−0.02	0.044	1	0.023	0.08
	−0.08	0.044	1	0.015	0.10

The likelihood of a spurious interaction between predictability and context frequency increases with increasing size of the predictability effect: the larger the main effect of predictability, the larger the difference in apparent effect size between frequent and rare contexts. The random effect of word effectively keeps this false alarm rate on interactions between predictability and context frequency under control. The presence of a true informativity effect reduces the likelihood of finding a spurious interaction between predictability and context frequency.

Overall, however, we can note that there is a high risk of an interaction between context frequency and informativity such that the informativity effect increases in rare contexts being spurious, due simply to noise in predictability estimates. All numbers in the right column are above the nominal level of 0.025.

Since both the spurious effect of informativity and its interaction with context frequency are attributable to partial pooling, the two effects should show up together in our simulations. This is shown in Table 6 for the mixed-effects model: when we detect a spurious informativity effect, we find that it gets smaller in frequent contexts 89–99% of the time, compared to the chance level of 50%. When a significant spurious interaction in this direction is detected (the informativity effect reduces in frequent contexts), a spurious simple effect is always detected as well. Thus, a spurious interaction with context frequency goes together with a spurious main effect of informativity. When informativity simply helps explain the variance that is really due to predictability the effect of informativity will almost always decrease in frequent contexts.

Table 6. The probability of detecting a spurious but significant effect of informativity in the expected direction (second column), the probability of the interaction between informativity and context frequency being in the expected direction given that a significant spurious main effect of informativity is detected (third column) and the probability of a main effect of informativity being in the expected direction given that the interaction is falsely detected (fourth column) as a function of the size of the true predictability effect.

b (Pred)	Probability of a False Alarm on Info	Probability of Info:CxFreq in the Expected Direction Given a False Alarm on Info	Probability of Info in the Expected Direction Given a False Alarm on Info:CxFreq
−0.02	0.50	0.99	1
−0.08	0.96	0.89	1

However, as shown in Figure 7, the simple effect of informativity is stronger than expected if all of the variance shared between informativity and predictability were attributable to predictability. Thus, again some of the informativity effect appears to not be spurious. For the interactions, there is more uncertainty that they are not due simply to predictability estimates being noisier in rare contexts. The interactions are significant at the 0.05 level (as shown by the fact the dashed line is farther from the middle of the distribution than the solid line in the middle and right panels). However, the interactions are not significant at the 0.001 level, even though $p < 0.0001$ in Table 3. This is because a spurious effect of informativity in this direction tends to come with interactions in this direction. That is, when the data are generated by a constant predictability effect that is independent of context frequency, a spurious informativity effect will appear to increase in rare contexts while the predictability effect will appear to decrease in rare contexts. The simulations in Figure 7 allow one to estimate whether this is a plausible explanation for the main effect of informativity and the interactions between predictability or informativity with context frequency observed in one's dataset.

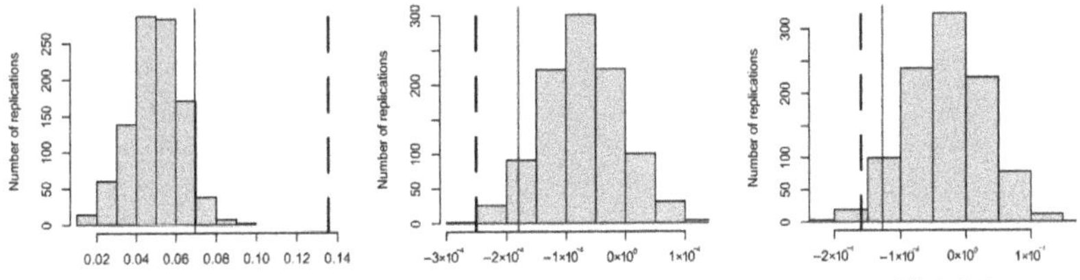

Figure 7. Fitting the full model in Table 3 to data generated without any true effect of informativity and no speaker/corpus probability mismatch. Dashed lines show observed coefficients from Table 3. Thin solid lines show the critical value of the coefficient in the same direction. Simple effect of informativity (**left**), and interactions of context frequency with informativity (**middle**) and predictability (**right**).

4. Discussion

Informative words have been repeatedly observed to be longer than less informative words. Furthermore, the effect of informativity has been observed to hold even if predictability given the local context is included in the regression model of word duration [1–11]. The present paper pointed out that informativity could improve local predictability estimates because of sampling noise in the latter, especially in rare contexts. This means that one would expect some effect of informativity to come out in a regression model that includes predictability even if reduction were driven by predictability alone. The same reasoning predicts that informativity effects should be stronger in rare contexts where the context-specific predictability estimates are noisiest. We showed this to be the case in two datasets at the opposite ends of the predictability, informativity and duration continua.

We then examined by simulation whether an effect of informativity observed in a real dataset is small enough to have emerged from sampling noise, in a world where differences in word duration were entirely due to predictability in the local context. We have shown that there is bias in informativity coefficient estimates: a spurious effect of informativity does come out significant and in the expected direction (showing an effect in the opposite direction of the reductive effect of predictability) more often than the p value in a regression model that includes both predictability and informativity would lead us to expect. This is also true for the interactions between context frequency and either predictability or informativity. However, the effect of informativity is smaller than the one observed in the real data. Therefore, we can conclude that the effect of informativity in the real data is unlikely to be entirely attributable to sampling noise in predictability estimates.

We showed that predictability and informativity interact with context frequency as one would expect from the predictability estimates being less reliable when they are conditioned on rare contexts. The interaction of predictability and context frequency has previously been observed in [26], whereas the interaction of informativity with context frequency is a new finding. This finding is predicted by the idea of informativity effects being due to adaptive partial pooling of probabilities across contexts. Our simulations show that the interaction of predictability with context frequency observed in the real data might be attributable to this effect of the Law of Large Numbers (i.e., $p > 0.05$). However, some of the interaction of informativity with context frequency may not be attributable to there being more noise in predictability estimates for informativity to compensate for in rare contexts. So, overall, not all of the effect of informativity is attributable to predictability, and the effect of informativity may genuinely be stronger in rare contexts.

This result extends prior simulation work in [24] that argued that the segmental informativity effect observed in [32] (e.g., [t] is less informative and more likely to be deleted than [k] in English) could not be attributed to noise in local predictability estimates. Like [24], we observe that most of the effect of informativity in the real data is not entirely due to noise in predictability estimates. However, this result is more surprising on the lexical level examined here than on the segmental level examined in [24], because (as pointed out there) segments are abundantly frequent and few in number, and therefore sampling noise in predictability is rather low at the segmental level. In contrast, word frequency distributions are Zipfian, with half the words occurring only once in the corpus, resulting in many contexts where estimating local predictability is virtually impossible. This concern was previously acknowledged in [24], which reported also running simulations at the lexical level, but this is the first time lexical simulation results are reported in the literature.

Unlike [24], which simulated how strongly true and estimated predictability and informativity values correlate depending on sample size, we took the approach of simulating the dependent variable, duration. This is likely because [24] focused on early studies that included only predictability or only informativity in their models, whereas we focus on the studies that included both predictors in the same model [1–11]. In the context of a model that includes both predictability and informativity, as in [1–11], a strong correlation between informativity and predictability is not a guarantee that an effect of predictability would often be falsely attributed to informativity. For example, in the post-disfluency dataset, informativity and predictability correlate at $|r| > 0.91$. Nonetheless, a true effect of predictability is seldom falsely attributed to informativity in its entirety—predictability is almost always significant in our simulations: despite the strong correlation, the sample size is large enough to detect a real effect of predictability with informativity in the model.

The advantage of simulating duration is also that we can determine how large an effect of informativity is expected to be if it were entirely spurious, and predictability had the same bivariate relationship with duration as in the real data. This is important because the likelihood of observing a spurious informativity effect and its size depend strongly on the size of the predictability effect on the dependent variable. Furthermore, by simulating the dependent variable we are able to estimate whether all of the observed informativity effect is likely to be attributable to noise in the predictability estimates.

Of course, showing that an informativity effect cannot be explained away by local predictability or by random differences between words leaves open the possibility that other stable characteristics of words that do not change across contexts and correlate with informativity would explain it away. We do not consider these alternatives here because such alternative explanations are adequately addressed by standard multiple regression methods, as in [1–11]. However, the present simulations show that some of the effect of informativity in these previous studies could be due to noise in predictability estimates. This raises the possibility that the remainder of the informativity effect might still be explained away by other stable, context-independent but word-specific factors (e.g., [44]). The methodology proposed here can incorporate such additional predictors, by creating simulated data in which all predictors but informativity have the effects they do in the real data, fitting a model with all predictors, including informativity, and testing how often we obtain an informativity effect of the observed magnitude and direction. A central point of this paper is that this remains an important direction for future work: the collinearity between informativity and predictability means that the effect of informativity is on looser ground than it appears to be.

The simulations above assume that corpus frequencies and probabilities are unbiased estimators of the probabilities (or activations, in a neural network framework) that drive reductions. While this is the standard simplifying assumption of all corpus studies, it may not be justified; e.g., [45]. For example, words of zero frequency in a large corpus nonetheless vary in the probability of a speaker knowing each word, and these probability differences affect lexical processing variables like reaction time in lexical decision [46] and likely also affect production measures like duration. The present simulations assume that words of frequency zero in the corpus have a zero probability of occurrence, which is of course not true for many such words. It does not appear to be a dangerous assumption because we do not have duration measurements for such words in any case so they would not be able to influence our estimate of the relationships between duration, predictability and informativity. However, the general possibility that the mismatch between predictability in a corpus and predictability in a speaker's experience might distort corpus predictability estimates for rare items and in rare contexts beyond simple sampling noise means that we might be underestimating noise in predictability estimates in such situations. If so, we may need to be particularly cautious in arguing that informativity effects found with such materials cannot be due to predictability.

However, let us assume that some of an effect of informativity is genuine. How then should we interpret it? Here, the notion of adaptive partial pooling also provides a novel perspective. A genuine effect of informativity and its interaction with context frequency can be due to speakers themselves representing their linguistic experience in a way that results in adaptive partial pooling, as argued in [17,47]. Adaptive partial pooling is directly predicted by probabilistic models of the mind [17,47,48], which propose that the language learner engages in hierarchical inference, building rich structured representations and apportioning credit for the observations among them. However, it is also compatible with other "rich memory" models in which experienced observations are not discarded even if the learner generalizes over them. All such rich memory models can be understood as performing adaptive partial pooling at the time of production. For example, an exemplar/instance-based model stores specific production exemplars/instances tagged with contextual features [49,50]. It then produces a weighted average of the stored exemplars, with the weight reflecting overlap between the tags associated with each exemplar and the current context. The average is naturally dominated by exemplars that match the context exactly when such exemplars are numerous and falls back on more distant exemplars when nearby exemplars are few. Modern deep neural networks also appear to be rich memory systems that implement a kind of hierarchical inference when generating text, in that best performance is achieved when the network is more than big enough to memorize the entire training corpus [51,52]. Increasing the size of the network beyond the point at which it can memorize everything appears to benefit the network specifically when the context frequency distribution is Zipfian [53,54], by allowing it to retrieve memorized examples when the context is familiar and to fall back on generalizing over similar contexts when it is not. The existence of implicit partial pooling in such models may be why attempts to add explicit partial pooling to them to improve generalization to novel contexts [55] have led to limited improvements in performance.

Within the usage-based phonology literature, informativity effects have been considered to be one of a number of effects that go by the name of "frequency in favorable contexts" (FFC). In general, the larger the proportion of a word's tokens that occur in contexts that favor reduction, the more it will be reduced in other contexts [18–23]. Informative words can be seen as one type of low-FFC word, because (by definition of informativity) they tend to have low predictability given the contexts in which they occur. In this literature, informativity effects have been argued to be due to storage of phonetically detailed pronunciation tokens in the memory representations of words [18,19] and are therefore often thought to be categorically distinct from the online, in-the-moment reduction effects driven by predictability in the current context.

However, the distinction between informativity and local predictability may also be an artifact of how we traditionally operationalize the notion of "current context". In reality, a word is never in exactly the same context twice. All contexts have a frequency of 1 if you look closely enough, and predictability of a word in a context always involves some generalization over contexts. That is, true predictability must always lie between predictability given the current context and average predictability across contexts (i.e., the inverse of informativity). The present paper treated contexts as being the same if they end in the same word. This is motivated by findings that the width of the context window does not appear to matter for predicting word duration (and other measures of word accessibility) from contextual predictability [35,56]. That is, language models with very narrow (one-word) and very broad windows generate upcoming word probabilities that are equally good at predicting word duration and fluency [35,56,57], suggesting that it is largely predictability given the preceding word that matters. However, [56] nonetheless shows that a large language model (with a short context window) outperforms an n-gram model in yielding predictability estimates predictive of word durations. It therefore seems likely that the notion of "current context" in our studies is too narrow and speakers actually generalize over the contexts that we considered distinct. For example, even if it is only the preceding word that matters, distributionally similar preceding words might behave alike, which a large language model would pick up on by using distributed semantics word representations (also known as embeddings) instead of the local codes of an n-gram model. Thus, another direction for future work is to try to determine what makes contexts alike in conditioning reduction.

Funding: This research received no external funding.

Institutional Review Board Statement: Not applicable.

Data Availability Statement: Code for replicating the simulations and running your own is available at https://osf.io/c8aws, accessed on 4 July 2025.

Acknowledgments: This work was presented at the 2024 meeting of the Association for Laboratory Phonology in Seoul, Korea, and the 2024 annual meeting of the Cognitive Science Society in Rotterdam, the Netherlands. Many thanks to T. Florian Jaeger for extremely helpful comments and providing a second dataset to apply the method to, two anonymous reviewers and the audience members for helpful discussion.

Conflicts of Interest: The author declares no conflicts of interest.

References

1. Seyfarth, S. Word informativity influences acoustic duration: Effects of contextual predictability on lexical representation. *Cognition* **2014**, *133*, 140–155. [CrossRef]
2. Cohen Priva, U. Informativity affects consonant duration and deletion rates. *Lab. Phonol.* **2015**, *6*, 243–278. [CrossRef]
3. Sóskuthy, M.; Hay, J. Changing word usage predicts changing word durations in New Zealand English. *Cognition* **2017**, *166*, 298–313. [CrossRef] [PubMed]
4. Kanwal, J.K. Word Length and the Principle of Least Effort: Language as an Evolving, Efficient Code for Information Transfer. Ph.D. Thesis, UCSD, La Jolla, CA, USA, 2018.
5. Barth, D. Effects of average and specific context probability on reduction of function words BE and HAVE. *Linguist. Vanguard* **2019**, *5*, 20180055. [CrossRef]
6. Hashimoto, D. Probabilistic reduction and mental accumulation in Japanese: Frequency, contextual predictability, and average predictability. *J. Phon.* **2021**, *87*, 101061. [CrossRef]
7. Tang, K.; Shaw, J.A. Prosody leaks into the memories of words. *Cognition* **2021**, *210*, 104601. [CrossRef]
8. Wang, S.F. The interaction between predictability and pre-boundary lengthening on syllable duration in Taiwan Southern Min. *Phonetica* **2022**, *79*, 315–352. [CrossRef]
9. Wang, S.F. Word-level and syllable-level predictability effects on syllable duration in Taiwan Mandarin. In Proceedings of the 20th International Congress of Phonetic Sciences, Prague, Czech Republic, 7–11 August 2023; pp. 1350–1354.
10. Wang, S.F. Contrast and predictability in the variability of tonal realizations in Taiwan Southern Min. In Proceedings of the Speech Prosody 2024, Leiden, The Netherlands, 2–5 July 2024; pp. 542–546.
11. Kaźmierski, K. The role of informativity and frequency in shaping word durations in English and in Polish. *Speech Commun.* **2025**, *171*, 103239. [CrossRef]
12. Jaeger, T.F.; Buz, E. Signal reduction and linguistic encoding. In *The Handbook of Psycholinguistics*; Fernández, E.M., Cairns, H.S., Eds.; Wiley: Malden, MA, USA, 2017; pp. 38–81.
13. Bybee, J. *Phonology and Language Use*; Cambridge University Press: Cambridge, UK, 2001.
14. Goldinger, S.D. Echoes of echoes? An episodic theory of lexical access. *Psychol. Rev.* **1998**, *105*, 251–279. [CrossRef]
15. Port, R.F.; Leary, A.P. Against formal phonology. *Language* **2005**, *81*, 927–964. [CrossRef]
16. Hall, K.C.; Hume, E.; Jaeger, T.F.; Wedel, A. The role of predictability in shaping phonological patterns. *Linguist. Vanguard* **2018**, *4*, 20170027. [CrossRef]
17. Kapatsinski, V. Hierarchical inference in sound change: Words, sounds, and frequency of use. *Front. Psychol.* **2021**, *12*, 652664. [CrossRef]
18. Brown, E. The long-term accrual in memory of contextual conditioning effects. In *The Handbook of Usage-Based Linguistics*; Díaz-Campos, M., Balasch, S., Eds.; Wiley: Malden, MA, USA, 2023; pp. 179–195.

19. Kapatsinski, V. Understanding type and token frequency effects in usage-based linguistics. In *The Handbook of Usage-Based Linguistics*; Díaz-Campos, M., Balasch, S., Eds.; Wiley: Malden, MA, USA, 2023; pp. 91–106.
20. Brown, E.K. The effect of forms' ratio of conditioning on word-final/s/voicing in Mexican Spanish. *Languages* **2020**, *5*, 61. [CrossRef]
21. Brown, E.L.; Raymond, W.D.; Brown, E.K.; File-Muriel, R.J. Lexically specific accumulation in memory of word and segment speech rates. *Corpus Linguist. Linguist. Theory* **2021**, *17*, 625–651. [CrossRef]
22. Raymond, W.D.; Brown, E.L.; Healy, A.F. Cumulative context effects and variant lexical representations: Word use and English final t/d deletion. *Lang. Var. Change* **2016**, *28*, 175–202. [CrossRef]
23. Brown, E.K. Cumulative exposure to fast speech conditions duration of content words in English. *Lang. Var. Change* **2023**, *35*, 153–173. [CrossRef]
24. Cohen Priva, U.; Jaeger, T.F. The interdependence of frequency, predictability, and informativity in the segmental domain. *Linguist. Vanguard* **2018**, *4*, 20170028. [CrossRef]
25. Katz, S. Estimation of probabilities from sparse data for the language model component of a speech recognizer. *IEEE Trans. Acoust. Speech Signal Process.* **1987**, *35*, 400–401. [CrossRef]
26. Jaeger, T.F. Redundancy and Syntactic Reduction in Spontaneous Speech. Ph.D. Thesis, Stanford University, Stanford, CA, USA, 2006.
27. Gelman, A.; Hill, J. *Data Analysis Using Regression and Multilevel/Hierarchical Models*; Cambridge University Press: Cambridge, UK, 2007.
28. Barth, D.; Kapatsinski, V. Evaluating logistic mixed-effects models of corpus-linguistic data in light of lexical diffusion. In *Mixed-Effects Regression Models in Linguistics*; Speelman, D., Heylen, K., Geeraerts, D., Eds.; Springer: Cham, Switzerland, 2018; pp. 99–116.
29. Collins, M.; Brooks, J. Prepositional phrase attachment through a backed-off model. *arXiv* **1995**, arXiv:cmp-lg/9506021.
30. Baayen, R.H. *Word Frequency Distributions*; Springer Science & Business Media: Cham, Switzerland, 2001.
31. Ellis, N.C.; O'Donnell, M.B.; Römer, U. Does language Zipf right along? Investigating robustness in the latent structure of usage and acquisition. In *Measured Language: Quantitative Approaches to Acquisition, Assessment and Variation*; Connor-Linton, J., Wander Amoroso, L., Eds.; Georgetown University Press: Washington, DC, USA, 2016; pp. 33–50.
32. Cohen Priva, U. Informativity and the actuation of lenition. *Language* **2017**, *93*, 569–597. [CrossRef]
33. Godfrey, J.J.; Holliman, E.C.; McDaniel, J. SWITCHBOARD: Telephone speech corpus for research and development. In Proceedings of the International Conference on Acoustics, Speech, & Signal Processing (ICASSP) '92, San Francisco, CA, USA, 23–26 March 1992; pp. 517–520.
34. Deshmukh, N.; Ganapathiraju, A.; Gleeson, A.; Hamaker, J.; Picone, J. Resegmentation of SWITCHBOARD. In Proceedings of the 5th International Conference on Spoken Language Processing (ICSLP), Sydney, Australia, 30 November–4 December 1998; pp. 1543–1546.
35. Harmon, Z.; Kapatsinski, V. A theory of repetition and retrieval in language production. *Psychol. Rev.* **2021**, *128*, 1112–1144. [CrossRef] [PubMed]
36. Hochreiter, S.; Schmidhuber, J. Long short-term memory. *Neural Comput.* **1997**, *9*, 1735–1780. [CrossRef]
37. Elman, J.L. Finding structure in time. *Cogn. Sci.* **1990**, *14*, 179–211. [CrossRef]
38. Dammalapati, S.; Rajkumar, R.; Ranjan, S.; Agarwal, S. Effects of duration, locality, and surprisal in speech disfluency prediction in English spontaneous speech. In Proceedings of the Society for Computation in Linguistics 2021, Online, 14–19 February 2021; pp. 91–101.
39. R Core Team. *R: A Language and Environment for Statistical Computing*; R Foundation for Statistical Computing: Vienna, Austria, 2024.
40. Bates, D.; Mächler, M.; Bolker, B.; Walker, S. Fitting linear mixed-effects models using lme4. *J. Stat. Softw.* **2015**, *67*, 1–48. [CrossRef]
41. Kuznetsova, A.; Brockhoff, P.B.; Christensen, R.H. lmerTest package: Tests in linear mixed effects models. *J. Stat. Softw.* **2017**, *82*, 1–26. [CrossRef]
42. Barton, K. MuMIn: Multi-Model Inference. R Package Version 1.48.11. 2025. Available online: https://CRAN.R-project.org/package=MuMIn (accessed on 4 July 2025).
43. Lüdecke, D. sjPlot: Data Visualization for Statistics in Social Science. R Package. Version 2.8.16. 2024. Available online: https://CRAN.R-project.org/package=sjPlot (accessed on 4 July 2025).
44. Bybee, J.; Brown, E.K. The role of constructions in understanding predictability measures and their correspondence to word duration. *Cogn. Linguist.* **2024**, *35*, 377–406. [CrossRef]
45. Shadrova, A. No three productions alike: Lexical variability, situated dynamics, and path dependence in task-based corpora. *Open Linguist.* **2025**, *11*, 20240036. [CrossRef]

46. Brysbaert, M.; Mandera, P.; Keuleers, E. The word frequency effect in word processing: An updated review. *Curr. Dir. Psychol. Sci.* **2018**, *27*, 45–50. [CrossRef]
47. Kleinschmidt, D.F.; Jaeger, T.F. Robust speech perception: Recognize the familiar, generalize to the similar, and adapt to the novel. *Psychol. Rev.* **2015**, *122*, 148–203. [CrossRef]
48. O'Donnell, T.J. *Productivity and Reuse in Language: A Theory of Linguistic Computation and Storage*; MIT Press: Cambridge, MA, USA, 2015.
49. Pierrehumbert, J.B. Word-specific phonetics. In *Laboratory Phonology 7*; Gussenhoven, C., Warner, N., Eds.; Mouton de Gruyter: Berlin, Germany, 2002; pp. 101–140.
50. Jamieson, R.K.; Johns, B.T.; Vokey, J.R.; Jones, M.N. Instance theory as a domain-general framework for cognitive psychology. *Nat. Rev. Psychol.* **2022**, *1*, 174–183. [CrossRef]
51. Belkin, M.; Hsu, D.; Ma, S.; Mandal, S. Reconciling modern machine-learning practice and the classical bias–variance trade-off. *Proc. Nat. Acad. Sci. USA* **2019**, *116*, 15849–15854. [CrossRef] [PubMed]
52. Zhang, C.; Bengio, S.; Hardt, M.; Recht, B.; Vinyals, O. Understanding deep learning (still) requires rethinking generalization. *Commun. ACM* **2021**, *64*, 107–115. [CrossRef]
53. Feldman, V. Does learning require memorization? A short tale about a long tail. In Proceedings of the 52nd Annual ACM SIGACT Symposium on Theory of Computing, Chicago, IL, USA, 22–26 June 2020; pp. 954–959.
54. Feldman, V.; Zhang, C. What neural networks memorize and why: Discovering the long tail via influence estimation. *Adv. Neural Inf. Process. Syst.* **2020**, *33*, 2881–2891.
55. White, J.; Goodman, N.; Hawkins, R. Mixed-effects transformers for hierarchical adaptation. *arXiv* **2022**, arXiv:2205.01749.
56. Clark, T.H.; Poliak, M.; Regev, T.; Haskins, A.J.; Gibson, E.; Robertson, C. The Relationship Between Surprisal, Prosody, and Backchannels in Conversation Reflects Intelligibility-Oriented Pressures. Ms. 2025. Available online: https://osf.io/preprints/psyarxiv/uydmx_v1 (accessed on 1 June 2025).
57. Regev, T.I.; Ohams, C.; Xie, S.; Wolf, L.; Fedorenko, E.; Warstadt, A.; Wilcox, E.G.; Pimentel, T. The time scale of redundancy between prosody and linguistic context. *arXiv* **2025**, arXiv:2503.11630.

entropy

MDPI

Article

Challenges and Opportunities in Causality Analysis Using Large Language Models

Wlodek W. Zadrozny

Computer Science & Data Science, University of North Carolina Charlotte, Charlotte, NC 28223, USA; wzadrozn@charlotte.edu

Abstract: This article examines the challenges and opportunities in extracting causal information from text with Large Language Models (LLMs). It first establishes the importance of causality extraction and then explores different views on causality, including common sense ideas informing different data annotation schemes, Aristotle's Four Causes, and Pearl's Ladder of Causation. The paper notes the relevance of this conceptual variety for the task. The text reviews datasets and work related to finding causal expressions, both using traditional machine learning methods and LLMs. Although the known limitations of LLMs—hallucinations and lack of common sense—affect the reliability of causal findings, GPT and Gemini models (GPT-5 and Gemini 2.5 Pro and others) show the ability to conduct causality analysis; moreover, they can even apply different perspectives, such as counterfactual and Aristotelian. They are also capable of explaining and critiquing causal analyses: we report an experiment showing that in addition to largely flawless analyses, the newer models exhibit very high agreement of 88–91% on causal relationships between events—much higher than the typically reported inter-annotator agreement of 30–70%. The article concludes with a discussion of the lessons learned about these challenges and questions how LLMs might help address them in the future. For example, LLMs should help address the sparsity of annotated data. Moreover, LLMs point to a future where causality analysis in texts focuses not on annotations but on understanding, as causality is about semantics and not word spans. The Appendices and shared data show examples of LLM outputs on tasks involving causal reasoning and causal information extraction, demonstrating the models' current abilities and limits.

Keywords: causality extraction; large language models; LLM; NLP; causality; GPT; Gemini; causality datasets; CNC corpus; causality analysis

1. Introduction

This article provides *an* overview of applications of Large Language Models (LLMs) to the "causality extraction" problem in Natural Language Processing (NLP). It aims to be conceptually comprehensive but it may be incomplete because the field is changing. It is also, in its conclusion, a position paper about how the field can change: with the appearance of new, very capable language models, the current focus on the *annotations* of causal segments can (should?) be replaced with *semantic analysis* and explainability, and employ different models and different viewpoints.

Large Language Models (LLMs) have been the dominant research theme in Natural Language Processing (NLP) since 2018, when BERT [1] was made available by Google. BERT gave rise to several related models, e.g., SciBERT, DistillBERT, BioBERT, and others.

Later OpenAI introduced GPT-2 and with GPT-3.5, we observed the emergence of "intelligent behavior" and the adoption of the ChatGPT app skyrocketed. Since then, we have seen great improvements in several versions of GPT: GPT-4o, o3 (with reasoning+agents) [2,3], and GPT-5 [4] and many interesting research results using publicly available Large Language Models (LLMs) such as LLAMA, Gemma, Gemini 2.5 Pro, Grok, Falcon, Claude, Mixtral, and others [5].

Due to the generative nature of LLMs, they often "hallucinate" the locations of relevant action, cause, or effect passages (discussed in more detail in a subsequent section). Additionally, their relatively shallow contextual understanding can hinder their performance in generating accurate interpretations, particularly for complex domains like political news, e.g., [6,7]. Thus, on the one hand we are seeing tremendous progress in the quality of LLMs, but on the other hand, we cannot be sure of the quality of produced results and their inexplicable occasional total failures. Therefore, we need to address the question of the reliability of LLMs for the causality analysis of texts, which this article realizes.

Causality extraction traditionally involves two key tasks: identifying phrases and clauses that describe causes, actions, and effects and situating these elements within broader contexts. Examples include improving a patient's condition or explaining the political implications of an event and its causes. In other cases, the task might involve finding causal relations between several related events, described in a newswire article, or diagnosing why a product is not reliable. For illustration, consider the following examples of recognized causal relations:

Example 1.

(1) *It was recognized that any* `<C>` *assessment of glycemia in early pregnancy* `</C>` *would also result in* `<E>` *detection of milder degrees of hyperglycemia short of overt diabetes* `</E>`.

(2) `<causal-relation>` *When a* `<cause>` *policyholder or insured person becomes sick or hurt* `</cause>`*, the company* `<trigger>` *pays* `</trigger>` `<outcome>` *cash benefits fairly and promptly for eligible claims* `</outcome>` `</causal-relation>`.

(3) *If he had* [*reduced his sugar intake*]$_{CO}$, *he would be* [*free from diabetes*]$_{E}$.

(4) *[[A] Timely referral] is indicated] if [[CO] chronic or recurrent symptoms severely affect the patient's productivity or quality of life]].*

The differences in notation are apparent and Sections 4, 6, and 7 will return to this problem. In the meantime, we can guess that in (1) `<C>` stands for "cause" and `<E>` for "effect"; and in (4), `[CO]` refers to a condition potentially leading to an action `[A]` (4) and (2) has `<outcome>`, leading to a question of whether it might be the same as `<E>`. Notable is also the fact that causal relations in (1) and (2) do not include any verbs unlike (4). Also note a counterfactual (3). Upon reflection, the vocabulary used in these examples can have multiple meanings. What **are** causes, effects, conditions, triggers, actions, and outcomes? Are effects and outcomes the same thing? What about conditions, e.g., is any vital sign of the patient part of a condition? In other words, where are the boundaries of these concepts drawn? We will return to this issue later in Sections 3 and 7.

Another example shows causality between events from the FinCausal corpus [8], which will be discussed later in Section 4. In this notation, events marked as $< e1 >$ are causes and $< e2 >$ are their effects.

Example 2.

$< e2 >$ *It was found that total U.S. healthcare spending would be about USD 3.9 trillion under Medicare for All in 2019, compared with about USD 3.8 trillion under the status quo.* $< /e2 >$

$< e1 >$ *Part of the reason is that Medicare for All would offer generous benefits with no copays and deductibles, except limited cost-sharing for certain medications.*$< /e1 >$

Overview of the article: The article presents an overview of the field, and then showcases recent findings up to early Fall 2025. It is a personal view informed by the author's experience and the findings of his students. (The overview part of the article is based on the author's tutorial at FLAIRS-38, May 2025 [9].) It complements other surveys, e.g., [10–13] (with still maintained pointers to the relevant articles https://github.com/zhijing-jin/CausalNLP_Papers (accessed on 10 October 2025)), as well as another repository on causality extraction [14] https://github.com/causaltext/causal-text-papers (accessed on 10 October 2025).

The present article highlights challenges in causality extraction using LLMs and suggests strategies to make it better. The next section briefly motivates the extraction of causality-related information from text. This is followed by pointing to the diversity of views on causality (Section 3), which as we show later is reflected in capabilities of LLMs. Section 4 grounds the article in an overview of datasets for causality extraction and provides a summary of the recent research on this topic. Then Section 5 introduces LLMs and discusses their growing capabilities, the lack of common sense, and the inevitability of hallucinations in generative models. Then, Section 6 and Appendix B show examples of successful causality analysis with several GPT and Gemini models, using different views of causality including Aristotelian and counterfactual. We conclude with a discussion of how LLMs can help improve the datasets for causality extraction and perhaps change our views of what the problems in understanding causality in text are.

2. Why Care About Extracting Causality-Related Information?

We care about finding causal dependencies for several reasons: On the practical side, because human understanding is often based on "mechanical," action–effect, or cause–effect models. In simpler cases, such as "smoking causes lung cancer," such relationships are apparent to a human and can also be extracted with high accuracy from simple statements. In more realistic analysis, causal relationships emerge from many factors and typically require a deep knowledge of the domain. For example, to understand and reason about the causes of production defects in industry, or a defeat in a game [15], one can use Ishikawa diagrams [16] to illustrate how the interaction of multiple causes in several categories produces an effect. The overall schema is shown in Figure 1.

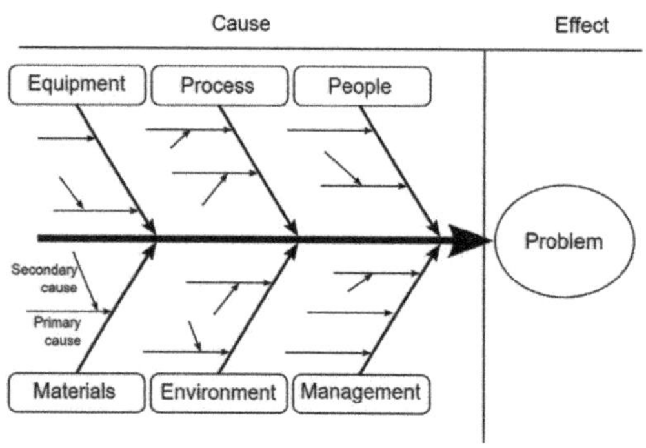

Figure 1. Ishikawa diagrams can be used for cause–effect analysis in complex environments. Source: Wikipedia [16].

There are many—often overlapping—reasons for extracting causality-related information from texts. The general theme seems to be explaining the behavior of complex systems. And an incomplete list would include making informed decisions, e.g., in medicine and business, and predicting outcomes, e.g., in market research and policy making. Such knowledge might also help in avoiding errors from data mining: correlation $\not\Rightarrow$ causation; for example, understanding the differences between "mediators" of causality and "confounders", by building causal graphs $X_i \longrightarrow X_j$, as in Figure 2, where arrows indicate causality and can be grouped into taxonomies as in Figure 1.

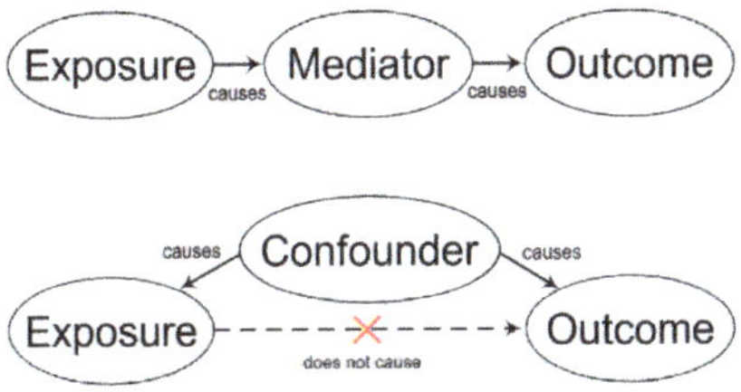

Figure 2. Causal graphs might also help in avoiding errors from data mining by visualizing "mediators" and "confounders." Source: Wikipedia [16].

Another list of reasons comes from AI itself: Natural Language Understanding (NLU) is not credible without question answering (QA), which itself cannot be conducted in any generality without an account for causality in texts (and other media). Building explainable AI, improving machine learning (including LLMs), and integration with formal reasoning systems could be added to this list. Finally, it is likely a requirement for AGI, that is Artificial General Intelligence [17], which used to be a topic rarely discussed by the general public and became mainstream after the wide adoption of ChatGPT.

3. Views on Causality

The topic of causality is vast, e.g., "cause effect relationship" appears about 8 M times in Google Scholar data, and "causality" about 2.8 M times. On the other hand, we obtain "causality extraction" about 900 times, of which 600 after 2021. (This makes this overview somewhat feasible.)

We observed earlier (Section 1) the diversity of annotation schemes in datasets for causality extraction, which presumably reflects the diversity of views of the authors on causality and related terms such as actions, effects, time, etc. (And also reflects the constraints, mainly the cost of annotations.)

To deepen our discussion and better understand the challenges, we need to broaden (slightly) our knowledge of what is or was meant by "causality" by different people and in different contexts. We will do so in the following order:

1. **NLP** adheres to a common sense view of causality as expressed in dictionaries. Tests for causality, administered by text annotators, are used to prepare data for information extraction, question answering, reasoning, and other AI tasks.
2. **Aristotle** introduced the "Four Types of Causes" and argued that all four are *necessary and sufficient* for explanations [18].
3. **Pearl** is mostly concerned with probabilistic causation and is the inventor of the Ladder of Causation (Associational, Interventional, counterfactual) [19,20]. This work is important for reasoning and the quality of explanations.
4. **Other views** will be alluded to mostly through references. This includes "Ontology of Causation" and seven views of causality in medical literature [21] .

3.1. Common Sense View of Causality in NLP

This exposition treats "causality extraction" as a separate domain of research in NLP, based on common sense (i.e., dictionary) views of cause and effect in texts. This is more operationally expressed as instructions for annotators, e.g., the "Five Tests for Causality" [22].

1. Why: The example is not causal if the reader is unable to construct a "Why" question regarding the effect.
2. Temporal order: The example is not causal if the cause does not precede the effect in time.
3. Counterfactual: The example is not causal if the effect is equally likely to occur or not occur without the cause.
4. Ontological asymmetry: The example is not causal if the reader can readily swap the cause and effect claims in place.
5. Linguistic: The example is likely to be causal if it can be rephrased into "X causes Y" or "Due to X, Y."

However, even with these relatively clear instructions, the inter-annotator agreement is between about 30% and 50% [22]. And as we discuss later, in Sections 6 and 7, one can ask the question whether the tests capture all aspects of interest.

3.2. Aristotle's Four Causes

According to Aristotle, to fully understand an object or event, one must account for the Four Causes: (1) the material cause, i.e., the material of an object or elements of an event; (2) the formal cause—its form or essence (e.g., the shape or design of the statue); (3) the efficient cause—the agent or process that brings; and (4) the final cause—its purpose. These are exemplified in Table 1 and Figure 3.

We first note that the efficient cause can be abstract ("carpentry") or concrete ("[t]he agent"). Second, the existence of final causes contradicts Test 2. That is, final causes precede their effects. Such motivating causes can appear in event descriptions, e.g., about a politician starting a smear campaign against the rival "because of the upcoming election." Thus, future events can influence current events. Of course, we could try to sidestep this problem by talking about a mental representation of "the upcoming election," but no one in practice would do it when creating annotations. Furthermore, to our knowledge, LLMs have not been trained to reason about mental representations and possible worlds (even though when asked by a series of prompts, they can create, imperfectly, such models). And finally, we see the philosophical (metaphysical) distinction between the form and the matter. We note that philosophy is rarely explicit in NLP, since the focus is almost always on "processing."

Table 1. The Four Causes with examples [18].

Cause Type	Description
Material	That from which something is generated and out of which it is made, e.g., the bronze of a statue.
Formal	The structure which the matter realizes and in terms of which it comes to be something determinate, e.g., the shape of the president, in virtue of which this quantity of bronze is said to be a statue of a president.
Efficient	The agent responsible for a quantity of matters coming to be informed, e.g., the sculptor who shaped the quantity of bronze into its current shape, the shape of the president.
Final	The purpose or goal of the compound of form and matter, e.g., the statue was created for the purpose of honoring the president.

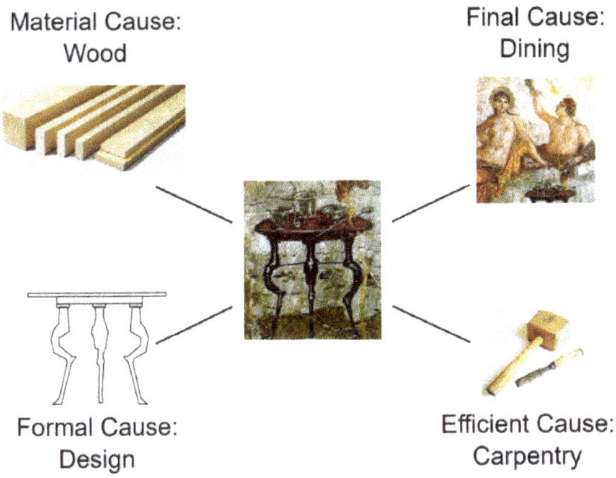

Figure 3. The Four Causes of Aristotle. Source: [23].

3.3. Pearl's Ladder of Causation

A methodological introduction to Pearl's view of causation can be found in [19]; this work is expanded to focus on measuring the influence of random variables on possible outcomes within a Bayesian framework [24,25]. Here, its main ideas are presented in Table 2. It is distinguished from the common sense (NLP) and from the Aristotelian views of causality by adding measurements, possibility of experimenting, and reasoning about counterfactuals using a set of previously identified variables.

Table 2. Ladder of Causation: Pearl's three-level Causal Hierarchy (PCH) with representative questions, examples, and ML paradigms. Reproduced from [20].

	Layer (Symbolic)	Typical Activity	Typical Question	Example	Machine Learning
L_1	Associational $P(y \mid x)$	Seeing	What is? How would seeing X change my belief in Y?	What does a symptom tell us about the disease?	Supervised and Unsupervised Learning
L_2	Interventional $P(y \mid do(x), c)$	Doing	What if? What if I do X?	What if I take aspirin, will my headache be cured?	Reinforcement Learning
L_3	Counterfactual $P(y_x \mid x', y')$	Imagining	Why? What if I had acted differently?	Was it the aspirin that stopped my headache?	

3.4. Other Views of Causation

Given the vastness of the literature on causality/causation, we limit ourselves to the ones with direct, in our view, relevance to the topic "LLMs and causality extraction." Therefore, we skip discussions of older views on causality (e.g., Hume [26]). We also do not venture into modern methods of understanding causality through data analysis of mediation and interaction of variables, as well as of causality in social networks. For an in-depth treatment of the former, we refer the reader to [27,28]. The last reference also discusses aspects of causality in social networks (spillovers) and has a short discussion of philosophical issues.

However, because of the diversity of annotation schemes for causality in text, we want to mention an article that attempts to organize views of causality in the medical literature into an ontology [21], as represented in the list below (this is our summary of the list from the article). The article also reports that *many studies used the terms "causality," "causation," and "cause and effect" as synonyms. However, those who use them may have different meanings and concepts in mind. Many studies appear to define these terms inconsistently or ignore their definition entirely.*

1. **Association**: Causality implied through correlation or co-occurrence, not mechanisms.
   Example: Genetic factors are associated with disease susceptibility.
2. **Determinism**: Causation as universal law: if A, then B.
   Example: Radiation causes cancer.
3. **Temporal order**: Cause must precede effect in time sequence.
   Example: Exposure to a toxin precedes the onset of symptoms.
4. **Disposition**: Objects have inherent powers to manifest effects.
   Example: A patient's disposition to experience adverse effects from a drug.
5. **Causal chain**: Events cause subsequent events in a linear sequence.
   Example: A pathogen triggers inflammation, which leads to organ damage.
6. **Influence**: One process modifies the likelihood of another occurring.
   Example: Nicotine withdrawal positively influences smoking relapse.
7. **Production**: Cause directly generates or triggers the effect.
   Example: A virus produces respiratory symptoms.

3.5. Why Care About This Diversity of Views?

As mentioned earlier, causality extraction is not an end in itself; rather, it is a means to improve machine understanding of texts, which again is a means to create better models of reality and, e.g., enable better decision making.

Since LLMs are trained on texts in which all senses of "cause" appear, they will also rely on all these senses when answering questions or performing formal analyses of texts. These views are partly overlapping, but are also distinct. Properties 1–5 do not naturally map into the Aristotle types of causation, nor Pearl's approach, nor other taxonomies of causal relations, e.g., the seven views of causality in the medical literature [21]. For example, future causes are admissible by Aristotle, but not by NLP annotators, even though we find plenty of news stories explaining how upcoming elections are a cause for dirty political campaigns now. Similarly, prevention—in the case of rare events—contradicts Rule 3 (counterfactual) of Section 3.1. Yet we commonly find texts with examples where skipping an early morning swim prevented being attacked by a shark or taking a child to school prevented being a victim of a terrorist.

With LLMs' growing importance and their use in both data preparation and causal analysis, we gain, as shown in Section 6, an ability to analyze text documents directly, without the intermediate steps of causality extraction. Moreover, LLMs can answer questions about counterfactuals and Aristotelian causes, even if they are not trained on annotated data for this task. Annotated datasets would still be of value, for example, to check for LLM hallucinations or adjust directions of analysis via better prompting. In addition, numbers (such as probabilities and correlations) are seldom part of annotated data; LLMs could add numerical information to the extracted data and help perform quantitative analysis, including perhaps the interaction of causes as it is performed, e.g., in epidemiology [27] and in other types of research (e.g., [29]).

4. Finding Causal Expressions: Datasets and Results

In this section, we provide an overview of causality datasets and published results on causality extraction from texts. As suggested by the preceding examples, this extraction process can operate on different levels: words, phrases, and sentences. Later we are going to see how it can be applied to documents. Furthermore, to be practical, it is necessary to situate the causal elements within broad contexts such as improving patients' conditions, explaining the political implications of events, or answering questions about a document.

4.1. Finding Causal Expressions Within Sentences

Finding "causal" expressions can be viewed as a subspecialty of "relation extraction" in NLP, as shown in sentences in Example 3, taken from the SemEval 2010 Task 8 dataset [30]. Multiple types of relations between entities are of interest and cause–effect is one of them. A recent example survey of "relation extraction" can be found in [31].

Example 3.

7800 "He was a < e1 > trouble < /e1 >< e2 > maker < /e2 > then, a leader in SDS and a trouble maker he remains."— `Cause-Effect(e1,e2)`

7795 "Interspersed in the design are < e1 > patterns < /e1 > derived from a variety of < e2 > textiles < /e2 >."—`Entity-Origin(e1,e2)`

However, this article focuses not on all relations, but narrowly on causality. Typically, the first step of this process is finding causal sentences (e.g., [32]). Then, within these sentences, as in all examples above (and below), different methods including LLMs can be used to mark causality expressions by identifying phrases and clauses describing causes, actions, and effects. Less frequently other characteristics, such as percentages or probabilities, are also extracted. This is followed by evaluations on one or more datasets.

We also need to mention causality within the context of multimodal analysis and reasoning, an emerging important sub-field of NLP, with applications to question answering [33] (but this is not a theme of this article).

4.2. Charlotte NLP Lab Work on Causality Extraction

The present article is informed to a considerable extent by a decade of research by the Charlotte NLP Lab, that is, my students and collaborators. Our research is summarized in this section and links are provided to the articles, programs, and datasets. There are two streams of this work: (1) causality extraction from medical guidelines; and (2) causality extraction from business texts. They are joined in the reported experiments on transfer learning. Later, Section 5 describes several experiments with LLMs performed in 2024 that will be relevant in the discussion of capabilities of the more recent language models. The next subsection (Section 4.3) attends to other datasets and summarizes other work on causality extraction.

(1) Causality extraction from medical guidelines: Clinical Practice Guidelines (CPGs) can contain complex concepts expressed in a complex language. Our work started by extending the results of [34], both by creating new public datasets covering several hundred examples of sentences from hypertension, rhinosinusitis, and asthma guidelines (https://github.com/hematialam (accessed on 15 May 2025)) and annotated primarily for medical conditions, actions, and consequences. Several models were used for causality extraction and example scores include an F1 of 60–70% and accuracy of 85–90+% for Logistic Regression, and for BioBERT [35], the F1 is about 85% and the accuracy is 90–95%. The details of dozens of experiments are given in [36–38].

The second dataset in this space is an extract from multiple Gestational Diabetes Guidelines containing about 290 items, which are available at https://github.com/gseetha04 (accessed on 15 May 2025). Here, example results also show the high performance of BERT (various versions) (over 90% accuracy on identifying conditional sentences), which was (as of Fall 2024) better than Llama 3.5 7B and GPT 3.5 and 4; details are provided in [39,40].

(2) Causality extraction from business texts: The ORG Dataset (Organizational Behavior) was developed with business school collaborators and has over 14K examples for fine-tuning and over 2.2K for training and testing annotated datasets(in the BIO style) and is available at https://github.com/GoPeaks-AI/text2causalgraph (accessed on 15 May 2025). The dataset comes with a taxonomy, which helps with causality extraction [29,41]. The extraction results are comparable with the ones for the medical guidelines. Namely, 85–91% F1 dataset with BERT (various versions), which were better (as of Fall 2024) than Llama 3.5 7B and GPT-3.5.

Additionally, the results on transfer learning show that when trying to predict performance based on pre-training on different datasets, it turns out that only K-L divergence, measuring the difference in tokens' distributions, is of value, while—surprisingly—the Kolmogorov–Smirnov test and Wasserstein distance are not predictive. For details, see [39,40,42].

4.3. Other Datasets and Studies on Causality Extraction

The experiments reported in this article use a sample of CNC, Chinese News Causality Dataset. This is perhaps the largest causality dataset https://github.com/twinkle121/CNC (accessed on 15 May 2025) and contains 25,629 event mentions and 5569 causal event pairs. Example annotations are shown in Figure 4, and an example result is [43] with an F1 score of 58–82% on the Event Story Line and CNC event causality corpus (however, only 44% on unseen predicates); their architecture incorporates BERT [1]. The mentioned Event Story Line dataset [44] contains 258 documents with temporal relations and causality identification https://github.com/tommasoc80/EventStoryLine (accessed on 15 May 2025). An example annotation in the ESL dataset is shown in Figure 5.

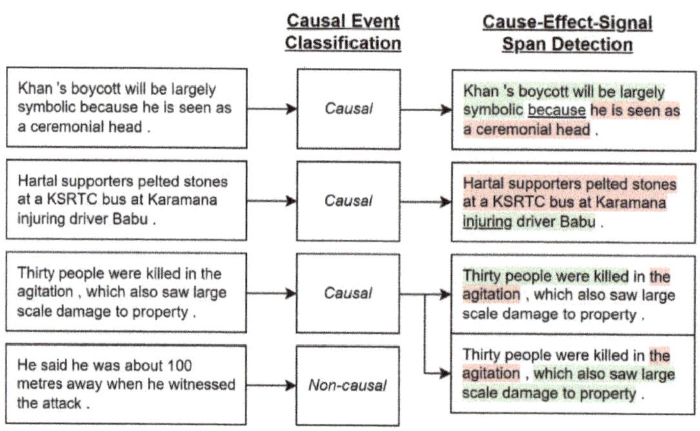

Figure 4. For figure source and an introduction to the **event causality extraction** problem and the CNC dataset, see [45].

<TIME_DATE>/</TIME_DATE> <TIME_DATE>2013</TIME_DATE>

<HUMAN_PART_PER>Lindsay</HUMAN_PART_PER>
<HUMAN_PART_PER>Lohan</HUMAN_PART_PER> has
<ACTION_OCCURRENCE>checked</ACTION_OCCURRENCE>
<ACTION_OCCURRENCE>into</ACTION_OCCURRENCE> <LOC_FAC>a</LOC_FAC>
<LOC_FAC>rehab</LOC_FAC> <LOC_FAC>facility</LOC_FAC> to
<ACTION_ASPECTUAL>begin</ACTION_ASPECTUAL> <TIME_DURATION>a</TIME_DURATION>
<TIME_DURATION>90</TIME_DURATION> <TIME_DURATION>-</TIME_DURATION>
<TIME_DURATION>day</TIME_DURATION> court - ordered
<ACTION_OCCURRENCE>stint</ACTION_OCCURRENCE> ,
<ACTION_OCCURRENCE>handed</ACTION_OCCURRENCE>
<ACTION_OCCURRENCE>down</ACTION_OCCURRENCE> as part of a
<ACTION_OCCURRENCE>sentence</ACTION_OCCURRENCE> for <TIME_DATE>a</TIME_DATE>
<TIME_DATE>2012</TIME_DATE> car <ACTION_OCCURRENCE>crash</ACTION_OCCURRENCE>
<ACTION_OCCURRENCE>case</ACTION_OCCURRENCE> .

However , it <ACTION_ASPECTUAL>remains</ACTION_ASPECTUAL>

Figure 5. An example of Event Story Line annotations [44] converted by GPT from complex XML to inline annotations for readability's sake (showing only a small segment of the document).

CNC can be compared to the SemEval 2010 Task 8 dataset (mentioned earlier) with 5236 sentences containing 64 embedded causality triplets (see https://huggingface.co/datasets/SemEvalWorkshop/sem_eval_2010_task_8 (accessed on 15 May 2025)). Note that, as shown in Example 3, only entities are being annotated. A recent work involving this dataset is [46], which explores multiple types of neural network architectures, and where yet again, BERT helps produce an F1 score of about 85% . The article also exemplifies multiple types of causality relations, as shown in Figure 6 and earlier in Figure 4.

Finally, we want to mention ExpliCa [47] (https://github.com/Unipisa/explica (accessed on 20 October 2025)), which can be used to evaluate LLMs in commonsense causal reasoning through Pairwise Causal Discovery tasks, i.e., potential causes and effects in pairs of sentences, e.g., as follows:

> *"The girl skipped school., The girl had not done her homework."*

A recent article [47] reports GPT-4o performance of 50–80% depending on a task.

4.4. Overall Impression

The easiest thing to notice in the annotated datasets is how different the assignments of causes and effects are. These multiple annotation schemes are not necessarily compatible. Second, the performance of LLMs is stuck at about 85% with various BERT versions often outperforming much larger LLMs (as of April 2025). Third, the annotated datasets are small in comparison with image datasets, which may have one or two orders of magnitude more of examples than the total for all textual causality datasets.

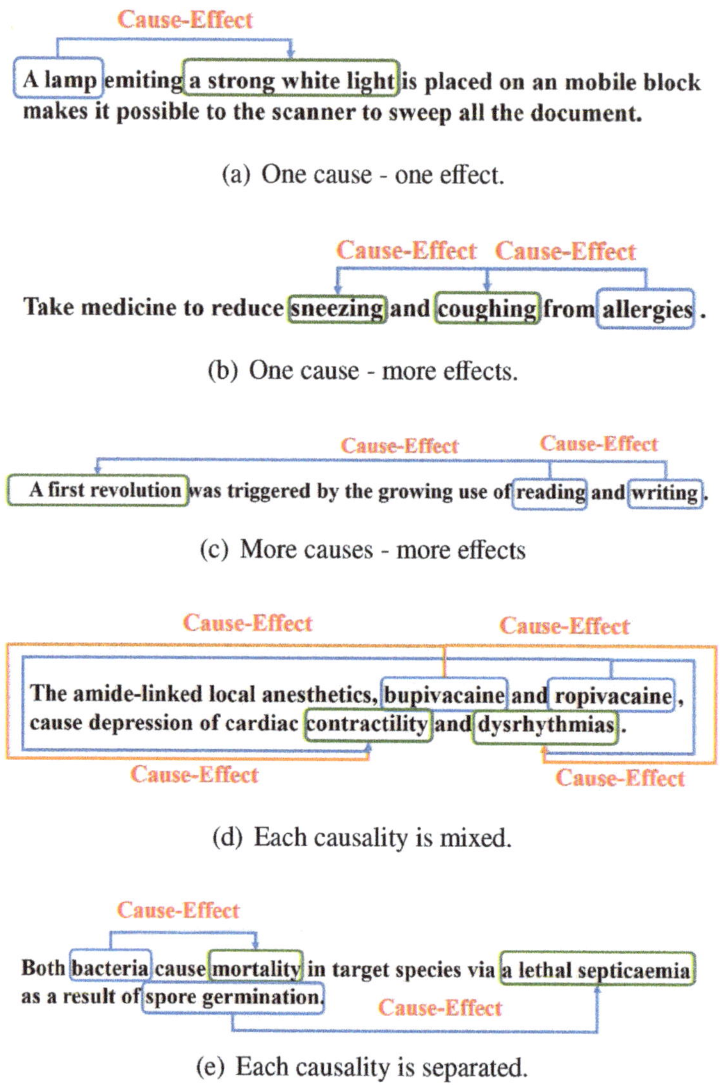

Figure 6. [Wu+24] exemplifies different types of causal relation on the SemEval dataset.

5. LLMs, Hallucinations, and Causality

LLMs are subject to many popular controversies. These include discussions on how intelligent they really are, what will be their impact on labor markets, or whether an AGI will appear in 2027 due to super-exponential growth in capabilities of AI [48]. The underlying question in these discussions is to what extent can we extrapolate from the progress of the last few years, e.g., measured by the performance of various models on several tasks, as illustrated by Figure 7.

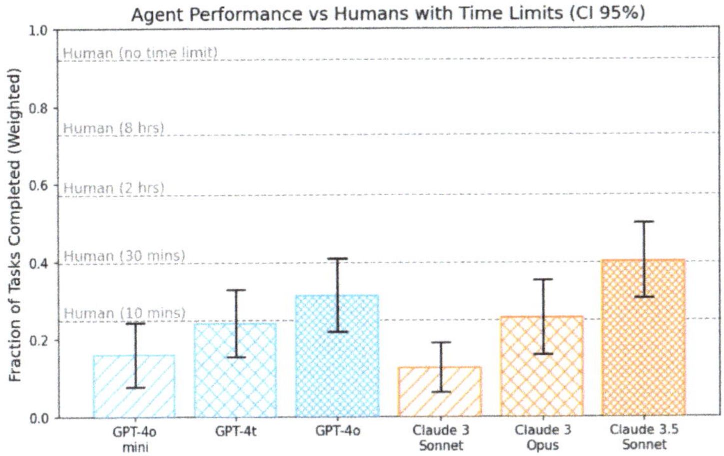

Figure 7. LLMs already help humans on time-consuming tasks. Their performance is changing on a weekly basis (image source: OpenAI [49]). Updates at https://llm-stats.com/.

There is also quite a bit of skepticism, e.g., performance growth seems to have flattened in 2025 as compared to 2024 https://llm-stats.com/, as measured by the GPQA: A Graduate-Level Google-Proof Q&A Benchmark" [50]. And notwithstanding the progress, the models lack common sense [51].

5.1. LLMs Limitations: Hallucinations Occur and Are Unavoidable

The term "hallucinations" in the context of LLMs seems to have entered the vernacular—it is estimated to appear (as of May 2025) at least three million times in the Google search results. However, the label "hallucination" is being applied to many LLM phenomena: disfluency, non-factuality, false confidence, contradicting provided or retrieved (RAG) sources, ignoring prompts, and others; ref. [52] presents a survey and a catalog of problems to which we apply this label. Many empirical methods try to mitigate or investigate hallucinations (surveyed in [53]). Often "fine-tuning" methods are used to reduce hallucinations (improve accuracy). However, such methods can be expensive (e.g., reinforcement fine-tuning [54]).

Importantly, it turns out that current generative LLMs "must hallucinate." There are several mathematical proofs that LLMs "must hallucinate" [55–60]. Intuitively, the space of "suggested facts" (from co-occurrences of phrases) is much larger than the space of established facts. And even though hallucinations can be made statistically negligible [58], the bad news is that larger and more accurate models hallucinate more [3]. Paradoxically, "models can hallucinate with high certainty even when they have the correct knowledge" [59]. And this, of course, can be a serious problem in applications.

5.2. Hallucinations and Causality Extraction

For causality extraction, a comparison of fine-tuning and prompting for BERT(s), Llama2, and GPT-4 can be found in [42], e.g., GPT-4 obtains an F1 score of 0.63 with fine-tuning and 0.60 with prompting on the gestational diabetes dataset mentioned earlier. A more recent article [61] discusses causal question answering (QA) and hallucinations. It shows a relatively high accuracy of GPT-4o in English and Spanish on the FinCausal dataset [8], introduced earlier in Example 2. It also offers a few insights: (1) *Prompt optimization and few-shot learning offer some improvements.* (2) *They were insufficient for consistently outperforming extractive methods in FinCausal, suffering from hallucinations.* (3) *In contrast, **fine-tuning generative models** was shown to be essential for minimizing hallucinations and achieving superior performance.* (4) "Both extractive models and generative models struggled at times to extract the correct answer in implicit causal relationships, where explicit causal markers (e.g., "because," "due to") were absent." Perhaps fine-tuning helps LLMs better "understand" what we are looking for by analyzing examples.

6. LLMs for Causality Analysis

Given the hallucination problems with LLMs and the accuracy limitations in causality *extraction*, we end on a positive note, showing the promise of current LLMs in causality *analysis*. The presented experiments show that LLMs are capable of sophisticated reasoning about causality.

Probing LLMs' Understanding of Causality

Initial experiments: Several small-scale experiments were performed by this author in Spring of 2025. A text from the CNC corpus was used (3463_ecbplus.xml.xml—full text is given in Appendix A). It concerned a landslide that struck a quarry, burying a 2000-square-meter plant. Seventeen workers were on site: fourteen were safely evacuated and three remained missing. Over 300 rescuers and multiple teams from various departments were mobilized for large-scale search and rescue operations. Possibly, the landslide was triggered by prolonged dry weather that loosened the mountain slope.

The experiment, with GPT-4o and Gemini Advanced 2.5 Pro, showed the potential of LLMs to "understand" causality. For example, both were able to create causal graphs and reason about indirect and direct causes, as well as perform a counterfactual and Aristotelian analysis. The respective causal graphs are shown in Figures 8 and 9. Note that the graphs are relatively correct (except for Node 10 in Figure 8) and informative. The details are given in Appendix A.

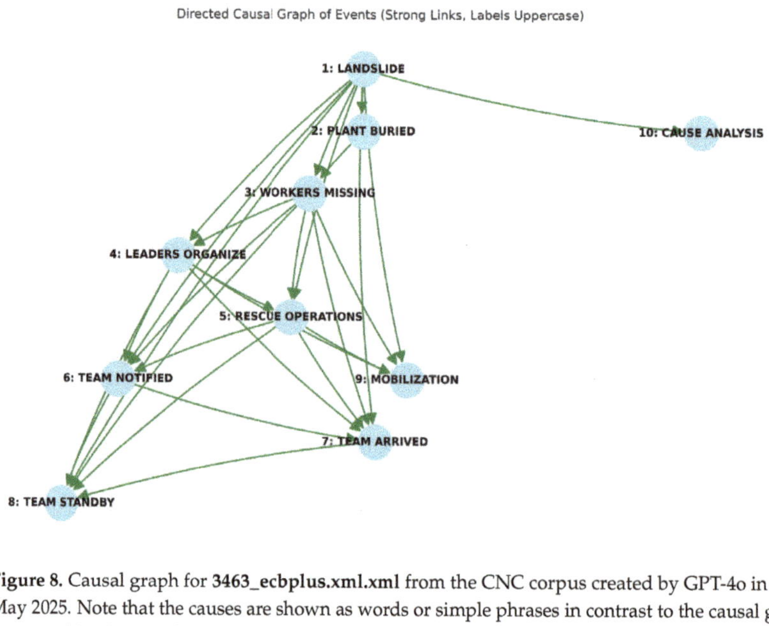

Figure 8. Causal graph for **3463_ecbplus.xml.xml** from the CNC corpus created by GPT-4o in early May 2025. Note that the causes are shown as words or simple phrases in contrast to the causal graph generated by Gemini shown in Figure 9.

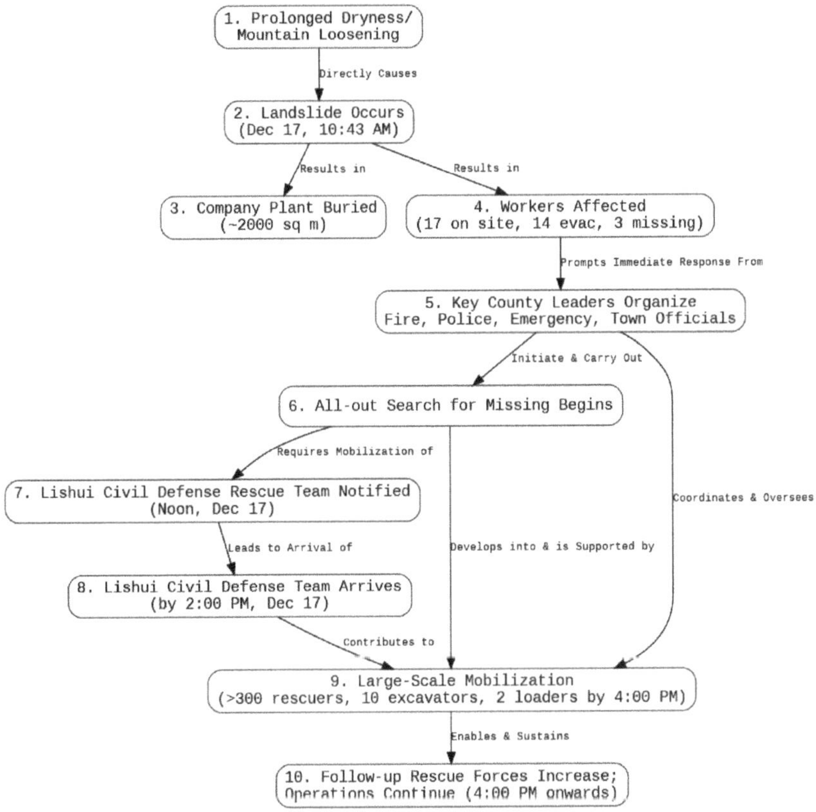

Figure 9. Graph of CNC corpus example **3463_ecbplus.xml.xml** using Gemini Advanced 2.5 Pro; early May 2025. Note that connections between states and events have their causes and relationships described, in contrast to Figure 8 generated by GPT.

Substantive experiments: In the Fall of 2025, a larger experiment on a random sample of 25 texts from the earlier-mentioned CNC corpus was performed by this author. The texts were translated into English (by DeepSeek). GPT-5 and Gemini 2.5 Pro were prompted to extract pairs of events and establish causal relations between them. The experiments were conducted using Google Colab Pro and the respective APIs, with default temperature settings. The prompts asked for extracting ten events from each text, ordering them chronologically, and finding direct or indirect causal relations between them. After finding causal relationships, the models were asked to perform a critique of the reasoning of the other LLM.

Some of the produced reports were slightly transformed using a few regular expressions (manually, using Notepad++) because the outputs were not always given in the requested format. Only formatting was affected and not any of the texts, detected causal pairs, or critiques. Such post-processing is not surprising, as online LLMs operate in parallel, possibly direct our requests to different versions of the generative models, and do not always provide consistent outputs. This phenomenon is more problematic for extracting causal relationships via annotation, but seems to have no effect on the semantic analysis in this experiment.

The results are interesting and show very high agreement between the LLMs in their causal analysis. (As contrasted with the relatively low inter-annotator agreement on the annotation task mentioned earlier in Section 3.1.) Gemini 2.5 Pro claimed 19 pairs to be incorrect out of 213 pairs analyzed by GPT-5, i.e., disagreed in about 9% of the cases. GPT-5, in analyzing Gemini 2.5 Pro, viewed 28 of 225 pairs as incorrect, i.e., disagreed in about 12% of the cases. Overall, the quality of the analysis was very high. In manual analysis, no hallucination events or clearly missing causal pairs were detected.

The number of errors is smaller than the number of disagreements. The majority reflected the underlying differences in reasoning. Thus, of the 28 suspect pairs detected by GPT-5, the manual analysis confirmed 13 errors; the other cases were borderline (10 relations) or likely correct (5 relations). For example, in a discussion about a roof collapse accident, GPT critiques Gemini:

```
"(e4, e5): Incorrect. The response was initiated because of
 the incident, not because of the confirmation of 14 trapped."
```

However, one could argue that the complex response described in the text was due to people being trapped, because a simple roof collapse, with no victims, would not generate such a response.

Notice that not only do we obtain high-quality causal analyses of events and their critiques, but also the arguments/reasoning supporting the detected causes and effects. In other words, the models give us explainability, which would be lacking if the programs were restricting themselves to creating annotations. In the created analyses, one can see a high degree of "common sense knowledge" (at least for the domain of news reports, which seems to follow common templates).

7. Concluding Discussion: Overcoming Challenges in Causal Analysis

In this section, we summarize our preceding observations as a list of challenges for causality analysis and causality extraction. Then, using what we have learned about LLMs in Section 6 (and earlier), we propose a few ways of addressing them (which we plan to do in followup work). Of course, this discussion must be hedged by the limitations of the experiments: one domain, one language, two dozen examples, and only about 450 causal pairs. Nevertheless, the results are strongly suggestive of the emerging capabilities of LLMs in understanding causal relations expressed in text.

7.1. Challenges in Causality Extraction—What Have We Learned?

The first challenge in causality analysis and causality extraction lies in *conceptualization*, that is, there are many views of causality and the examples and references provided in this article (mainly in Section 3) are by no means exhaustive. The question arises whether different domains (e.g., news events vs. medical) require different ontologies for expressing causation, which seems to follow from some cognitive science experiments [62]. Or, even more, whether they should be different for different sets of experimental data, as seems to be the case in medicine [21].

This is further illustrated by the many datasets for causality extraction and their incompatible representations of causality. These datasets follow different notational conventions and, more importantly, have different (implicit) ontologies.

Furthermore, the causality datasets are small because the manual annotations are expensive to create and high inter-annotator agreements are difficult to achieve. And this is yet another, important, challenge. A third—and related—challenge is the low inter-annotator agreement, and our lack of the ability to question the decisions made by the annotators once the datasets are published.

A fourth challenge has to do with LLMs having difficulty in keeping track of the sequences of annotations, and " hallucinating" beginnings and ends of annotations. In all interactions with LLMs, one should be aware of hallucinations and keep in mind cautionary examples (e.g., [63]) pointing to the fact that LLMs operate on "well-worn mental shortcuts" and can produce incorrect results, even if challenged. One might hope that symbolic co-processing would eventually help, but at this point, there are no guarantees of correctness.

7.2. Can Causality Analysis with LLMs Help Address These Four Challenges?

Based on the example results discussed in the previous section, Appendix B, and in the experiment data made available online, the answer is a tentative "Yes". Yet in the absence of large-scale analyses, this conclusion is still tentative, although informed by the author's experience with LLMs and causality extraction [29,38,40,42,64,65].

To start with, different conceptualizations can be used with LLMs, for example, their common sense ones are different, and they approach Aristotelian point of view differently. Still, experiments to confirm or disprove this possibility are yet to be performed.

Can the ontologies which are implicit in annotation instructions, or explicit ones, be used to guide LLMs towards a particular type of analysis? This article is unable to provide an answer. Thus, the first challenge remains open.

Next, let us discuss the challenge of data paucity. In the best case, when the data is unambiguous and follows a pattern, one can fine-tune LLMs to good performance. The creators of datasets often attempt to follow patterns and avoid ambiguity. This suggests that LLM-assisted translation between different dataset formats should be possible. For example, in Figures 8 and 9, we see that causal dependencies are expressed both as short phrases and as events connected by specific actions (based on the common text).

Furthermore, given the high-quality machine translation, it is possible to translate datasets between languages, e.g., from Chinese to English, as we have conducted in our analyzed example. These two steps would help increase the amounts of available data and perhaps enable additional experiments in other languages, e.g., for which no causality data exists.

The paucity of data can also be addressed by using LLMs as annotators (this is not a novel idea). If performed at scale, we could create large datasets by keeping only the items on which they agree, and perhaps analyzing the disagreements, either manually or automatically, for conceptual disagreements. At the minimum, these steps would give us (somewhat?) interoperable datasets and address the third problem: low inter-annotator agreement. Notice that even if the quality of causal analysis by LLMs is better than now, we will still need datasets to measure what LLMs do right and where they fail. As suggested by our examples, with proper setup (good data preparation, precise prompts, etc.), they can reason about causes and effects and find regions of agreement.

In the context of multiple annotation schemes and multiple views of causality, likely producing lower inter-annotator agreements, an obvious question arises as to whether forcing one or more ontologies on LLMs would improve the quality of the new data, i.e., do we need standards for causality analysis that differ for different domains (politics vs. medicine)? Perhaps they should also differ according to applications or goal, e.g., in medicine, often there is no agreement on causes, actions, and effects, resulting in different guidelines [37,42,66,67].

Such standards can be made explicit. For example, causal analysis with LLMs has to be different when used to create treatment guidelines vs. as a tool in patient–physician decision making. The former depends on the perspective of the medical organization creating the recommendations, while the latter happens in the context of existing recommendations, the patient's situation, and the physician's knowledge and diagnostic intuitions. For this,

LLMs are improving in understanding contexts of requests or analyses. In addition, in these and many other cases, numbers are important, e.g., for Pearl-style analysis mentioned in Section 3.

To summarize, we have presented an overview of challenges of causality extraction from text (lack of agreement on the concept of causation and its representation and the paucity and quality of available data) and suggested how, controlling for hallucinations, Large Language Models can help address the more general problem of causal analysis of text documents and help address these challenges.

Regarding the fourth challenge (i.e., producing correct annotations), one can expect further improvements in the abilities of LLMs. However, one can also ask whether in most practical cases, we can skip causality extraction and simply proceed with causality analysis, as shown in this article.

Funding: This research received no external funding.

Institutional Review Board Statement: Not applicable.

Data Availability Statement: Pointers to available datasets are given in the text of the article. The code and data used in the experiments are available at https://drive.google.com/drive/folders/1VujYfpQ-6iDuGJut81dpHXnlpqGuHYkP (accessed on 10 December 2025).

Acknowledgments: The author would like to thank the reviewers for their comments and suggested improvements.

Conflicts of Interest: The author declares no conflicts of interest.

Abbreviations

The following abbreviations are used in this manuscript:

LLM(s)	Large Language Model(s)
NLP	Natural Language Processing

Appendix A. Working with GPT-4o on a Longer Text

We use a random example from the CNC corpus mentioned earlier (**3463_ecbplus.xml.xml**). The example is in Chinese; therefore, it needs to be translated into English, which is followed by a causality analysis of English text in the steps below, of which we show only results for steps 1, 4, and 6. The whole conversation is available at https://chatgpt.com/share/685d81d7-0260-800a-a2a2-9ac21226761a (accessed on 10 October 2025).

1. Translation of the text from Chinese to English (here using GPT, earlier in the article DeepSeek was used, to separate translation from interpretation).
2. Semantic translation including annotations.
3. Finding causal verbs.
4. Listing the numbered points of what happened, in temporal order.
5. Causal analysis for each pair of the points.
6. Creation of the causal graph (several attempts).

GPT Translation: *On 17 December in Lishui, reporter Zhou Yulong learned from the Publicity Department of Songyang, Zhejiang, that as of 4:00 p.m. on the 17th, more than 300 rescuers, 10 excavators and 2 loaders had been mobilized to the scene to carry out full-scale search and rescue operations. Follow-up rescue forces were still being increased. Pan Saiqiu and other members of the Lishui Civil Defense Rescue Team received the notice at noon and arrived at the site by 2:00 p.m. "Some of our team members have already entered the site," said Pan Saiqiu. "Because there's* *too much rescue equipment, those of us who haven't entered the site are standing by at any time." Currently, more than ten rescue teams—including public security, firefighting, emergency response, militia, medical, electrical, telecommunications and civil rescue forces—are carrying out rescue work at the scene. Around 10:43 a.m. on 17 December, a landslide occurred at the quarry of Anluyuan Mechanized Sand Co., Ltd. in Xinxing Town, Songyang County, Zhejiang Province. Initial checks identified 17 workers on site; three were missing, while the others had been safely evacuated. The company's plant, covering about 2000 square meters, was buried. After the incident, key leaders of the county's party committee and government immediately organized firefighting, public security, emergency management and Xinxing Town officials to rush to the scene and carry out an all-out search for the missing persons. Rescue operations are still underway. Preliminary analysis suggests the landslide was caused by prolonged abnormal dryness of the weather, leading to loosening of the mountain.*

Prompt: `take this paragraph and tell me in 10 numbered points what happened, in temporal order.`

Response:

1. On the morning of 17 December at around 10:43 a.m. a landslide occurred at the quarry of Anluyuan Mechanized Sand Co., Ltd. in Xinxing Town, Songyang County, Zhejiang Province.
2. The landslide buried the company's plant, covering approximately 2000 m^2.
3. An initial investigation identified 17 workers present at the scene; three workers were missing, while the remaining workers were safely evacuated.
4. Immediately after the incident, county party-committee and government leaders organized firefighting, public-security, emergency-management and township officials to rush to the scene and search for the missing persons.
5. Rescue operations were launched with combined forces from public security, firefighting, emergency response, militia, medical, electrical, telecommunications and civil-rescue teams.
6. By noon on 17 December, Pan Saiqiu and other members of the Lishui Civil Defense Rescue Team had been notified of the incident.
7. By 2:00 p.m., Pan Saiqiu's team arrived at the scene to assist in the rescue efforts.
8. Some team members entered the site, but owing to the large amount of equipment already present, others remained on standby outside.
9. By 4:00 p.m. on 17 December, more than 300 rescuers, ten excavators and two loaders had been mobilized to continue full-scale search and rescue operations.
10. A preliminary analysis concluded that the landslide was likely caused by prolonged abnormal dryness of the weather, which loosened the mountain slope.

Causality Analysis with Gemini Advanced 2.5 Pro; Early May 2025

We analyzed the same English version of the CNC text using Gemini Advanced 2.5 Pro in early May 2025 [68] https://ai.google.dev/gemini-api/docs/quickstart?lang=python (accessed on 10 October 2025). We present the prompts and the informative parts of the responses, slightly reformatted, with deletions marked by (...). The transcript of whole session is available at https://g.co/gemini/share/cd52a6147a54 (accessed on 10 October 2025).

Prompt: `analyze this text:` *On 17 December in Lishui, reporter Zhou Yulong learned from the Publicity Department of Songyang, (...).*

Response:

This text describes an emergency incident and the subsequent rescue efforts. Here's an analysis:

- **Key Event:** A landslide occurred at the quarry of Anluyuan Mechanized Sand Co., Ltd. **Location:** Xinxing Town, Songyang County, Lishui, Zhejiang Province, China.
- **Date and Time of Incident:** 17 December, around 10:43 a.m. (The year is not specified in the text, but the reporting is from 17 December of that unspecified year).
- **Impact:** *Human Casualties:* 17 workers were on site. 3 are missing. The other 14 were safely evacuated. *Property Damage:* The company's plant, covering approximately 2000 square meters, was buried.
- **Response Efforts (as of 4:00 p.m. on 17 December):** *Mobilization:* Over 300 rescuers. 10 excavators. 2 loaders. Follow-up rescue forces were still being increased. *Participating Agencies/Groups:* A multi-agency response is underway, including: (. . .)
- **Quoted Individual:** *Pan Saiqiu (Lishui Civil Defense Rescue Team):* Confirms his team's arrival by 2:00 p.m. and mentions that some members have entered the site, while others are on standby due to the amount of rescue equipment.
- **Source of Information for the Report:** Reporter Zhou Yulong learned the details from the Publicity Department of Songyang, Zhejiang.
- **Preliminary Cause of Landslide:** Prolonged abnormal dryness of the weather, leading to the loosening of the mountain.
- **Overall Tone:** The tone is factual and urgent, focusing on the details of the incident and the scale of the rescue operation. It highlights the immediate and comprehensive response from various authorities and volunteer groups.
- **Missing Information:** The specific year of the incident is not mentioned. The condition of the three missing workers. Specific details about the "prolonged abnormal dryness" (e.g., duration).
- **In summary**, the text provides a snapshot of an active emergency situation—a landslide at a quarry—detailing the immediate aftermath, the significant rescue operation launched, the human impact and the preliminary suspected cause. It emphasizes a coordinated, large-scale response from multiple official and civil organizations.

Prompt: `Now extract 10 key events and put them in chronological order`
Response: (We skip it, since Figure 9 adequately represents the extracted events)

Prompt: `take each pair of bullets and tell me if bullet i caused bullet j and why.`

Response (Segment of analysis)

(. . .)

2. Did Bullet 2 (Landslide) cause Bullet 3 (Plant buried)?

Yes. Why: The landslide is the physical event that directly resulted in the company's plant being covered by debris.

(. . .)

Prompt: `--what about indirect causation e.g., 1 to 10 could you investigate these too.`

Prompt: `--now create a directed labeled graph of these causal relations` (Note. Three attempts to create the graph did not work properly and no .png file was produced).

Prompt: `--can you create python code to generate this graph`
Response: Note: The last prompt resulted in an informative graph shown in Figure 9.

We observe that, with no training examples, both analyses provide a reasonable analysis of causes and effects in the text and both systems can explain their reasoning. No hallucinations are apparent in the event-by-event causality analysis. These analyses are not identical, e.g., the cause of the landslide is provided by Gemini but not by GPT. The graphs in Figures 8 and 9 are different, but so are the representations of causality described earlier: in Example 3 and Figure 6, causal relationships are between words or phrases, while in Figure 4, they are between clauses or events. And given both the lack of a unified perspective on causality and the relatively low inter-annotator agreement mentioned earlier in Section 3.1, using LLMs directly for causality analysis—without the causality extraction—might give us better ways of analyzing and extracting information from text. In particular, as shown in the next subsection, we can ask for analyses with different perspectives. The old and new annotated datasets would still be of value, as a check for LLMs hallucinations, or to guide directions of analysis via better prompting.

Appendix B. Four Causes and Counterfactuals

In these, very preliminary, experiments both Gemini and GPT were asked to provide both a counterfactual causal analysis and an analysis using the Aristotelian categories. All analyses by GPT and Gemini are given in the shared data (https://drive.google.com/drive/folders/1VujYfpQ-6iDuGJut81dpHXnlpqGuHYkP (accessed on 10 December 2025)) and they are easily followed in the chats at these links (https://g.co/gemini/share/cd52a6147a54 (accessed on 10 December 2025), https://chatgpt.com/share/685d81d7-0260-800a-a2a2-9ac21226761a (accessed on 10 December 2025)). A comparison of Aristotelian and counterfactual analysis will be given in a future work.. In this Appendix, we provide a summary illustrating the spirit of the analysis.

Let us start with Aristotle. Since Final Causes are not given in news reports, should we regard the answers below as unjustified speculations ("hallucinations") or an indication of plausible, common sense reasoning?

GPT speculated that the Final Cause(s) possibly were as follows:

> "Economic aim: extract construction aggregate at lowest cost; Corporate mandate: maximize output, minimize downtime; Local demand for sand/concrete in Zhejiang's building sector." And it commented "Teleologically, the quarry wall was shaped the way it was so that large volumes of rock could be removed rapidly and cheaply. Reduced catch-bench spacing or steeper faces keep haul distances short, satisfying the final cause (profit), but at the expense of higher geotechnical risk."

On the other hand, Gemini made the following warning:

> "The final cause, or *telos*, refers to the end, purpose, or the sake for which an event occurs. This is challenging for natural events which don't have intention."

Appendix B.1. FourCauses Analysis by Gemini 2.5 Pro

The Event: A landslide at the quarry of Anluyuan Mechanized Sand Co., Ltd. in Xinxing Town, Songyang County, Zhejiang Province, on 17 December.

1. **Material Cause (What is it made of?)**
   The material cause refers to the underlying matter from which the event or object is constituted. For the landslide:
   - **The Mountain Itself:** This includes the rock, soil, earth and any other geological materials that comprised the slope before it failed.
   - **Water (or lack thereof, then its presence indirectly):** While the *prolonged dryness* is cited as a factor for loosening, the actual movement in many landslides involves the interaction of water with the soil/rock. In this specific case, the dryness led to "loosening," implying a change in the material properties that made it susceptible to movement.
   - **The Company's Plant and Equipment:** These became part of the debris and the consequence of the landslide, but the primary material of the *landslide itself* is the earth material.

   Essentially, the material cause is the physical "stuff" that moved and was affected.
2. **Formal Cause (What is its form or essence?)**
   The formal cause is the pattern, model, or structure that defines the event; what makes it recognizable as a "landslide."
   - **The Definition of a Landslide:** The form is the rapid movement of a mass of rock, debris, or earth down a slope.
   - **The Specific Characteristics of *This* Landslide:**
     - Occurred at a quarry.
     - Buried a plant of approximately 2000 square meters.
     - Involved a specific volume and type of material (though not detailed in the text beyond "mountain").
     - Had a specific failure mechanism (e.g., a slump, a debris flow, a rockfall–the text doesn't specify, but the "loosening" suggests a less cohesive failure).
   - **The Shape and Trajectory of the Moving Mass:** The way the earth detached, moved and came to rest.

   The formal cause is what allows us to identify this event as a "landslide" rather than, say, an earthquake or a flood.
3. **Efficient Cause (What brought it into being?)**
   The efficient (or moving) cause is the agent or process that initiated the event. This is often what we think of as the "cause" in modern scientific terms.
   - **Primary Efficient Cause (as per text):** "Prolonged abnormal dryness of the weather, leading to loosening of the mountain."
   - **Contributing/Underlying Efficient Causes (Implied or General):**
     - **Gravity:** The constant force pulling the loosened material downwards.
     - **Geological Conditions:** The inherent properties of the rock and soil, slope angle and geological structure of the mountain at the quarry site.
     - **Quarrying Activities (Potentially):** The landslide occurred *at a quarry*. This human activity could be a significant contributing efficient cause.
     - **The specific moment of failure:** Some final, perhaps imperceptible, trigger that caused the already loosened material to give way.

   The efficient cause is the "why it happened now" and "what made it happen."

4. **Final Cause (Telos: What is its purpose or end?)**
   The final cause, or *telos*, refers to the end, purpose, or the sake for which an event occurs. This is challenging for natural events which don't have intention.
   - **In a Purely Naturalistic Sense:**
     - **Achieving a More Stable State:** From a physics perspective, the materials moved to reach a lower energy state, a more stable equilibrium under the force of gravity.
     - **The Natural Process of Erosion:** Landslides are a natural geological process that shapes the Earth's surface over time.
   - **From a Human Perspective (Consequences, not "Purpose'):**
     - The text describes the *consequences*: workers missing, a plant buried, a large-scale rescue operation.
     - It's important not to anthropomorphize the landslide as having an intention to cause harm.

   The human response (rescue) has its own final causes (saving lives, understanding the event).

Appendix B.2. Causal Analysis Using Aristotle's Four Causes—GPT-o3

1. **Material + Formal Causes:** Supplied the latent vulnerability:
   - Weak rock.
   - Over-steep configuration.
2. **Efficient Cause:** Acted on 17 December to actualise that vulnerability:
   - Crack propagation → block detachment.
3. **Final Cause:** Explains why the slope was engineered so aggressively:
   - Economic rationale shaped both formal geometry and tolerance for risk.

   Aristotle's framework clarifies that addressing only the efficient cause (e.g., blaming drought) is insufficient.
4. **Preventive Interventions by Causal Type:**
   - **Material fixes:** rock bolting, shotcrete, drainage to improve inherent strength.
   - **Formal fixes:** redesign bench angles, add catch benches, flatten overall slope.
   - **Efficient controls:** drought monitoring, blast vibration limits, continuous slope radar.
   - **Final-cause alignment:** integrate safety margin into the quarry's business objectives so the purpose itself no longer incentivises risky geometry.
5. **Conclusion:** By tackling all four causal dimensions, future failures of this type become far less probable.

References

1. Devlin, J. BERT: Pre-training of deep bidirectional transformers for language understanding. *arXiv* **2018**, arXiv:1810.04805.
2. Achiam, J.; Adler, S.; Agarwal, S.; Ahmad, L.; Akkaya, I.; Aleman, F.L.; Almeida, D.; Altenschmidt, J.; Altman, S.; Anadkat, S.; et al. Gpt-4 technical report. *arXiv* **2023**, arXiv:2303.08774. [CrossRef]
3. OpenAI. *OpenAI o3 and o4-Mini System Card*; Technical Report; OpenAI: San Francisco, CA, USA, 2025.
4. OpenAI. *GPT-5 System Card*; System Card; OpenAI: San Francisco, CA, USA, 2025.
5. Contributors, W. List of Large Language Models—Wikipedia, The Free Encyclopedia. 2025. Available online: https://en.wikipedia.org/wiki/List_of_large_language_models (accessed on 17 May 2025).
6. Rettenberger, L.; Reischl, M.; Schutera, M. Assessing political bias in large language models. *J. Comput. Soc. Sci.* **2025**, *8*, 1–17. [CrossRef]

7. Bang, Y.; Chen, D.; Lee, N.; Fung, P. Measuring Political Bias in Large Language Models: What Is Said and How It Is Said. In Proceedings of the 62nd Annual Meeting of the Association for Computational Linguistics (Volume 1: Long Papers), Bangkok, Thailand, 11–16 August 2024; Ku, L.W., Martins, A., Srikumar, V., Eds.; Association for Computational Linguistics: Stroudsburg, PA, USA, 2024; pp. 11142–11159. [CrossRef]
8. Mariko, D.; Abi Akl, H.; Trottier, K.; El-Haj, M. The financial causality extraction shared task (FinCausal 2022). In Proceedings of the 4th Financial Narrative Processing Workshop@ LREC2022, Marseille, France, 15–24 June 2022; pp. 105–107.
9. Zadrozny, W. Large Language Models (LLMs) and Causality Extraction from Text: Tutorial at FLAIRS-38. In Proceedings of the Florida Artificial Intelligence Research Society Conference (FLAIRS-38), LibraryPress@UF, Daytona Beach, FL, USA, 20–23 May 2025; Volume 38. [CrossRef]
10. Li, X.; Cai, Z.; Wang, S.; Yu, K.; Chen, F. A Survey on Enhancing Causal Reasoning Ability of Large Language Models. *arXiv* **2025**, arXiv:2503.09326. [CrossRef]
11. Liu, X.; Xu, P.; Wu, J.; Yuan, J.; Yang, Y.; Zhou, Y.; Liu, F.; Guan, T.; Wang, H.; Yu, T.; et al. Large Language Models and Causal Inference in Collaboration: A Comprehensive Survey. In Proceedings of the Findings of the Association for Computational Linguistics: NAACL 2025, Albuquerque, Mexico, 29 April–4 May 2025; Association for Computational Linguistics: Stroudsburg, PA, USA, 2025; pp. 7668–7684.
12. Wang, L.; Shen, Y. Evaluating Causal Reasoning Capabilities of Large Language Models: A Systematic Analysis Across Three Scenarios. *Electronics* **2024**, *13*, 4584. [CrossRef]
13. Jin, Z. Causality for NLP Reading List. 2021. Available online: https://github.com/zhijing-jin/Causality4NLP_Papers (accessed on 10 December 2025).
14. Feder, A.; Keith, K.A.; Manzoor, E.; Pryzant, R.; Sridhar, D.; Wood-Doughty, Z.; Eisenstein, J.; Grimmer, J.; Reichart, R.; Roberts, M.E.; et al. Causal Inference in Natural Language Processing: Estimation, Prediction, Interpretation and Beyond. *Trans. Assoc. Comput. Linguist.* **2022**, *10*, 1138–1158. [CrossRef]
15. THORS eLearning Solutions. 7QC Tools: Fishbone Diagrams. 2025. Available online: https://thors.com/manufacturing-courses/manufacturing-fundamentals/quality/7qc-tools-fishbone-diagrams/ (accessed on 26 May 2025).
16. Causality. Causality—Wikipedia, The Free Encyclopedia, 2022. Available online: https://en.wikipedia.org/wiki/Causality (accessed on 10 December 2025).
17. Wang, P. Introduction to Artificial General Intelligence (AGI). 2023. Available online: https://cis.temple.edu/~pwang/AGI-Intro.html (accessed on 27 June 2025)
18. Shields, C. Aristotle. 2022. Available online: https://plato.stanford.edu/archives/fall2022/entries/aristotle/#FouCauAccExpAde (accessed on 14 December 2025).
19. Halpern, J.Y.; Pearl, J. Causes and explanations: A structural-model approach. Part I: Causes. *Br. J. Philos. Sci.* **2005**. Available online: https://www.journals.uchicago.edu/doi/abs/10.1093/bjps/axi147 (accessed on 14 December 2025). [CrossRef]
20. Bareinboim, E.; Correa, J.D.; Ibeling, D.; Icard, T.F. On Pearl's Hierarchy and the Foundations of Causal Inference. In *Probabilistic and Causal Inference*; Association for Computing Machinery: New York, NY, USA, 2022.
21. Sawesi, S.; Rashrash, M.; Dammann, O. The Representation of Causality and Causation with Ontologies: A Systematic Literature Review. *Online J. Public Health Inform.* **2022**, *14*, e4. [CrossRef]
22. Tan, F.A.; Hürriyetoğlu, A.; Caselli, T.; Oostdijk, N.; Nomoto, T.; Hettiarachchi, H.; Ameer, I.; Uca, O.; Liza, F.F.; Hu, T. The Causal News Corpus: Annotating Causal Relations in Event Sentences from News. In Proceedings of the Thirteenth Language Resources and Evaluation Conference, Marseille, France, 20–25 June 2022; pp. 2298–2310.
23. Wikipedia Contributors. Four Causes. 2024. Available online: https://en.wikipedia.org/wiki/Four_causes (accessed on 5 February 2024).
24. Pearl, J. *Models, Reasoning and Inference*, 2nd ed.; Cambridge University Press: Cambridge, UK, 2009.
25. Pearl, J. The seven tools of causal inference, with reflections on machine learning. *Commun. ACM* **2019**, *62*, 54–60. [CrossRef]
26. Lorkowski, C.M. Hume, David: Causation. *Internet Encyclopedia of Philosophy* **2011**. Available online: https://iep.utm.edu/hume-causation/ accessed on 1 June 2025).
27. VanderWeele, T.J.; Knol, M.J. A tutorial on interaction. *Epidemiol. Methods* **2014**, *3*, 33–72. [CrossRef]
28. VanderWeele, T. *Explanation in Causal Inference: Methods for Mediation and Interaction*; Oxford University Press: Oxford, UK, 2015.
29. Gopalakrishnan, S.; Chen, V.Z.; Dou, W.; Hahn-Powell, G.; Nedunuri, S.; Zadrozny, W. Text to Causal Knowledge Graph: A Framework to Synthesize Knowledge from Unstructured Business Texts into Causal Graphs. *Information* **2023**, *14*, 367. [CrossRef]
30. Hendrickx, I.; Kim, S.N.; Kozareva, Z.; Nakov, P.; Séaghdha, D.O.; Padó, S.; Pennacchiotti, M.; Romano, L.; Szpakowicz, S. Semeval-2010 task 8: Multi-way classification of semantic relations between pairs of nominals. *arXiv* **2019**, arXiv:1911.10422.
31. Diaz-Garcia, J.A.; Lopez, J.A.D. A survey on cutting-edge relation extraction techniques based on language models. *arXiv* **2024**, arXiv:2411.18157. [CrossRef]

32. Norouzi, R.; Kleinberg, B.; Vermunt, J.K.; van Lissa, C.J. Capturing causal claims: A fine-tuned text mining model for extracting causal sentences from social science papers. *Res. Synth. Methods* **2025**, *16*, 139–156. [CrossRef]
33. Demirhan, H.; Zadrozny, W. Survey of multimodal medical question answering. *BioMedInformatics* **2024**, *4*, 50–74. [CrossRef]
34. Wenzina, R.; Kaiser, K. Identifying condition-action sentences using a heuristic-based information extraction method. In *Process Support and Knowledge Representation in Health Care*; Springer: Cham, Switzerland, 2013; pp. 26–38.
35. Lee, J.; Yoon, W.; Kim, S.; Kim, D.; Kim, S.; So, C.H.; Kang, J. BioBERT: A pre-trained biomedical language representation model for biomedical text mining. *Bioinformatics* **2020**, *36*, 1234–1240. [CrossRef]
36. Hematialam, H.; Zadrozny, W. Identifying condition-action statements in medical guidelines using domain-independent features. *arXiv* **2017**, arXiv:1706.04206.
37. Hematialam, H. Knowledge Extraction and Analysis of Medical Text with Particular Emphasis on Medical Guidelines. Ph.D. Thesis, The University of North Carolina at Charlotte, Charlotte, NC, USA, 2021.
38. Hematialam, H.; Zadrozny, W. Identifying Condition-action Statements in Medical Guidelines: Three Studies using Machine Learning and Domain Adaptation. 2021. Available online: https://www.researchsquare.com/article/rs-500521/v1 (accessed on 20 December 2025).
39. Gopalakrishnan, S. Building Computational Representations of Medical Guidelines Using Large Language Models and Transfer Learning. Ph.D. Thesis, The University of North Carolina at Charlotte, Charlotte, NC, USA, 2023.
40. Gopalakrishnan, S.; Chen, V.Z.; Dou, W.; Zadrozny, W. On the relation between K–L divergence and transfer learning performance on causality extraction tasks. *Nat. Lang. Process. J.* **2024**, *6*, 100055. [CrossRef]
41. Chen, V.Z.; Montano-Campos, F.; Zadrozny, W.; Canfield, E. Machine reading of hypotheses for organizational research reviews and pre-trained models via R shiny app for non-programmers. *arXiv* **2021**, arXiv:2106.16102. [CrossRef]
42. Gopalakrishnan, S.; Garbayo, L.; Zadrozny, W. Causality extraction from medical text using large language models (LLMs). *Information* **2024**, *16*, 13. [CrossRef]
43. Liu, X.; Yang, W.; Wei, F.; Wu, Z. Semantic aware enhanced event causality identification. *(Nat.) Sci. Rep.* **2024**, *14*, 32078. [CrossRef]
44. Caselli, T.; Vossen, P. The Event StoryLine Corpus: A New Benchmark for Causal and Temporal Relation Extraction. In Proceedings of the Events and Stories in the News Workshop, Vancouver, BC, Canada, 4 August 2017; pp. 77–86. [CrossRef]
45. Tan, F.A.; Hettiarachchi, H.; Hürriyetoğlu, A.; Caselli, T.; Uca, O.; Liza, F.F.; Oostdijk, N. Event Causality Identification with Causal News Corpus–Shared Task 3, CASE 2022. *arXiv* **2022**, arXiv:2211.12154.
46. Wu, M.; Zhang, Q.; Wu, C.; Wang, G. End-to-end multi-granulation causality extraction model. *Digit. Commun. Netw.* **2024**, *10*, 1864–1873. [CrossRef]
47. Miliani, M.; Auriemma, S.; Bondielli, A.; Chersoni, E.; Passaro, L.; Sucameli, I.; Lenci, A. ExpliCa: Evaluating Explicit Causal Reasoning in Large Language Models. *arXiv* **2025**, arXiv:2502.15487. [CrossRef]
48. AI Futures Project (Daniel Kokotajlo, Eli Lifland, Thomas Larsen, Romeo Dean). AI 2027: A Research-Backed AI Scenario Forecast. 2025. Available online: https://ai-2027.com/ (accessed on 15 June 2025).
49. OPEN AI. Gpt-4o system card. *arXiv* **2024**, arXiv:2410.21276. [CrossRef]
50. Rein, D.; Hou, B.L.; Stickland, A.C.; Petty, J.; Pang, R.Y.; Dirani, J.; Michael, J.; Bowman, S.R. GPQA: A Graduate-Level Google-Proof Q&A Benchmark. In Proceedings of the First Conference on Language Modeling, Philadelphia, PA, USA, 7–9 October 2024.
51. Marcus, G. The Latest AI Scaling Graph, and Why It Should Worry You. 2024. Available online: https://garymarcus.substack.com/p/the-latest-ai-scaling-graph-and-why (accessed on 27 May 2025).
52. Venkit, P.N.; Chakravorti, T.; Gupta, V.; Biggs, H.; Srinath, M.; Goswami, K.; Rajtmajer, S.; Wilson, S. An Audit on the Perspectives and Challenges of Hallucinations in NLP. In Proceedings of the 2024 Conference on Empirical Methods in Natural Language Processing, Miami, FL, USA, 12–16 November 2024; pp. 6528–6548.
53. Huang, L.; Yu, W.; Ma, W.; Zhong, W.; Feng, Z.; Wang, H.; Chen, Q.; Peng, W.; Feng, X.; Qin, B.; et al. A survey on hallucination in large language models: Principles, taxonomy, challenges, and open questions. *ACM Trans. Inf. Syst.* **2025**, *43*, 1–55. [CrossRef]
54. OpenAI. Fine-Tuning Guide. 2025. Available online: https://platform.openai.com/docs/guides/fine-tuning (accessed on 22 May 2025).
55. Xu, Z.; Jain, S.; Kankanhalli, M. Hallucination is inevitable: An innate limitation of large language models. *arXiv* **2024**, arXiv:2401.11817. [CrossRef]
56. Kalai, A.T.; Vempala, S.S. Calibrated language models must hallucinate. In Proceedings of the 56th Annual ACM Symposium on Theory of Computing, Vancouver, BC, Canada, 24–28 June 2024; pp. 160–171.
57. Wu, C.; Grama, A.; Szpankowski, W. No Free Lunch: Fundamental Limits of Learning Non-Hallucinating Generative Models. *arXiv* **2024**, arXiv:2410.19217. [CrossRef]

58. Suzuki, A.; He, Y.; Tian, F.; Wang, Z. Hallucinations are inevitable but statistically negligible. *arXiv* **2025**, arXiv:2502.12187. [CrossRef]
59. Simhi, A.; Itzhak, I.; Barez, F.; Stanovsky, G.; Belinkov, Y. Trust Me, I'm Wrong: High-Certainty Hallucinations in LLMs. *arXiv* **2025**, arXiv:2502.12964.
60. Zubić, N.; Soldá, F.; Sulser, A.; Scaramuzza, D. Limits of deep learning: Sequence modeling through the lens of complexity theory. *arXiv* **2024**, arXiv:2405.16674. [CrossRef]
61. Niess, G.; Razouk, H.; Mandic, S.; Kern, R. Addressing Hallucination in Causal Q&A: The Efficacy of Fine-tuning over Prompting in LLMs. In Proceedings of the Joint Workshop of the 9th Financial Technology and Natural Language Processing (FinNLP), the 6th Financial Narrative Processing (FNP), and the 1st Workshop on Large Language Models for Finance and Legal (LLMFinLegal), Abu Dhabi, United Arab Emirates, 19–20 January 2025; pp. 253–258.
62. Dündar-Coecke, S.; Goldin, G.; Sloman, S.A. Causal reasoning without mechanism. *PLoS ONE* **2022**, *17*, e0268219. [CrossRef] [PubMed]
63. Mollick, E. On Jagged AGI: o3, Gemini 2.5, and Everthing After. 2025. Available online: https://www.oneusefulthing.org/p/on-jagged-agi-o3-gemini-25-and-everything (accessed on 25 May 2025).
64. Gopalakrishnan, S.; Chen, V.; Hahn-Powell, G.; Tirunagar, B. Computer-assisted construct classification of organizational performance concerning different stakeholder groups. *arXiv* **2021**, arXiv:2107.05133. [CrossRef]
65. Gopalakrishnan, S.; Padithala, S.; Demirhan, H.; Zadrozny, W. MDS_UNCC Question Answering System for Biomedical Data with Preliminary Error Analysis. In Proceedings of the CLEF (Working Notes), Bucharest, Romania, 21–24 September 2021; pp. 231–240.
66. Zadrozny, W.; Garbayo, L. A Sheaf Model of Contradictions and Disagreements. Preliminary Report and Discussion. *arXiv* **2018**, arXiv:1801.09036. [CrossRef]
67. Hematialam, H.; Garbayo, L.; Gopalakrishnan, S.; Zadrozny, W.W. A Method for Computing Conceptual Distances between Medical Recommendations: Experiments in Modeling Medical Disagreement. *Appl. Sci.* **2021**, *11*, 2045. [CrossRef]
68. Team, G. Gemini 1.5: Unlocking multimodal understanding across millions of tokens of context. *arXiv* **2024**, arXiv:2403.05530. [CrossRef]

MDPI AG
Grosspeteranlage 5
4052 Basel
Switzerland
Tel.: +41 61 683 77 34

Entropy Editorial Office
E-mail: entropy@mdpi.com
www.mdpi.com/journal/entropy